Textile Fibre

Textile Fibre

Ajay Shrivastava

RANDOM

RANDOM PUBLICATIONS

NEW DELHI (INDIA)

Textile Fibre

ISBN 978-93-5111-646-2

Published in 2015 in India by

RANDOM PUBLICATIONS

4376-A/4B, Gali Murari Lal, Ansari Road
New Delhi-110 002
Phone : +9111-43580356, 011-23289044, 011-43142548
e-mail: sales@randompublications.com,
info@randompublications.com, randomexports@gmail.com

Reprinted 2026

Type Setting by : Friends Media, Delhi-110089

Preface

It is defined as one of the delicate, hair portions of the tissues of a plant or animal or other substances that are very small in diameter in relation to there length. A fiber is a material which is several hundred times as long as its thick. Textile fiber has some characteristics which differ between fibers to Textile fiber. Textile fiber can be spun into a yarn or made into a fabric by various methods including weaving, knitting, and braiding, felting, and twisting. The essential requirements for fibers to be spun into yarn include a length of at least 5 millimeters, flexibility, cohesiveness, and sufficient strength. Other important properties include elasticity, fineness, uniformity, durability, and luster. Banana fiber is one kind of fiber but it is not a textile fiber. So we can say that all fiber are not textile fiber. Natural fibers include those produced by plants, animals, and geological processes. They are biodegradable over time. They can be classified according to their origin. It is also known as manufactured fiber. Synthetic or man-made fibers generally come from synthetic materials such as petrochemicals. But some types of synthetic fibers are manufactured from natural cellulose; including rayon, modal, and the more recently developed Lyocell.

I would like to thank my team for standing beside me throughout my career and writing this book. My special thanks go to "Random Publications" who have published the book.

– Ajay Shrivastava

Contents

1

Introduction

WHAT IS A FIBRE?

FIBRES ARE FUNDAMENTAL

Most basic of all in selecting fabrics is the Behaviour of the Fibres, the smallest visible components of most fabrics. The two main divisions in textile Fibres are natural and manufactured. Natural Fibres are those found in nature already usable to make fabrics. The two main sources are plants (with the fibers called cellulosic from the main building material, cellulose), and animals (with the fibers called protein from the main building material). Manufactured (MF) fibers (formerly termed "man-made") are formed from a suitable raw material as a thick, sticky liquid, which is "spun" or extruded through spinneret holes, forming streams that are solidified into fibers. The raw material for MF fibers may be itself a natural substance, or it may be synthetic (synthesized from basic chemical units), but it is converted into textile fibers by a manufacturing process. While there are MF fibers made of natural rubber (as well as of synthetic rubber), there is no such thing as a natural rubber Fibre. Similarly, Tencel® lyocell is not a natural Fibre; it is an MF Fibre made of a natural material, cellulose. Textile fibers may be staple or filament. Staple fibers are relatively short, measured in millimeters or inches. Filament fibers are relatively long, measured in meters or yards. Most natural fibers are staple; the only natural filament Fibre is reeled or cultivated silk. On the other hand, all MF fibers can be staple or filament; they begin as filament, and in this form can give silky or (reeled) silk-like fabrics. They can also be cut or broken into staple to give fabrics that look and feel more like wool, cotton, or linen.

TEXTILE FIBERS ARE SPECIAL

Many fibrous materials are not suitable to make into fabrics, *e.g.*, corn silk or wood slivers. Textile fibers must have certain properties: flexible, thin (but not too thin), long (enough), cohesive, and strong (enough). Textile fibers must be flexible. Wood fibers (unless processed, as into pulp) do not bend easily—

you cannot make fabrics from slivers! Textile fibers are also very thin—long in relation to diameter. To be mechanically spun into yarn—drawn out and twisted—staple fibers must have sufficient length, strength, and cohesiveness (Fibre-to-Fibre friction). Many seed fibers are too short, weak, and slippery to spin into yarn; kapok, for instance, can be used only for stuffing. Of course, for a Fibre to be used a great deal, there must be a reasonable supply and price.

WHAT'S IN A NAME? FIBRE GENERIC NAMES

Fibers are classified into groups, each of which is given a generic name or "generic." For natural fibers, this generic is the name of the plant (cotton) or animal (camel) from which the Fibre comes, or an accepted name, such as wool or mohair. For MF fibers, the generic name is given to a family of fibers of similar chemical makeup and is defined in textile labeling regulations in the United States, Canada, and many other countries. Only a few of these are used extensively by consumers; these *major generics* as a "pruned" textile Fibre family tree. All are summarized and discussed in this section. Note that some generic names legitimate in Europe or Canada are not recognized by the United States in its Textile Fibre Products Identification Act, although they are allowed to be used if recognized by the International Organization for Standardization. Some of these are: *alginate, carbon, chlorofibre, cupro, elastane, elastodiene, fluorofibre, metal fibre, modal, polyamide, polyethylene, polyimide, polypropylene, vinylal, viscose*. Conversely, ISO rejected *rayon* as a generic name in 1971, although it remains a legitimate generic name in the United States and Canada. Some names are not official in the United States or Canada:

- *polynosic* (a term for one type of High Wet
- Modulus [HWM] viscose rayon)
- *fibranne* (a term for viscose rayon used in France)
- *PVC* (an acronym for polyvinyl chloride, United States generic vinyon)
- *PVA* (an acronym for polyvinyl alcohol—vinal)
- *polyacrylonitrile* (PAN) or polyacryl (terms for acrylic)

FIBRE CONTENT LABELING

In the United States, the Textile Fibre Products Identification Act2 (TFPIA), enforced by the Federal Trade Commission, specifies requirements for labels, invoices, or advertising for most textile Fibre consumer products (there are some exemptions); in Canada, the Textile Labelling and Advertising Regulations made under the Textile Labelling Act, administered by Industry Canada, give similar specifications.3 Labels should state in English (Canada: English and French) the content by generic name (not capitalized) of each Fibre present in an amount of 5 percent or more, listed in descending order of percentage. A Fibre present as less than 5 percent, unless it clearly has functional significance, and excluding decoration, is labeled "other Fibre."

Generic names are defined in the regulations. A trademark name may be added in immediate conjunction with and in type of roughly equal size and prominence to the generic. When the Fibre content cannot be determined, terms such as "unknown fibers" and their origin, such as "waste materials" are allowable. A company name or registered identification number must also appear on the label, and each act details declaration of the country where the textile product was processed or made.

WHAT IS A TRADEMARK NAME?

With many fashion-forward, "high tech," or "Space Age" articles, consumers are more familiar with trademark names than those of the maker, in many cases, or the generic type of a Fibre. A few for examples: Fortrel®, a polyester Fibre; Supplex®, a textured yarn; Ultrasuede® or Gore-Tex®, fabrics; and Scotchgard™, a finish. Trademarks (brand, house, or trade names) are names or symbols given to a product by a company or association that has registered and owns the name. Many names we use indiscriminately for any of similar products are actually trademarks of wellknown products, such as Coke, Kodak, and Kleenex. These are powerful promotional tools, especially when they include or use a memorable *logo* (word in distinctive type) or symbol (such as the interlocking rings of the Olympic games). The Seal of Cotton symbol (a stylized cotton boll) plus the logo used by the trade association Cotton Incorporated.

DuPont, a company that owned some of the most famous trademarks in textiles, guided by market research, in 2000 "restructured" the definitions of some of its brands, including one of the best-known in the world, Lycra®, which originally meant the first spandex (elastane) Fibre. Because of that restructuring, the name Lycra, now owned by Invista, does not necessarily connote a spandex Fibre, but is attached to fabrics promising freedom of movement in clothing, no matter how achieved. Some others no longer associated with a particular generic, type of yarn, or finish are: Coolmax®, Supplex®, and Thermolite®, all now owned by Invista. Considerable emphasis in *Fabric Reference* has formerly been placed on trademark names as carriers of specific information, with the generic or chemical makeup most basic of all, but this shift has brought listing of all names into question, and in this edition, *many* trademark names have been removed, and virtually all information on names that are obsolete. A great deal of money can be invested in establishing a trademark with positive connotations; manufacturers naturally wish these to remain distinctive but part of their corporate image, rather than drifting into the dreaded area of a generic name, as *aspirin* has done. In this text, every effort has been made to add the appropriate symbol and give the name of the company that owns a trademark in brackets immediately after, at least for the first use on a page; in advertisements you will usually find a footnote used to denote ownership, such as "Abcd is the Registered Trademark of Wxyz Company." Such well-known trademarks carry a definite promise of quality to the consumer; see "Private Sector Labeling with Implications for Care" in Section Six.

Fig. The "Seal of Cotton" is a registered servicemark/ Trademark of Cotton Incorpoarted.

Fibers or fabrics sold without a trademark name or end product certification are termed *commodity fibers* or *commodity goods*.

LEGAL FRAMEWORK OF TRADEMARKS

The symbol for a registered trademark is ® (™ if pending), and for a certification mark, ©. Such a trademark is owned as "intellectual property" and may legally be used exclusively by the company that applied for it. The owner may license others to use the mark, within provisions of legislation in each country. In the United States, this falls within the jurisdiction of the Department of Commerce, through the Patent, Trademark and Copyright Information Office. A trademark, if for interstate commerce, may be registered with the federal Patent and Trademark Office. In Canada, the governing legislation is the Trade Marks Act. Since trademarks or brands are such an important factor in advertising and hence in competition, governments handle registrations but leave policing of the system to those who own the trademarks. Infringement of trademark rights is pursued through the courts. A few words and symbols cannot be used (*e.g.*, *United Nations;* national, territorial, or civic flags; and the Red Cross symbol).

2

Fibre to Fabric

You know that fibres are what fabrics are made of. Just look at the fabric of the dress you are wearing. It is made of the thread prepared from small fibres. Just pull out a thread from a piece of fabric and open it up by untwisting. You will see those small fibres. You do know that there are a large number of fibres available to us and from which a variety of fabrics are made. This is why in any cloth shop you find such a variety of fabrics on the shelves of cloth shops in the market. In this lesson we want you to get all the knowledge and acquire skills needed to recognize the fabric you want to buy by associating it with its specific properties and assessing its suitability to specific uses. You are familiar with the variety of uses various fabrics have in our daily life. The study of fibres and fabrics gives us the complete knowledge of properties and uses of various fabrics available in the market. This knowledge will help us to choose wisely the fabric we require and make the best use of it after purchase.

FUNCTION OF CLOTHING AND HOUSEHOLD USES OF FABRICS

Why do we wear clothes? Our clothes are rightly recognized as "the second skin." At every stage of life and in every ceremony, clothes play an important role. They cover our body and protect it from adverse climatic conditions. Clothes enhance our personality. These are made from various kinds of materials and sewn in several forms called dresses or garments. A well-fitted dress, worn according to the occasion and time speaks a lot about the person's habits, taste, social status, behaviour and many other traits. Generally, men, women and children dress up differently and they also have sets of dresses for different purposes and occasions, for example, dresses for formal, casual occasions, sports and nightwear. Colour of our clothing, in our country, is decided according to season, climate, age, occasion, marital status, gender, community, happiness and sorrow. The religion and occupation of a person can also be judged by the clothes a person wears. Lastly and most importantly, clothes also tell us about the habits of the wearer and his/ her approach to hygiene. If the clothes are dirty and crushed we tend to think the person is careless, shabby and poor. Crisp, freshly washed and well ironed clothes lend a well groomed appearance to the wearer. You have just learnt about the functions of clothes. Two examples are given below. List three more functions of clothes.

I	To look smart.
ii.	To show individuality.
iii	
iv	
v	

OTHER USES OF FABRICS

Think and write some of the other uses of fabrics in your house. Besides clothing, you are using fabrics in the house for bed linen, curtains, upholstery, cushion covers, etc. In the kitchen, you need dusters, mops, covers, etc. which are also produced from fibres. You use old cloth for dusting, wiping and covering, etc. So you see you have varied uses of fabrics in the house. Generally cotton fabric is suitable for most of the purposes. Once you learn about the properties of various fibres available, you will know why cotton is preferred to other fibres.

FIBRES AND THEIR SOURCES

We have talked about fibre as the basic unit of fabric. Do you know what a fibre looks like? To understand, this take a small amount of cotton and pull out the smallest part from it. Study carefully. These may be long or short strands with a smooth structure that looks like white hair. Actually, these are fibres. A single fibre of cotton is difficult to locate but can be easily recognized in a mass of cotton. Now, consider the structure of wool fibres. You know these are hair of animals like sheep, goat, rabbit, camel etc. Sweaters, socks, gloves, scarves, shawls and coats etc., are made from these hair/fibres. You can check this by opening a thread pulled out from a pure woolen fabric, or knitting wool. What do you see? Similarly, there are many other fibres available to us, which can be used to make cloth and we will learn about these in the next section. A fibre is a fine hair like strand and is the basic unit of textiles from which we make yarns and then the fabric.

SOURCES OF FIBRES AND THEIR CLASSIFICATION

There are many different sources from which we can obtain fibres and therefore, we classify them accordingly.

- Fibres from Natural Sources: All the fibres obtained from nature, *i.e.*, plants or animals, are known as natural fibres. *e.g.* cotton, wool, linen, silk, etc. Fibres obtained from plant source are called cellulosic fibre *e.g.*, cotton and linen. The fibres that come from animal sources are also known as protein fibres, *e.g.*, wool and silk.

- Man-Made fibres: The fibres which are made in laboratories using chemicals are known as man-made fibres and these are of following two types:
- Regenerated fibres – These fibres are made from extremely small cotton fibres or any other fibre source such as wood pulp, milk protein, etc. Chemicals are used to dissolve these and the solution is then converted into solid fibres. Examples are rayon (cellulose out of viscose/acetate/triacetate) of different types, casein fibre (from milk) and soya bean fibre.
- Synthetic fibres - These are made using various petrochemical products. Nylon, acrylic and polyester are all synthetic fibres.

It is advisable to use garments made of natural fibres which are eco-friendly in nature. Sometimes synthetic fibres may cause allergies if worn next to skin. Sources of commonly used natural and man made fibres are presented in Table.

Table. Commonly used natural man-made fibres

Type of fibres	Name of fibre	Sources
NATURAL FIBRES		
Plants (cellulosic)	cotton	cotton ball
	linen	bark of flax stalk
Animal (protein)	wool	hair of sheep, goat, rabbit, llama etc.
	silk	silk worm
MAN-MADE FIBRES / MANUFACTURED		
Regenerated	rayon (viscose, acetate)	cotton linters or wood pulp + chemicals
Synthetic	nylon	chemicals
	polyester	chemicals
	acrylic	chemicals

CLASSIFICATION ACCORDING TO THE LENGTH OF FIBRES

The fibres we have listed above are short or long. The short length fibres are called staple fibres and are measured in inches or centimeters, *e.g.*, cotton, wool and linen. The long fibres are known as filaments and are measured in yards/ meters, *e.g.*, silk and all man-made fibres.

	Length of fibre	Class	Appearance	Unit of measurement
FIBRE	long	filament fibres		yards / meters
LENGTH	short	staple fibres		inches/ centimeters

COMMON CHARACTERISTICS OF DIFFERENT FIBRES

COTTON

Cotton fibre is the smallest of all the textile fibres. They are white, cream or light brown in colour and fine and strong. These are absorbent, porous and cool and allow the body heat to go out.

Hence, fabrics made out of it are used as summer wear as cotton wrinkles very easily. Fabrics made from cotton are strong, durable and easy to wash and are used to make towels, sheets, pillow covers, etc., that require frequent washing.

FLAX

It is a 'bast fibre' and fabric made from it is called linen. It is a staple fibre though its length (20-30 inches) is more than the other staple fibres available. Linen fabric is shiny, smooth, durable and easy to wash. Like cotton, it wrinkles very easily, is cool, absorbent and is suitable for summer wear.

JUTE

Like flax jute is also a bast fibre. Maximum production of jute is in India. The fibres are short and lusturous but weaker than flax. The fibres are hairy and generally rough. It is used for making gunny bags and cords.

WOOL

It is obtained from the fleece of domestic goats, sheep, rabbits, etc. The colour of wool fibres may vary from off-white to light cream. Fabrics made from wool are soft, smooth, absorbent and do not wrinkle easily. These do not allow the body heat to go out and act as insulators. This is why the fabric made out of these fibres is used as winter wear. Wool is a weak fibre and is easily affected by common washing soaps, powders and friction.

SILK

Silk is a natural, protein filament produced by silk worm. Fabrics made from silk are soft, fine, smooth, lustrous, warm and stronger than wool. It is called 'Queen of the Fibres' and is used for formal wear.

RAYON

It is a man-made filament fibre which is lusturous, smooth, cool and absorbent but is weak in nature. It wrinkles very easily. Because of its close resemblance to silk, rayon is also called 'artificial silk' or 'art silk.' It is used as a summer wear. These fibres are thermoplastic in nature *i.e.*, they are heat sensitive and soften and melt on application of heat.

SYNTHETIC FIBRES

Synthetic fibres are made from petroleum products. Nylon, polyester, acrylic, etc., are the examples of synthetic fibres. Like rayon these are also thermoplastic fibres. Since these fibres catch fire easily and can stick to the body, they should not be worn while working in kitchen and near a flame. Synthetics do not wrinkle and can be made dull or shiny. They have good strength and are easy to wash and dry quickly. In other words, these fabrics are easy to care and maintain.

IDENTIFICATION OF FIBRE THROUGH NON–TECHNICAL TEST

dentification of fibres by visual test and feel of the fabric By now, you know the names and some basic properties of fibres obtained from different sources. On the basis of this knowledge can you identify the fabric you are buying? Yes, to some extent you can. If you remember the characteristics of a fibre then you will also know the properties of the fabric made out of it because the fabric will have the same properties. Look for these properties in the fabric and add to this your personal experiences like touch, feel and visual inspection of the fabric. Chances are that you will be able to name the fabric. In the following Table we are presenting to you some of the typical characteristics of different fabrics. If you examine visually, these will help you to recognize the fabric/ fibres.

Table. Characteristics of Different Fibrics whick help Identification.

Fibres	*Appearance*	*Touch*	*Feel*	*Care required*
Cotton	dull in appearance but lustrous when starched	feels smooth and soft to touch	gives a cool feeling	wrinkles easily more if it is starched
Linen	low to medium luster	soft and smooth texture	gives a warm feeling	wrinkles easily
Jute	dull	Rough and hairy texture	gives a warm and rough feeling	does not wrinkle easily
Wool	medium to low luster; poor quality has no luster	soft, smooth and absorbent; also bulky to look at	warm to touch	does not wrinkle easily
Silk	delicate looking and lustrous	smooth, soft and light	warm to touch	does not wrinkle easily
Rayon	can be lustrous or without it	soft and shiny but heavier than silk	gives cool feeling	wrinkles easily
Synthetic fibres	can be dull or semi dull or lustrous acrylic fibres look like wool	heat sensitive soften and melt on application of heat	most fabrics feel warm	able to withstand friction and do not wrinkle hence easy to care

To test the information given in table, select various items of clothing you are using and observe the nature of the fibre in terms of appearance, touch and feel.

IDENTIFICATION OF FIBRES USING BURNING TEST

Burning test tells us about the composition of fibres. *i.e.*, whether the fibres of a fabric are from a plant/ animal source or are man-made. Follow the steps mentioned below to conduct the burning test:

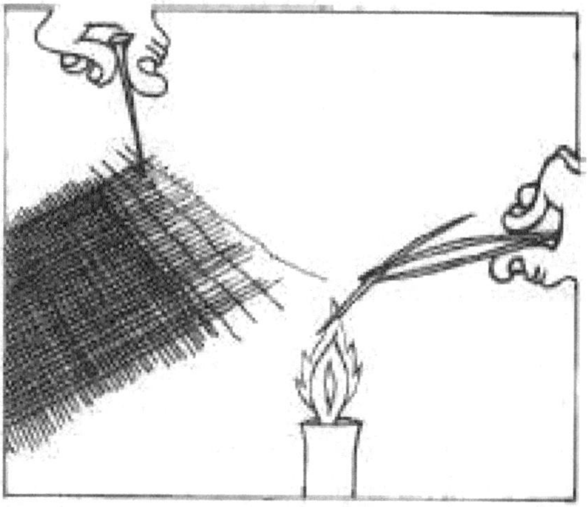

Fig. Burning Test.

Take out a few strands of fibres from a yarn or a fabric and then burn them with the help of a candle flame or a match stick. Observe the following points and record your observations:

The behaviour of the fibres on approaching the flame, in the flame, on burning, and the residue left after burning. Since different types of fibres have specific burning pattern, one can recognize them accordingly.

Table. Describes the Burning tests for identification of fibres.

Fibre	*Near flame*	*Type of burning / flame*	*Odour of burning*	*Residue*
cellulosic fibres – cotton, linen, jute, rayon, etc.	catches fire easily	continue to burn with a bright flame; have an afterglow	burning paper like smell	light, feathery, grayish /black smooth ash
protein fibres – wool, silk	smolder and burn	slow flickering flame; sizzle and curl	Burning hair or feathers like smell	silk-crisp dark ash; wool- dark, irregular, crushable bead

Limitations of the burning test

The result of the burning can be confusing if the fabric is made by mixing two or more types of fibres or yarns

YARNS, YARN MAKING AND FABRIC CONSTRUCTION

A yarn is a long continuous length of interlocked fibres. Strands of fibres are brought closer to each other by twisting. Twists impart strength to the fibre strand which is then termed as a yarn. It is suitable for the production of fabrics, thread for sewing, crocheting, knitting, embroidery and/or rope making. A thread is a highly twisted and smooth strand of fibre. It is used for sewing, embroidery, etc.

PROCESS OF YARN MAKING

Spinning of yarns

Do you know how a yarn is made? Yes, you guessed it right, a yarn is spun. Spinning is the process by which a group of fibres is pulled, drawn and twisted together to make a yarn. Do you remember Mahatma Gandhi and his *charkha* or the spinning wheel? Mahatma Gandhi, would daily take a hand full of cotton and spin it into a yarn on his *charkha*. He promoted *charkha* during India's freedom struggle as a symbol of self-reliance and a source of income. A *charkha* is for hand spinning. The yarn spun on a *charkha* can have different thickness. Thick yarn is used for floor coverings, medium thickness for upholstery items and fine quality yarn is used for making dress material. Different types of fibres - cotton, wool, hemp and silk are spun on *charkha* in the villages in many states of India. Twists given to fibre strands for formation of a yarn can be either 'S-twist' (clockwise) or 'Z-twist' (anticlockwise). The quality and strength of yarn is affected by the number of twists in inch. Lesser the number of twists per inch, bulkier and less strong is the yarn and more the number of twists, finer and stronger is the yarn.

Spinning by Machine

Both, the fibres as well as filaments are spun into yarns that are then used for different end uses. Fibres available in the filament form are first cut into short lengths and then made into yarns called spun yarns. Various steps followed for making yarns are:

- Cleaning: When the natural fibres are harvested or collected, these contain dry leaves, stems, seeds, dirt and unwanted materials that are removed during cleaning.
- Carding: The fibres sometimes get matted and stick to each other. Carding machine opens and arranges the fibres in a parallel manner. The carded web of fibres is turned into a soft rope called sliver.

- Combing: It is an optional step used for making fine quality yarn. Carded slivers are combed to separate long and short fibres with the help of series of combs. Cleaning, carding and combing steps are omitted while making spun yarns from cut filaments of synthetic fibres. For these synthetic fibres only spinning and winding is done.
- Spinning: Carded and combed slivers are further drawn and spun into yarns. The yarn is a single strand but may be plied into several strands:
- Winding: The yarn is wound into various packages according to the weight or length of the yarn and its end use. Some of the common yarn packages for fabric construction are- ball (yarns for hand knitting), reels or bobbins for sewing; embroidery and hanks, cones, etc.

After spinning, a specific length of yarn is wound in the form of packages called balls, reels, hanks, cones, etc., depending on the weight or length of yarn and its end use.

FABRIC

Fabric is a pliable, strong sheet made from fibres or yarns. You must have heard names such as poplin, ***khadder***, ***mulmul***, denim, rubia, terricot, etc. All these are fabrics are prepared by weaving the yarn. Human beings learnt to weave by taking inspiration from nature by observing the nests of birds and entangled branches of trees. Fabrics are manufactured by many techniques such as weaving, knitting, felting, nets, etc. However, weaving and knitting, the two most popular methods of fabric construction have been discussed in detail here.

WEAVING

Weaving is interlacing of two sets of yarns –warp and weft at 90o angles to each other. Straight yarns in fabric are known as warp yarns. Horizontal yarns are known as weft yarns. Along the length of the woven fabric, on both sides, end yarns are woven very densely and the portion is named as selvedge. It does not allow the fabric yarns to come out from the lengthwise edge. The portion between the two selvedges is the body of the fabric.

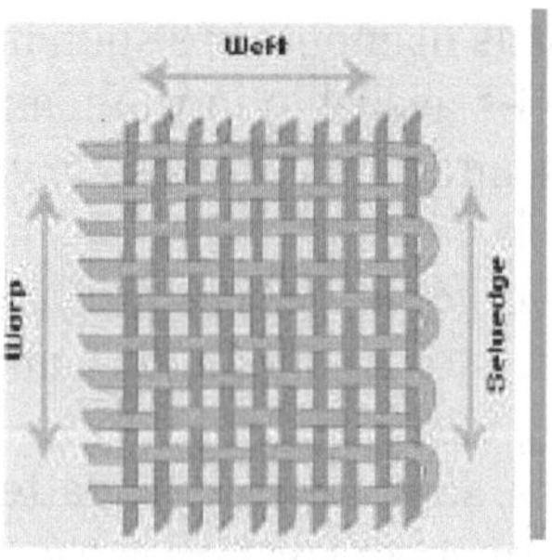

Fig. Woven Fabric

Merits of weaving

- Weaving gives a firm fabric.
- Woven fabrics do not stretch and are thus easy to handle.
- Woven fabrics are easy to print and embroider.
- Weaving is most commonly used method of fabric construction.

BASIC WEAVES

Weaves are broadly classified as basic and novelty weaves. Most of fabrics are produced in basic weaves, which are of three types- plain, twill and satin weave. Their construction is explained in figure

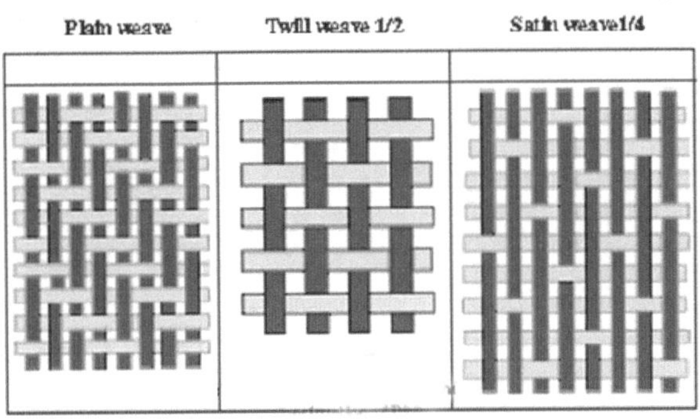

- Plain weave - Plain weave is also known as homespun, tabby or taffeta weave. It is the easiest to weave where one weft yarn alternatively moves over one and under another warp yarn. Maximum production of fabric is done in plain weave. It is inexpensive weave, most suitable for printing and embroideries. To see the variations of the weave, note the fabrics like muslin, cambric, hand spun and hand woven khaddar, organdy, poplin, voile, etc.
- Twill weave - It is woven on three to four harness loom. In this, one weft yarn moves over two and under one warp yarn. Twill woven fabric is distinguished by a continuous diagonal line called wale. Variation in diagonal lines produces various designs of twill. Twill weave is woven tightly, that is why it is suitable for work clothes and for men's clothes. Examples of Twill woven fabrics are gabardine, tweed, denim, jean, etc.
- Satin weave - It is woven on five to twelve harness loom. If woven on a five harness loom, one weft yarn passes under 4 warp yarns and goes over one warp yarn. It differs from Twill weave as it has long yarns floating on the surface. There is no design visible on the face of the fabric but it has a smooth and shiny surface. Satin fabric is an example of satin weave. Fabrics woven in this weave are suitable for making formal wear garments.

END USES OF DIFFERENT FABRICS

When you go to a shop, you give specification of the fabric you want, to the salesperson. Often you go to different shops to buy fabric or ready made garments. In other words, shops usually specialize in the type of items they sell. This way it is easy for you to find what you want and the shop can also stock good variety of related products. Since you have already studied about the properties of various types of fibres, yarns, fabrics and weaves, it will be easy for you to use this knowledge to recognize fabrics and choose them for the end use in your mind. Cotton fabrics are available in the form of muslin, khadi, poplin, rubia, organdy lawn and denim. Similarly, wool is available in the form of felt, knits and woven fabric silk fabric is available as raw silk, crepe and satin silk.

3

Fibre Theory, Formation, and Characterization

FIBRE THEORY AND FORMATION

The word "textile" was originally used to define a woven fabric and the processes involved in weaving. Over the years the term has taken on broad connotations, including the following:

- Staple filaments and fibers for use in yarns or preparation of woven, knitted, tufted or Non-woven fabrics,
- Yarns made from natural or man-made fibers,
- Fabrics and other products made from fibers or from yarns, and
- Apparel or other articles fabricated from the above which retain the flexibility and drape of the original fabrics.

This broad definition will generally cover all of the products produced by the textile industry intended for intermediate structures or final products. Textile fabrics are planar structures produced by interlacing or entangling yarns or fibers in some manner.

In turn, textile yarns are continuous strands made up of textile fibers, the basic physical structures or elements which makes up textile products.

Each individual Fibre is made up of millions of individual long molecular chains of discrete chemical structure.

The arrangement and orientation of these molecules within the individual Fibre, as well as the gross cross section and shape of the Fibre (morphology), will affect Fibre properties, but by far the molecular structure of the long molecular chains which make up the Fibre will determine its basic physical and chemical nature.

Usually, the polymeric molecular chains found in fibers have a definite chemical sequence which repeats itself along the length of the molecule. The total number of units which repeat themselves in a chain (n) varies from a few units to several hundred and is referred to as the degree of polymerization (DP) for molecules within that Fibre.

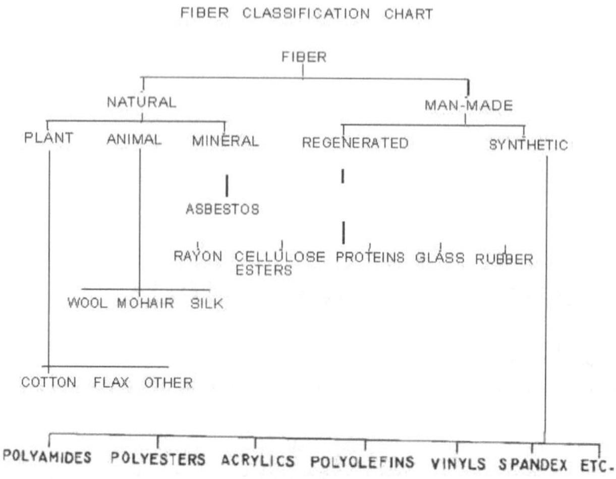

Fig. Classification of natural and man-made fibers.

FIBRE CLASSIFICATION

Textile fibers are normally broken down into two main classes, natural and man-made fibers. All fibers which come from natural sources (animals, plants, etc.) and do not require Fibre formation or reformation are classed as natural fibers. Natural fibers include the protein fibers such as wool and silk, the cellulose fibers such as cotton and linen, and the mineral Fibre asbestos. Man-made fibers are fibers in which either the basic chemical units have been formed by chemical synthesis followed by Fibre formation or the polymers from natural sources have been dissolved and regenerated after passage through a spinneret to form fibers. Those fibers made by chemical synthesis are often called synthetic fibers, while fibers regenerated from natural polymer sources are called regenerated fibers or natural polymer fibers. In other words, all synthetic fibers and regenerated fibers are man-made fibers, since man is involved in the actual Fibre formation process. In contrast, fibers from natural sources are provided by nature in ready-made form.

The synthetic man-made fibers include the polyamides (nylon), polyesters, acrylics, polyolefins, vinyls, and elastomeric fibers, while the regenerated fibers include rayon, the cellulose acetates, the regenerated proteins, glass and rubber fibers. Figure shows a classification chart for the major fibers. Another method of classifying fibers would be according to chemical structure without regard of the origin of the Fibre and its starting materials. In this manner all fibers of similar chemical structure would be classed together. The natural man-made Fibre classification given in Figure does this to a certain extent. In this way, all fibers having the basic cellulosic unit in their structures would be grouped together rather than separated into natural and man-made fibers. This book essentially presents the fibers in groups of similar basic chemical structure, with two exceptions. In one case the elastomeric fibers have been grouped together due to their exceptional physical property, high extensibility and recovery. In the other case, new fibers which do not properly "fit" into anyone category have been placed in a separate chapter. An outline for the arrangement for fibers by chemical class as presented in this sourcebook follows:

Cellulosic Fibers:
- Cotton
- Flax
- Other natural cellulosic
- fibers
- Rayon

Cellulosic Ester Fibers:
- Acetate
- Tri acetate

Protein (Natural Polyamide) Fibers
- Wool
- Silk
- Other natural and regenerated
- protein fibers

Polyamide (Nylon) Fibers
- Nylon 6 and 6,6
- Aramid
- Other nylon fibers

Polyester Fibers:
- Polyethylene terephthalate
- Poly-l,4-cyclohexylenedimethylene terephthalate
- Other polyester fibers

Acrylic and Modacrylic Fibers:
- Acrylic
- Modacrylic
- Other acrylics

Polyolefin Fibers:
- Polyethylene
- Polypropyl ene

Vi nyl Fibers:
- Vinyon
- Vinal
- Vinyon-vinal matrix
- Saran
- Polytetrafluoroethylene

Elastomeric Fibers
- Rubber

- Spandex
- Other elastomeric fibers

Mineral and Metallic Fibers

- Glass
- Inorganic
- Asbestos
- Metallic

Miscellaneous Fibers

- Novaloid
- Carbon
- Polyimide

FIBRE PROPERTIES

There are several primary properties necessary for a polymeric material to make an adequate Fibre: (1) Fibre length to width ratio, (2) Fibre uniformity, (3) Fibre strength and flexibility, (4) Fibre extensibility and elasticity, and (5) Fibre cohesiveness.

Certain other Fibre properties increase its value and desirability in its intended end-use but are not necessary properties essential to make a Fibre. Such secondary properties include moisture absorption characteristics, Fibre resiliency, abrasion resistance, density, luster, chemical resistance, thermal characteristics, and flammability. A more detailed description of both primary and secondary properties follows.

PRIMARY PROPERTIES

Fibre Length to Width Ratio:

Fibrous materials must have sufficient length so that they can be made into twisted yarns. In addition, the width of the Fibre (the diameter of the cross section) must be much less than the overall length of the Fibre, and usually the Fibre diameter should be 1/100 of the length of the Fibre. The Fibre may be "infinitely" long, as found with continuous filament fibers, or as short as 0.5 inches (1.3 em), as found in staple fibers. Most natural fibers are staple fibers, whereas man-made fibers come in either staple or filament form depending on processing prior to yarn formation.

Fibre Uniformity

Fibers suitable for processing into yarns and fabrics must be fairly uniform in shape and size. Without sufficient uniformity of dimensions and properties in a given set of fibers to be twisted into yarn. the actual formation of the yarn may be impossible or the resulting yarn may be weak, rough, and irregular in size and shape and unsuitable for textile usage. Natural fibers must be sorted and graded to assure Fibre uniformity, whereas synthetic fibers may be "tailored" by cutting into appropriate uniform lengths to give a proper degree of Fibre uniformity.

Fibre Strength and Flexibility

A Fibre or yarn made from the Fibre must possess sufficient strength to be processed into a textile fabric or other textile article. Following fabrication into a textile article. the resulting textile must have sufficient strength to provide adequate durability during end-use. Many experts consider a single Fibre strength of 5 grams per denier to be necessary for a Fibre suitable in most textile applications, although certain fibers with strengths as low as 1.0 gram per denier have been found suitable for some applications. The strength of a single Fibre is called the tenacity, defined as the force per unit linear density necessary to break a known unit of that Fibre. The breaking tenacity of a Fibre may be expressed in grams per denier (g/d) or grams per tex (g/tex). Both denier and tex are units of linear density (mass per unit of Fibre length) and are defined as the number of grams of Fibre measuring 9000 meters and 1000 meters, respectively.

As a result, the denier of a Fibre or yarn will always be 9 times the tex of the same Fibre. Since tenacities of fibers or yarns are obtained by dividing the force by denier or tex, the tenacity of a Fibre in grams per denier will be 1/9 that of the Fibre tenacity in grams per tex. As a result of the adaption of the International System of Units. ref erred to as SI. the appropriate length un it for breaking tenac ity becomes kilometer (km) of breaking length or Newtons per tex (N/tex) and will be equivalent in value to g/tex. The strength of a Fibre. yarn, or fabric can be expressed in terms of force per unit area, and when expressed in this way the term is tensile strength. The most common unit used in the past for tensile strength has been pounds force per square inch or grams force per square centimeter. In 51 units, the pounds force per square inch x 6.895 will become kilopascals (kPa) and grams force per square centimeter x 9.807 will become megapascals (MPa). A Fibre must be sufficiently flexible to go through repeated bending without significant strength deterioration or breakage of the Fibre. Without adequate flexibility, it would be impossible to convert fibers into yarns and fabrics, since flexing and bending of the individual fibers is a necessary part of this conversion. In addition, individual fibers in a textile will be subjected to considerable bending and flexing during enduse.

Fibre Extensibility and Elasticity

An individual Fibre must be able to undergo slight extensions in length (less than 5per cent) without breakage of the Fibre. At the same time the Fibre must be able to almost completely recover following slight Fibre deformation.

In other words, the extension deformation of the Fibre must be nearly elastic. These properties are important because the individual fibers in textiles are often subjected to sudden stresses, and the textile must be able to give and recover without significant overall deformation of the textile.

Fibre Cohesiveness

Fibers must be capable of adhering to one another when spun into a yarn. The cohesiveness of the Fibre may be due to the shape and contour of the individual Fibre s or the nature of the surface of the fibers. In addition, long-filament fibers by virtue of their length can be twisted together to give stability without true cohesion between fibers. Often the term "spinning quality" is used to state the overall attractiveness of fibers for one another.

SECONDARY PROPERTIES

Moisture Absorption and Desorption

Most fibers tend to absorb moisture (water vapor) when in contact with the atmosphere. The amount of water absorbed by the textile Fibre will depend on the chemical and physical structure and properties of the Fibre, as well as the temperature and humidity of the surroundings. The percentage absorption of water vapor by a Fibre is often expressed as its moisture regain. The regain is determined by weighing a dry Fibre, then placing it in a room set to standard temperature and humidity (21° ± 1°C and 65per cent relative humidity are commonly used). From these measurements, the percentage moisture regain of the Fibre is determined:

$$\text{Percentage regain} = \frac{\text{Conditioned weight} - \text{Dry weight}}{\text{Dry weight}} \times 100\%$$

Percentage moisture content of a Fibre is the percentage of the total weight of the Fibre which is due to the moisture present, and is obtained from the following formula:

$$\text{Percentage moisture content} \quad \frac{\text{Conditioned weight} - \text{Dry weight}}{\text{Conditioned weight}} \times 100\%$$

The percentage moisture content will always be the smaller of the two values. Fibers vary greatly in their regain, with hydrophobic (water-repelling) fibers having regains near zero and hydrophilic (water-seeking) fibers like cotton, rayon, and wool having regains as high as 15per cent at 21°C and 65per cent RH. The ability of fibers to absorb high regains of water affects the basic properties of the Fibre in end-use. Absorbent fibers are able to absorb large amounts of water before they feel wet, an important factor where absorption of perspiration is necessary. Fibers with high regains will be easier to process, finish, and dye in aqueous solutions, but will dry more slowly. The low regain found for many man-made fibers makes them quick drying, a distinct advantage in certain applications. Fibers with high regains are often desirable because they provide a "breathable" fabric which can conduct moisture from the body to the outside atmosphere readily, due to their favorable moisture absorption-desorption properties. The tensile properties of fibers as well as their dimensional properties are known to be affected by moisture.

Fibre Resiliency and Abrasion Resistance

The ability of a Fibre to absorb shock and recover from deformation and to be generally resistant to abrasion forces is important to its end-use and wear characteristics. In consumer use, fibers in fabrics are often placed under stress through compression, bending, and twisting (torsion) forces under a variety of temperature and humidity conditions. If the fibers within the fabric possess good elastic recovery properties from such deformative actions, the Fibre has good resiliency and better overall appearance in end-use. For example, cotton and wool show poor wrinkle recovery under hot moist conditions, whereas polyester exhibits good recovery from deformation as a result of its high resiliency.

Resistance of a Fibre to damage when mobile forces or stresses come in contact with Fibre structures is referred to as abrasion resistance. If a Fibre is able to effectively absorb and dissipate these forces without damage, the Fibre will show good abrasion resistance. The toughness and hardness of the Fibre is related to its chemical and physical structure and morphology of the Fibre and will influence the abrasion of the Fibre. A rigid, brittle Fibre such as glass, which is unable to dissipate the forces of abrasive action, results in Fibre damage and breakage, whereas a tough but more plastic Fibre such as polyester shows better resistance to abrasion forces. Finishes can affect Fibre properties including resiliency and abrasion resistance.

Luster

Luster refers to the degree of light that is reflected from the surface of *a* Fibre or the degree of gloss or sheen that the Fibre possesses. The inherent chemical and physical structure and shape of the Fibre *can* affect the relative luster of the Fibre. With natural fibers the luster of the Fibre is dependent on the morphological form that nature gives the Fibre, although the relative luster can be changed by chemical and/or physical treatment of the Fibre as found in processes such *as* mercerization of cotton. Man-made fibers can vary in luster from bright to dull depending on the amount of delusterant added to the Fibre. Oelusterants such *as* titanium dioxide tend to scatter and absorb 1ight, thereby making the Fibre appear duller. The desirability of luster for *a* given Fibre application will vary and is often dependent on the intended end-use of the Fibre in *a* fabric or garment form and on current fashion trends.

Resistance to Chemicals in the Environment

A textile Fibre to be useful must have reasonable resistance to chemicals it comes in contact with in its environment during use and maintenance. It should have resistance to oxidation by oxygen and other gases in the air, particularly in the presence of light, and be resistant to attack by microorganisms and other biological agents. Many fibers undergo light-induced reactions, and fibers from natural sources are susceptible to biological attack, but such deficiencies can be minimized by treatment with appropriate finishes. Textile fibers come in contact with a large range of chemical agents on laundering and dry cleaning and must be resistant from attack under such conditions.

Oensity

The density of a Fibre is related to its inherent chemical structure and the packing of the molecular chains within that structure. The density of a Fibre will have a noticeable effect on its aesthetic appeal and its usefulness in given applications. For example, glass and silk fabrics of the same denier would have noticeable differences in weight due to their broad differences in density. Fishnets of polypropylene fibers are of great utility because their density is less than that of water.

Oensities are usually expressed in units of grams per cubic centimeter, but in 51 units will be expressed as kilograms per cubic meter, which gives a value 1000 times larger.

Thermal and Flammability Characteristics

Fibers used in textiles must be resistant to wet and dry heat, must not ignite readily when coming in contact with a flame, and ideally should self-extinguish when the flame is removed. Heat stability is particularly important to a Fibre during dyeing and finishing of the textile and during cleaning and general maintenance by the consumer. Textile fibers for the most part are made up of organic polymeric materials containing carbon and burn on ignition from a flame or other propagating source. The chemical structure of a Fibre establishes its overall flammability characteristics, and appropriate textile finishes can reduce the degree of flammability. A number of Federal, state, and local statutes eliminate the most dangerous flammable fabrics from the marketplace.

PRIMARY FIBRE PROPERTIES FROM AN ENGINEERING PERSPECTIVE

The textile and polymer engineer must consider a number of criteria essential for formation, fabrication, and assembly of fibers into textile substrates. Often the criteria used will be similar to those set forth above concerning end-use properties. Ideally a textile Fibre should have the following properties:

- A melting and/or decomposition point above 220°C.
- A tensile strength of 5 g/denier or greater.
- Elongation at break above 10per cent with reversible elongation up to 5per cent strain.
- A moisture absorptivity of 2per cent-5per cent moisture uptake.
- Combined moisture regain and air entrapment capability.
- High abrasion resistance.
- Resistance to attack, swelling, or solution in solvents, acids, and bases.
- Self-extinguishing when removed from a flame.

FIBRE FORMATION AND MORPHOLOGY

Fibre morphology refers to the form and structure of a Fibre, including the molecular arrangement of individual molecules and groups of molecules within the Fibre. Most fibers are organic materials derived from carbon combined with other atoms such as oxygen, nitrogen, and halogens. The basic building blocks that organic materials form as covalently-bonded organic compounds are called monomers. Covalent bonds involve the sharing of electrons between adjacent atoms within the monomer, and the structure trons between adjacent atoms within the monomer, and the structure of the monomer is determined by the type, location, and nature of bonding of atoms within the monomer and by the nature of covalent bonding between atoms. Monomers react or condense to form long-chain molecules called polymers made up of a given number (n) of monomer units which are the basic building unit of fibers. On formation into fibers and orientation by natural or mechanical means the polymeric molecules possess ordered crystalline and Non-ordered amorphous areas, depending on the nature of the polymer and the relative packing of molecules within the Fibre. For a monomer A the sequence of events to Fibre formation and orientation would appear as shown in Figure.

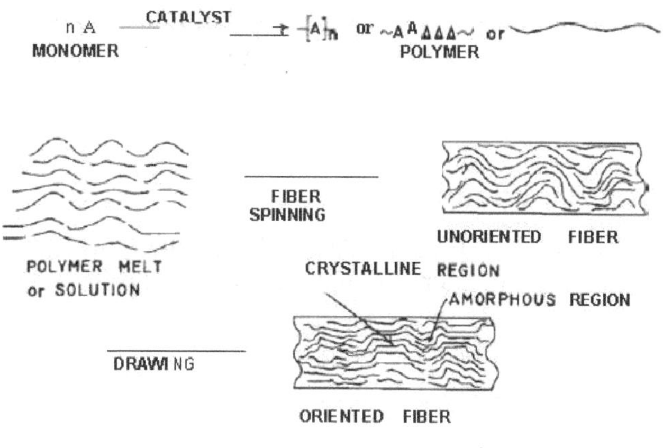

Fig. Polymerization sequence and Fibre formation

Polymers with repeating units of the same monomer (An) would be referred to as homopolymers. If a second unit B is introduced into the basic structure, structures which are copolymers are formed with structures as outlined in Figure.

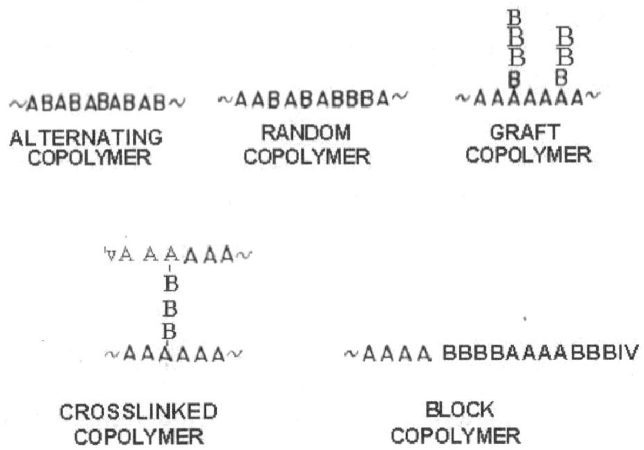

Fig. Copolymer structures.

POLYMER FORMATION

Synthetic polymers used to form fibers are often classified on the basis of their mechanism of polymerization—step growth (condensation) or chain growth (addition) polymerization. Step growth polymerization involves multifunctional monomers which undergo successive condensation with a second monomer or with itself to form a dimer, which in turn condenses with another dimer to form a tetramer, etc., usually with loss of a small molecule such as water.

Chain growth involves the instantaneous growth of a long molecular chain from unsaturated monomer units, followed by initiation of a second chain, etc. The two methods are outlined below schematically:

$$\text{Step growth: } nA \rightarrow \frac{n}{2} AA \rightarrow \frac{n}{4} AAAA \rightarrow$$

$$\text{Chain growth: } nA \rightarrow (A)n \quad nA \rightarrow (A)n \quad nA \rightarrow (A)n$$

The average number of monomer repeating units in a polymer chain (n) is often referred to also as the degree of polymerization, DP. The DP must exceed an average 20 units in most cases to give a polymer of s ufficient molecular size to have desirable Fibre-forming properties. The overall breadth of distribution of molecular chain lengths in the polymer will affect the ultimate properties of the fibers, with wide polymer size distributions leading to an overall reduction of Fibre properties. Although the polymers from natural fibers and regenerated natural fibers do not undergo polymerization by the mechanisms found for synthetic fibers, most natural polymers have characteristic repeating units and high degrees of polymerization and are related to step growth polymers.

FIBRE SPINNING

Although natural fibers come in a morphological form determined by nature, regenerated and synthetic man-made fibers can be "tailor-made" depending on the shape and dimensions of the orifice (spinning jet) that the polymer is forced through to form the Fibre. There are several methods used to spin a Fibre from its polymer, including melt, dry, wet, emulsion, and suspension spinning. Melt spinning is the least complex of the methods. The polymer from which the Fibre is made is melted and then forced through a spinneret and into air to cause solidification and Fibre formation.

Dry and wet spinning processes involve dissolving the Fibre-forming polymer in an appropriate solvent, followed by passing a concentrated solution (20per cent-50per cent polymer) through the spinneret and into dry air to evaporate the solvent in the case of dry spinning or into a coagulating bath to cause precipitation or regeneration of the polymer in Fibre form in the case of wet spinning. There is a net contraction of the spinning solution on loss of solvent. If a skin of polymer is formed on the Fibre followed by diffusion of the remainder of the solvent from the core of the forming Fibre, the cross section of the Fibre as it contracts may collapse to form an irregular popcornlike cross section. Emulsion spinning is used only for those Fibre-forming polymers that are insoluble. Polymer is mixed with a surface-active agent (detergent) and possibly a solvent and then mixed at high speeds with water to form an emulsion of the polymer. The polymer is passed through the spinneret and into a coagulating bath to form the Fibre. In suspension spinning, the polymer is swollen and suspended in a swelling solvent. The swollen suspended polymer is forced through the spinneret into dry hot air to drive off solvent or into a wet non-solvent bath to cause the Fibre to form through coagulation. The spinning process can be divided into three steps:

- flow of spinning fluid within and through the spinneret under high stress and sheer.
- exit of fluid from the spinneret with relief of stress and an increase in volume (ballooning of flow).
- elongation of the fluid jet as it is subjected to tensile force as it cools and solidifies with orientation of molecular structure within the Fibre.

Common cross sections of man-made fibers include round, trilobal, pentalobal, dog-bone, and crescent shapes. When two polymers are used in Fibre formation as in bicomponent or biconstituent fibers, the two components can be arranged in a matrix, side-by-side, or sheath-core configuration. Round cross sections are also found where skin formation has caused Fibre contraction and puckering (as with rayons) has occurred or where the spinneret shape has provided a hollow Fibre. Complex Fibre cross-sectional shapes with special properties are also used.

FIBRE DRAWING AND MORPHOLOGY

On drawing and orientation the man-made fibers become smaller in diameter and more crystalline, and imperfections in the Fibre morphology are improved somewhat. Side-by-side bi component or bi constituent fibers on drawing become wavy and bulky. In natural fibers the orientation of the molecules within the Fibre is determined by the biological source during the growth and maturity process of the Fibre. The form and structure of polymer molecules with relation to each other within the Fibre will depend on the relative alignment of the molecules in relationship to one another. Those areas where the polymer chains are closely aligned and packed close together are crystalline areas within the Fibre, whereas those areas where there is essentially no molecular alignment are referred to as amorphous areas. Dyes and finishes can penetrate the amorphous portion of the Fibre, but not the ordered crystalline portion. A number of theories exist concerning the arrangement of crystalline and amorphous areas within a Fibre. Individual crystal 1ine areas in a Fibre are often referred to as microfibrils. Microfibrils can associate into larger crystalline groups, which are called fibrils or micelles. Microfibrils are 30-100 A (lO-le meters) in length, whereas fibrils and micelles are usually 200-600 A in length. This compares to the individual molecular chains, which vary from 300 to 1500 A in length and which are usually part of both crystal Iine and amorphous areas of the polymer and therefore give continuity and association of the various crystaline and amorphous areas within the Fibre. A number of theories have been developed to explain the interconnection of crystal Iine and amorphous areas in the Fibre and include such concepts as fringed micelles or fringed fibrils, molecular chain foldings, and extended chain concepts. The amorphous areas within a Fibre will be relatively loosely packed and associated with each other, and spaces or voids will appear due to discontinuities within the structure. Figure outlines the various aspects of internal Fibre morphology with regard to polymer chains.

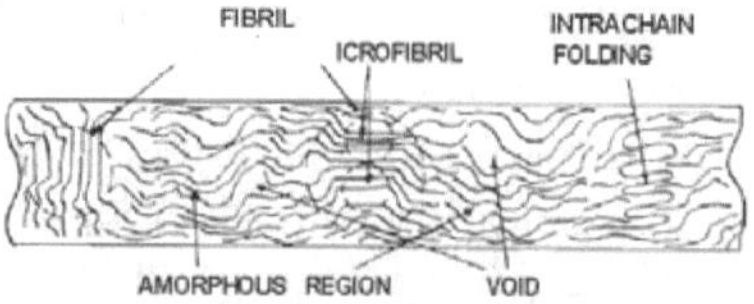

Fig. Aspects of internal Fibre morphology.

4

Cotton Fibre

Cotton is a soft, fluffy staple Fibre that grows in a boll, or protective capsule, around the seeds of cotton plants of the genus Gossypium in the family of Malvaceae. The Fibre is almost pure cellulose. Under natural conditions, the cotton bolls will tend to increase the dispersion of the seeds.

The plant is a shrub native to tropical and subtropical regions around the world, including the Americas, Africa, and India. The greatest diversity of wild cotton species is found in Mexico, followed by Australia and Africa. Cotton was independently domesticated in the Old and New Worlds. The English name derives from the Arabic (al) qum n which began to be used circa 1400 AD.

The Fibre is most often spun into yarn or thread and used to make a soft, breathable textile. The use of cotton for fabric is known to date to prehistoric times; fragments of cotton fabric dated from 5000 BC have been excavated in Mexico and the Indus Valley Civilization (modern-day Pakistan and some parts of India). Although cultivated since antiquity, it was the invention of the cotton gin that lowered the cost of production that led to its widespread use, and it is the most widely used natural Fibre cloth in clothing today.

Current estimates for world production are about 25 million tonnes or 110 million bales annually, accounting for 2.5per cent of the world's arable land. China is the world's largest producer of cotton, but most of this is used domestically. The United States has been the largest exporter for many years. In the United States, cotton is usually measured in bales, which measure approximately 0.48 cubic metres (17 cubic feet) and weigh 226.8 kilograms (500 pounds).

TYPES

There are four commercially grown species of cotton, all domesticated in antiquity:

- Gossypium hirsutum – upland cotton, native to Central America, Mexico, the Caribbean and southern Florida, (90per cent of world production)
- Gossypium barbadense – known as extra-long staple cotton, native to tropical South America (8per cent of world production)

- Gossypium arboreum – tree cotton, native to India and Pakistan (less than 2per cent)
- Gossypium herbaceum – Levant cotton, native to southern Africa and the Arabian Peninsula (less than 2per cent)

The two New World cotton species account for the vast majority of modern cotton production, but the two Old World species were widely used before the 1900s. While cotton fibers occur naturally in colors of white, brown, pink and green, fears of contaminating the genetics of white cotton have led many cotton-growing locations to ban the growing of colored cotton varieties, which remain a specialty product.

GOSSYPIUM HIRSUTUM

Gossypium hirsutum, also known as upland cotton or Mexican cotton, is the most widely planted species of cotton in the United States, constituting some 95per cent of all cotton production there. It is native to Mexico, the West Indies, northern South America, Central America and possibly tropical Florida. Worldwide, the figure is about 90per cent of all cotton production is of cultivars derived from this species.

Archeological evidence from the Tehuacan Valley in Mexico shows the cultivation of this species as long ago as 3,500 BC, although there is as yet no evidence as to exactly where it may have been first domesticated. This is the earliest evidence of cotton cultivation in the Americas found thus far.

Gossypium hirsutum includes a number of varieties or cross-bred cultivars with varying Fibre lengths and tolerances to a number of growing conditions. The longer length varieties are called "long staple upland" and the shorter length varieties are referred to as "short staple upland". The long staple varieties are the most widely cultivated in commercial production. Besides being fibre crops, Gossypium hirsutum and Gossypium herbaceum are the main species used to produce cottonseed oil. The Zuni people use this plant to make ceremonial garments, and the fuzz is made into cords and used ceremonially.

GOSSYPIUM BARBADENSE

Gossypium barbadense, also known as extra long staple (ELS) cotton as it generally has a staple of at least 1 3/8" or longer, is a species of cotton plant. Varieties of ELS cotton include American Pima, Egyptian Giza, Indian Suvin, Chinese Xinjiang, Sudanese Barakat, and Russian Tonkovoloknistyi. It is a tropical, frost-sensitive perennial plant that produces yellow flowers and has black seeds. It grows as a small, bushy tree and yields cotton with unusually long, silky fibers. To grow, it requires full sun and high humidity and rainfall.

This plant contains the chemical gossypol, which reduces its susceptibility to insect and fungal damage. In Suriname's traditional medicine, the leaves of G. barbadense are used to treat hypertension and delayed/irregular menstruation.

The name Pima was applied in Honour of the Pima Indians, who helped raise the cotton on USDA experimental farms in Arizona in the early 1900s. The first clear sign of domestication of this cotton species comes the Early Valdivia phase site of Real Alto on the coast of Ecuador (4400 BC) and from Ancon, a site on the Peruvian coast, where cotton bolls dating to 4200 BC were found. By 1000 BC, Peruvian cotton bolls were indistinguishable from modern cultivars of G. barbadense. Cotton growing became widespread in South America and spread to the West Indies, where Christopher Columbus encountered it. Cotton became a commercial plantation crop tended by slaves in the West Indies, so that by the 1650s, Barbados had become the first British West Indies colony to export cotton.In about 1786, planting of Sea Island cotton began in the former British North American colonies, on the Sea Islands of South Carolina and Georgia, when cotton planters were brought over from Barbados. Among the earliest planters of Sea Island cotton in America was an Englishman, Francis Levett, who later fled his Georgia plantation at the outbreak of the American Revolution and went to the Bahamas, where he attempted to introduce cotton production, but failed. Sea Island cotton commanded the highest price of all the cottons, due to its long staple (1.5 to 2.5 inches, 35 to 60 mm) and its silky texture; it was used for the finest cotton counts and often mixed with silk. It was also grown on the uplands of Georgia, where the quality was inferior, and was soon surpassed in commercial production by another native American species, upland cotton (Gossypium hirsutum), which today represents about 95per cent of U.S. production.The term Egyptian cotton is usually applied to the extra long staple cotton produced in Egypt and used by luxury and upmarket brands worldwide. It also has the most upper thread count.

American Pima accounts for less than 5per cent of U.S. cotton production. It is grown chiefly in California, with small acreages in West Texas, New Mexico and Arizona.

For purposes of federal support, the 2002 farm bill (P.L. 101-171, Sec. 1001) defined ELS cotton.

ELS cotton, like upland cotton, is eligible for marketing assistance loans and loan deficiency payments (LDPs). The national loan rate for ELS cotton under the 2002 farm bill was $0.7977 per pound. ELS cotton, in contrast to upland cotton, does not qualify for direct payments or counter-cyclical payments.

GOSSYPIUM ARBOREUM

Gossypium arboreum, commonly called tree cotton, is a species of cotton native to India, Pakistan and other tropical and subtropical regions of the Old World. There is evidence of its cultivation as long ago as the Harappan civilization of the Indus Valley for the production of cotton textiles. This species of cotton was also introduced into East Africa and was grown by the Meroe civilization in Nubia. The shrub was included in Linnaeus's Species Plantarum published in 1753. The holotype was also supplied by him, which is now in the Linnean Herbarium in the Swedish Museum of Natural History. Tree cotton is a shrub attaining heights of one to two metres. Its branches are covered with pubescence and are purple in colour. Stipules are present at the leaf base and they are linear to lanceolate in shape and sometimes falcate (*i.e.* sickle-shaped). The leaves are attached to the stem by a 1.5 to 10 cm petiole. The blades are ovate to orbicular in shape and have five to seven lobes, making them superficially resemble a maple leaf. The lobes are linear to lanceolate, and often a tooth is present in the sinus. Glands are present along the midrib or occasionally on the adjacent nerves. The leaves are glabrescent, meaning the pubescence is lost with age, but when it is present on young leaves, it is both stellate (*i.e.* star-shaped) and simple.

The flowers are set on short pedicels (*i.e.* flower stalks). An epicalyx is present, which is a series of subtending bracts that resemble sepals. Its large, ovate segments are dentate (*i.e.* toothed along the margins), though sometimes only very slightly so. They are cordate (*i.e.* heart-shaped) at the base and acute at the apex. The true calyx is small, measuring only about 5 mm in length. Its shape is cupular, and five subtle dentations are present. The corolla is a pale yellow on colour, sometimes with a purple centre, and occasionally entirely purple. It measures 3 to 4 cm in length. The staminal tube bears the anthers and is 1.5 to 2 cm in length. The fruit is a three- or four-celled capsule measuring 1.5 to 2.5 cm across. It is ovoid or oblong in shape and glabrous (*i.e.* hairless). The surface is pitted and a beak is present at the terminal end. The seeds within are globular and are covered in long white cotton.

GOSSYPIUM HERBACEUM

Gossypium herbaceum, commonly known as Levant cotton, is a species of cotton native to the semi-arid regions of sub-Saharan Africa and Arabia where it still grows in the wild as a perennial shrub. It is a sister-species of Gossypium arboreum.

It was first cultivated in Western Sudan, from there it spread to India, before being introduced to Egypt. It reached China in the 700 AD and was first cultivated from this period. It can also be found in Pakistan and almost all of the former Soviet Union areas.

A legend was perpetuated from a factual description of this plant by Greek historian Herodotus in the 5th century BC. Although his book, simply titled Histories, was an account of a war between the Persian Empire and the Greek city-states. It also contained descriptions of vast lands beyond the boundaries of the world known by the Greeks at that time. He wrote: "certain trees...bear forth their fruit fleeces surpassing those of sheep in beauty and excellence, and the natives clothe themselves in cloths made therefrom." From this description came the legend of the "vegetable lamb plant" which was said to be a real sheep. The tree would grow from a melon-like seed and grow into a lamb rooted to the earth by a stem from its navel. It was said to graze on the surrounding vegetation until the all greenery around it was devoured at which point it would wither and die. A 14th-century traveler by the name of Sir John Mandeville, professed to eating the flesh of this herbal beast. Although scientist tried to debunk this tale it was not officially labeled as a fable until 1887.

G. herbaceum has high stems that grow 2 feet (0.61 m) to 6 feet (1.8 m) high with wide, hairy leaves. Their flowers are small and yellow with a purple center. When ripe and in warm weather, the flower capsule will burst and expose the cotton surrounding the seeds firmly. The cotton produced by this plant is short, about 2 inches (5.1 cm) long and is firmly attached to the seed, which is covered in hairy down. An acre of cotton can be expected to produce about 300 pounds (140 kg).

Cotton is usually used as a textile while making clothing and can be made into yarns and sheets of fabric. In the Levant seeds are often used for as food. It is utilized so often because of its comfortable, breathable properties. It has been cultivated for women's menstrual cycle pains and irregular bleeding, and it also has been used after birth to expel placenta afterbirth and to increase lactation. Cotton has been used for gastrointestinal issues also, such as hemorrhages, nausea, and diarrhea, as well as fevers and headaches, especially in the southern United States. Levant cotton seed extract, gossypol, also has a potential use as a male contraceptive but can cause infertility after discontinuing. In lab rat studies, it has been able to stop pregnancies early. Orally administered ethyl ether and ethanol extracts of Gossypium herbaceum significantly decreased the blood glucose level. Gossypium herbaceum is not only lowered TC, TG, LDL, VLDL levels but also increased level of cardioprotective lipid HDL Therefore, Gossypium herbaceum has potential role to prevent formation of atherosclerosis and coronary heart disease. The study demonstrated that both above given extracts of Gossypium herbaceum could be useful in management of diabetes associated with abnormalities in lipid profiles.

HISTORY

Cotton was used in the Old World at least 7,000 years ago (5th millennium BC). Evidence of cotton use has been found at the site of Mehrgarh, where early cotton threads have been preserved in copper beads. Cotton cultivation became more widespread during the Indus Valley Civilization, which covered parts of modern eastern Pakistan and northwestern India. The Indus cotton industry was well developed and some methods used in cotton spinning and fabrication continued to be used until the industrialization of India. Between 2000 and 1000 BC cotton became widespread across much of India. For example, it has been found at the site of Hallus in Karnataka dating from around 1000 BC.

Cotton fabrics discovered in a cave near Tehuacán, Mexico have been dated to around 5800 BC, although it is difficult to know for certain due to Fibre decay. Other sources date the domestication of cotton in Mexico to approximately 5000 to 3000 BC.

The Greeks and the Arabs were not familiar with cotton until the Wars of Alexander the Great, as his contemporary Megasthenes told Seleucus I Nicator of "there being trees on which wool grows" in "Indica". This might actually be a reference to the 'tree cotton', Gossypium arboreum, which is a native of the Indian subcontinent.

According to the Columbia Encyclopaedia:

- Cotton has been spun, woven, and dyed since prehistoric times. It clothed the people of ancient India, Egypt, and China. Hundreds of years before the Christian era, cotton textiles were woven in India with matchless skill, and their use spread to the Mediterranean countries.

In Iran (Persia), the history of cotton dates back to the Achaemenid era (5th century BC); however, there are few sources about the planting of cotton in pre-Islamic Iran. The planting of cotton was common in Merv, Ray and Pars of Iran. In the poems of Persian poets, especially Ferdowsi's Shahname, there are references to cotton ("panbe" in Persian). Marco Polo (13th century) refers to the major products of Persia, including cotton. John Chardin, a French traveler of 17th century, who had visited the Safavid Persia, has approved the vast cotton farms of Persia.

During the Han dynasty, cotton was grown by Non- Chinese peoples in the southern Chinese province of Yunnan.

In Peru, cultivation of the indigenous cotton species Gossypium barbadense was the backbone of the development of coastal cultures, such as the Norte Chico, Moche and Nazca. Cotton was grown upriver, made into nets and traded with fishing villages along the coast for large supplies of fish. The Spanish who came to Mexico and Peru in the early 16th century found the people growing cotton and wearing clothing made of it.

During the late medieval period, cotton became known as an imported Fibre in northern Europe, without any knowledge of how it was derived, other than that it was a plant. Because Herodotus had written in his Histories, Book III, 106, that in India trees grew in the wild producing wool, it was assumed that the plant was a tree, rather than a shrub. This aspect is retained in the name for cotton in several Germanic languages, such as German Baumwolle, which translates as "tree wool" (Baum means "tree"; Wolle means "wool"). Noting its similarities to wool, people in the region could only imagine that cotton must be produced by plant-borne sheep. John Mandeville, writing in 1350, stated as fact the now-preposterous belief: "There grew there [India] a wonderful tree which bore tiny lambs on the endes of its branches. These branches were so pliable that they bent down to allow the lambs to feed when they are hungrie. By the end of the 16th century, cotton was cultivated throughout the warmer regions in Asia and the Americas.

India's cotton-processing sector gradually declined during British expansion in India and the establishment of colonial rule during the late 18th and early 19th centuries. This was largely due to aggressive colonialist mercantile policies of the British East India Company, which made cotton processing and manufacturing workshops in India uncompetitive. Indian markets were increasingly forced to supply only raw cotton and were forced, by British-imposed law, to purchase manufactured textiles from Britain.

INDUSTRIAL REVOLUTION IN BRITAIN

The advent of the Industrial Revolution in Britain provided a great boost to cotton manufacture, as textiles emerged as Britain's leading export. In 1738, Lewis Paul and John Wyatt, of Birmingham, England, patented the roller spinning machine, and the flyer-and-bobbin system for drawing cotton to a more even thickness using two sets of rollers that traveled at different speeds. Later, the invention of the James Hargreaves' spinning jenny in 1764, Richard Arkwright's spinning frame in 1769 and Samuel Crompton's spinning mule in 1775 enabled British spinners to produce cotton yarn at much higher rates. From the late 18th century on, the British city of Manchester acquired the nickname "Cottonopolis" due to the cotton industry's omnipresence within the city, and Manchester's role as the heart of the global cotton trade.

Production capacity in Britain and the United States was improved by the invention of the cotton gin by the American Eli Whitney in 1793. Before the development of cotton gins, the cotton fibers had to be pulled from the seeds tediously by hand. By the late 1700s a number of crude ginning machines had been developed. However, to produce a bale of cotton required over 600 hours of human Labour, making large scale production uneconomical in the United States, even with the use of humans as slave Labour. The gin that Whitney manufactured (the Holmes design) reduced the hours down to just a dozen or so per bale. Although Whitney patented his own design for a cotton gin, he manufactured a prior design from Henry Odgen Holmes, for which Holmes filed a patent in 1796. Improving technology and increasing control of world markets allowed British traders to develop a commercial chain in which raw cotton fibers were (at first) purchased from colonial plantations, processed into cotton cloth in the mills of Lancashire, and then exported on British ships to captive colonial markets in West Africa, India, and China (via Shanghai and Hong Kong).

By the 1840s, India was no longer capable of supplying the vast quantities of cotton fibers needed by mechanized British factories, while shipping bulky, low-price cotton from India to Britain was time-consuming and expensive. This, coupled with the emergence of American cotton as a superior type (due to the longer, stronger fibers of the two domesticated native American species, Gossypium hirsutum and Gossypium barbadense), encouraged British traders to purchase cotton from plantations in the United States and plantations in the Caribbean. By the mid-19th century, "King Cotton" had become the backbone of the southern American economy. In the United States, cultivating and harvesting cotton became the leading occupation of slaves.

During the American Civil War, American cotton exports slumped due to a Union blockade on Southern ports, and also because of a strategic decision by the Confederate government to cut exports, hoping to force Britain to recognize the Confederacy or enter the war. This prompted the main purchasers of cotton, Britain and France, to turn to Egyptian cotton. British and French traders invested heavily in cotton plantations.

The Egyptian government of Viceroy Isma'il took out substantial loans from European bankers and stock exchanges. After the American Civil War ended in 1865, British and French traders abandoned Egyptian cotton and returned to cheap American exports, sending Egypt into a deficit spiral that led to the country declaring bankruptcy in 1876, a key factor behind Egypt's occupation by the British Empire in 1882.

During this time, cotton cultivation in the British Empire, especially India, greatly increased to replace the lost production of the American South. Through tariffs and other restrictions, the British government discouraged the production of cotton cloth in India; rather, the raw Fibre was sent to England for processing. The Indian Mahatma Gandhi described the process:

1. English people buy Indian cotton in the field, picked by Indian Labour at seven cents a day, through an optional monopoly.
2. This cotton is shipped on British ships, a three-week journey across the Indian Ocean, down the Red Sea, across the Mediterranean, through Gibraltar, across the Bay of Biscay and the Atlantic Ocean to London. One hundred per cent profit on this freight is regarded as small.
3. The cotton is turned into cloth in Lancashire. You pay shilling wages instead of Indian pennies to your workers. The English worker not only has the advantage of better wages, but the steel companies of England get the profit of building the factories and machines. Wages; profits; all these are spent in England.
4. The finished product is sent back to India at European shipping rates, once again on British ships. The captains, officers, sailors of these ships, whose wages must be paid, are English. The only Indians who profit are a few lascars who do the dirty work on the boats for a few cents a day.

5. The cloth is finally sold back to the kings and landlords of India who got the money to buy this expensive cloth out of the poor peasants of India who worked at seven cents a day. (Fisher 1932 pp 154–156)

In the United States, Southern cotton provided capital for the continuing development of the North. The cotton produced by enslaved African Americans not only helped the South, but also enriched Northern merchants. Much of the Southern cotton was trans-shipped through northern ports.

Cotton remained a key crop in the Southern economy after emancipation and the end of the Civil War in 1865. Across the South, sharecropping evolved, in which free black farmers and landless white farmers worked on white-owned cotton plantations of the wealthy in return for a share of the profits. Cotton plantations required vast Labour forces to hand-pick cotton.

It was not until the 1950s that reliable harvesting machinery was introduced (prior to this, cotton-harvesting machinery had been too clumsy to pick cotton without shredding the fibers). During the first half of the 20th century, employment in the cotton industry fell, as machines began to replace laborers and the South's rural Labour force dwindled during the World Wars.

Cotton remains a major export of the southern United States, and a majority of the world's annual cotton crop is of the long-staple American variety.

CULTIVATION

Successful cultivation of cotton requires a long frost-free period, plenty of sunshine, and a moderate rainfall, usually from 600 to 1,200 mm (24 to 47 in). Soils usually need to be fairly heavy, although the level of nutrients does not need to be exceptional. In general, these conditions are met within the seasonally dry tropics and subtropics in the Northern and Southern hemispheres, but a large proportion of the cotton grown today is cultivated in areas with less rainfall that obtain the water from irrigation. Production of the crop for a given year usually starts soon after harvesting the preceding autumn. Planting time in spring in the Northern hemisphere varies from the beginning of February to the beginning of June. The area of the United States known as the South Plains is the largest contiguous cotton-growing region in the world. While dryland (non-irrigated) cotton is successfully grown in this region, consistent yields are only produced with heavy reliance on irrigation water drawn from the Ogallala Aquifer. Since cotton is somewhat salt and drought tolerant, this makes it an attractive crop for arid and semiarid regions. As water resources get tighter around the world, economies that rely on it face difficulties and conflict, as well as potential environmental problems. For example, improper cropping and irrigation practices have led to desertification in areas of Uzbekistan, where cotton is a major export. In the days of the Soviet Union, the Aral Sea was tapped for agricultural irrigation, largely of cotton, and now salination is widespread.

Cotton can also be cultivated to have colors other than the yellowish off-white typical of modern commercial cotton fibers. Naturally colored cotton can come in red, green, and several shades of brown.

GENETIC MODIFICATION

Genetically modified (GM) cotton was developed to reduce the heavy reliance on pesticides. The bacterium Bacillus thuringiensis (Bt) naturally produces a chemical harmful only to a small fraction of insects, most notably the larvae of moths and butterflies, beetles, and flies, and harmless to other forms of life. The gene coding for Bt toxin has been inserted into cotton, causing cotton, called Bt cotton, to produce this natural insecticide in its tissues. In many regions, the main pests in commercial cotton are lepidopteran larvae, which are killed by the Bt protein in the transgenic cotton they eat. This eliminates the need to use large amounts of broad-spectrum insecticides to kill lepidopteran pests (some of which have developed pyrethroid resistance). This spares natural insect predators in the farm ecology and further contributes to Non-insecticide pest management.

But cotton is ineffective against many cotton pests, however, such as plant bugs, stink bugs, and aphids; depending on circumstances it may still be desirable to use insecticides against these. A 2006 study done by Cornell researchers, the Center for Chinese Agricultural Policy and the Chinese Academy of Science on Bt cotton farming in China found that after seven years these secondary pests that were normally controlled by pesticide had increased, necessitating the use of pesticides at similar levels to non-Bt cotton and causing less profit for farmers because of the extra expense of GM seeds. However, a 2009 study by the Chinese Academy of Sciences, Stanford University and Rutgers University refuted this. They concluded that the GM cotton effectively controlled bollworm. The secondary pests were mostly miridae (plant bugs) whose increase was related to local temperature and rainfall and only continued to increase in half the villages studied. Moreover, the increase in insecticide use for the control of these secondary insects was far smaller than the reduction in total insecticide use due to Bt cotton adoption. A 2012 Chinese study concluded that Bt cotton halved the use of pesticides and doubled the level of ladybirds, lacewings and spiders. The International Service for the Acquisition of Agri-biotech Applications (ISAAA) said that, worldwide, GM cotton was planted on an area of 25 million hectares in 2011. This was 69per cent of the worldwide total area planted in cotton.

GM cotton acreage in India grew at a rapid rate, increasing from 50,000 hectares in 2002 to 10.6 million hectares in 2011. The total cotton area in India was 12.1 million hectares in 2011, so GM cotton was grown on 88per cent of the cotton area. This made India the country with the largest area of GM cotton in the world. A long-term study on the economic impacts of Bt cotton in India, published in the Journal PNAS in 2012, showed that Bt cotton has increased yields, profits, and living standards of smallholder farmers. The U.S. GM cotton crop was 4.0 million hectares in 2011 the second largest area in the world, the Chinese GM cotton crop was third largest by area with 3.9 million hectares and Pakistan had the fourth largest GM cotton crop area of 2.6 million hectares in 2011. The initial introduction of GM cotton proved to be a success in Australia – the yields were equivalent to the Non- transgenic varieties and the crop used much less pesticide to produce (85per cent reduction). The subsequent introduction of a second variety of GM cotton led to increases in GM cotton production until 95per cent of the Australian cotton crop was GM in 2009 making Australia the country with the fifth largest GM cotton crop in the world. Other GM cotton growing countries in 2011 were Argentina, Myanmar, Burkina Faso, Brazil, Mexico, Colombia, South Africa and Costa Rica.

Cotton has been genetically modified for resistance to glyphosate a broad-spectrum herbicide discovered by Monsanto which also sells some of the Bt cotton seeds to farmers. There are also a number of other cotton seed companies selling GM cotton around the world. About 62per cent of the GM cotton grown from 1996 to 2011 was insect resistant, 24per cent stacked product and 14per cent herbicide resistant.

Cotton has gossypol, a toxin that makes it inedible. However, scientists have silenced the gene that produces the toxin, making it a potential food crop.

ORGANIC PRODUCTION

Organic cotton is generally understood as cotton, from plants not genetically modified, that is certified to be grown without the use of any synthetic agricultural chemicals, such as fertilizers or pesticides. Its production also promotes and enhances biodiversity and biological cycles. United States cotton plantations are required to enforce the National Organic Programme (NOP). This institution determines the allowed practices for pest control, growing, fertilizing, and handling of organic crops. As of 2007, 265,517 bales of organic cotton were produced in 24 countries, and worldwide production was growing at a rate of more than 50per cent per year.

PESTS AND WEEDS

The cotton industry relies heavily on chemicals, such as herbicides, fertilizers and insecticides, although a very small number of farmers are moving Towards an organic model of production, and organic cotton products are now available for purchase at limited locations. These are popular for baby clothes and diapers. Under most definitions, organic products do not use genetic engineering. All natural cotton products are known to be both sustainable and hypoallergenic. Historically, in North America, one of the most economically destructive pests in cotton production has been the boll weevil. Due to the US

Department of Agriculture's highly successful Boll Weevil Eradication Programme (BWEP), this pest has been eliminated from cotton in most of the United States. This Programme, along with the introduction of genetically engineered Bt cotton (which contains a bacterial gene that codes for a plant-produced protein that is toxic to a number of pests such as cotton bollworm and pink bollworm), has allowed a reduction in the use of synthetic insecticides.

Other significant global pests of cotton include the pink bollworm, Pectinophora gossypiella; the chili thrips, Scirtothrips dorsalis; the cotton seed bug, Oxycarenus hyalinipennis; the tarnish plant bug, Lygus lineolaris; and the fall armyworm, Spodoptera frugiperda, Xanthomonas citri subsp. malvacearum.

HARVESTING

Most cotton in the United States, Europe, and Australia is harvested mechanically, either by a cotton picker, a machine that removes the cotton from the boll without damaging the cotton plant, or by a cotton stripper, which strips the entire boll off the plant. Cotton strippers are used in regions where it is too windy to grow picker varieties of cotton, and usually after application of a chemical defoliant or the natural defoliation that occurs after a freeze. Cotton is a perennial crop in the tropics, and without defoliation or freezing, the plant will continue to grow.

Cotton continues to be picked by hand in developing countries.

COMPETITION FROM SYNTHETIC FIBERS

The era of manufactured fibers began with the development of rayon in France in the 1890s. Rayon is derived from a natural cellulose and cannot be considered synthetic, but requires extensive processing in a manufacturing process, and led the less expensive replacement of more naturally derived materials. A succession of new synthetic fibers were introduced by the chemicals industry in the following decades. Acetate in Fibre form was developed in 1924. Nylon, the first Fibre synthesized entirely from petrochemicals, was introduced as a sewing thread by DuPont in 1936, followed by DuPont's acrylic in 1944. Some garments were created from fabrics based on these fibers, such as women's hosiery from nylon, but it was not until the introduction of polyester into the Fibre marketplace in the early 1950s that the market for cotton came under threat. The rapid uptake of polyester garments in the 1960s caused economic hardship in cotton-exporting economies, especially in Central American countries, such as Nicaragua, where cotton production had boomed tenfold between 1950 and 1965 with the advent of cheap chemical pesticides. Cotton production recovered in the 1970s, but crashed to pre-1960 levels in the early 1990s.

Beginning as a self-help Programme in the mid-1960s, the Cotton Research and Promotion Programme (CRPP) was organized by U.S. cotton producers in response to cotton's steady decline in market share. At that time, producers voted to set up a per-bale assessment system to fund the Programme, with built-in safeguards to protect their investments. With the passage of the Cotton Research and Promotion Act of 1966, the Programme joined forces and began battling synthetic competitors and re-establishing markets for cotton. Today, the success of this Programme has made cotton the best-selling Fibre in the U.S. and one of the best-selling fibers in the world.

Administered by the Cotton Board and conducted by Cotton Incorporated, the CRPP works to greatly increase the demand for and profitability of cotton through various research and promotion activities. It is funded by U.S. cotton producers and importers.

USES

Cotton is used to make a number of textile products. These include terrycloth for highly absorbent bath towels and robes; denim for blue jeans; cambric, popularly used in the manufacture of blue work shirts (from which we get the term "blue-collar"); and corduroy, seersucker, and cotton twill. Socks, underwear, and most T-shirts are made from cotton. Bed sheets often are made from cotton. Cotton also is used to make yarn used in crochet and knitting. Fabric also can be made from recycled or recovered cotton that otherwise would be thrown away during the spinning, weaving, or cutting process. While many fabrics are made completely of cotton, some materials blend cotton with other fibers, including rayon and synthetic fibers such as polyester. It can either be used in knitted or woven fabrics, as it can be blended with elastine to make a stretchier thread for knitted fabrics, and apparel such as stretch jeans.

In addition to the textile industry, cotton is used in fishing nets, coffee filters, tents, explosives manufacture (see nitrocellulose), cotton paper, and in bookbinding. The first Chinese paper was made of cotton Fibre. Fire hoses were once made of cotton.

The cottonseed which remains after the cotton is ginned is used to produce cottonseed oil, which, after refining, can be consumed by humans like any other vegetable oil. The cottonseed meal that is left generally is fed to ruminant livestock; the gossypol remaining in the meal is toxic to monogastric animals. Cottonseed hulls can be added to dairy cattle rations for roughage. During the American slavery period, cotton root bark was used in folk remedies as an abortifacient, that is, to induce a miscarriage. Gossypol was one of the many substances found in all parts of the cotton plant and it was described by the scientists as 'poisonous pigment'. It also appears to inhibit the development of sperm or even restrict the mobility of the sperm. Also, it is thought to interfere with the menstrual cycle by restricting the release of certain hormones.

Cotton linters are fine, silky fibers which adhere to the seeds of the cotton plant after ginning. These curly fibers typically are less than D_8 inch (3.2 mm) long. The term also may apply to the longer textile Fibre staple lint as well as the shorter fuzzy fibers from some upland species. Linters are traditionally used in the manufacture of paper and as a raw material in the manufacture of cellulose. In the UK, linters are referred to as "cotton wool". This can also be a refined product (absorbent cotton in U.S. usage) which has medical, cosmetic and many other practical uses. The first medical use of cotton wool was by Dr. Joseph Sampson Gamgee at the Queen's Hospital (later the General Hospital) in Birmingham, England.

Shiny cotton is a processed version of the Fibre that can be made into cloth resembling satin for shirts and suits. However, it is hydrophobic (does not absorb water easily), which makes it unfit for use in bath and dish towels (although examples of these made from shiny cotton are seen).

The name Egyptian cotton is broadly associated with quality products, however only a small percentage of Egyptian cotton production is actually of superior quality. Most products bearing the name are not made with the finest cottons from Egypt. Pima cotton is often compared to Egyptian cotton, as both are used in high quality bed sheets and other cotton products. It is even considered superior by some authorities. Pima cotton is grown in the American southwest.

INTERNATIONAL TRADE

The largest producers of cotton, currently (2009), are China and India, with annual production of about 34 million bales and 27 million bales, respectively; most of this production is consumed by their respective textile industries. The largest exporters of raw cotton are the United States, with sales of $4.9 billion, and Africa, with sales of $2.1 billion. The total international trade is estimated to be $12 billion. Africa's share of the cotton trade has doubled since 1980. Neither area has a significant domestic textile industry, textile manufacturing having moved to developing nations in Eastern and South Asia such as India and China. In Africa, cotton is grown by numerous small holders. Dunavant Enterprises, based in Memphis, Tennessee, is the leading cotton broker in Africa, with hundreds of purchasing agents. It operates cotton gins in Uganda, Mozambique, and Zambia. In Zambia, it often offers loans for seed and expenses to the 180,000 small farmers who grow cotton for it, as well as advice on farming methods. Cargill also purchases cotton in Africa for export.

The 25,000 cotton growers in the United States of America are heavily subsidized at the rate of $2 billion per year although China now provides the highest overall level of cotton sector support. The future of these subsidies is uncertain and has led to anticipatory expansion of cotton brokers' operations in Africa. Dunavant expanded in Africa by buying out local operations. This is only possible in former British colonies and Mozambique; former French colonies continue to maintain tight monopolies, inherited from their former colonialist masters, on cotton purchases at low fixed prices.

LEADING PRODUCER COUNTRIES

The five leading exporters of cotton in 2011 are (1) the United States, (2) India, (3) Brazil, (4) Australia, and (5) Uzbekistan. The largest Non-producing importers are Korea, Taiwan, Russia,, and Japan.In India, the states of Maharashtra (26.63per cent), Gujarat (17.96per cent) and Andhra Pradesh (13.75per cent) and also Madhya Pradesh are the leading cotton producing states, these states have a predominantly tropical wet and dry climate. In Pakistan, cotton is grown predominantly in the provinces of Punjab, and Sindh. The leading area of cotton production is the south Punjab, comprising the areas around Rahim Yar Khan, Bahawalpur, Bahawalnagar, Multan, Dera Ghazi Khan, Muzaffargarh, Vehari, and Khanewal. In Sindh Sanghar is the most important cotton producing district. Faisalabad is a leader in textiles within Pakistan. Punjab has a tropical wet and dry climate throughout the year therefore enhancing the growth of cotton. In the United States, the state of Texas led in total production as of 2004, while the state of California had the highest yield per acre.

FAIR TRADE

Cotton is an enormously important commodity throughout the world. However, many farmers in developing countries receive a low price for their produce, or find it difficult to compete with developed countries. This has led to an international dispute (see United States – Brazil cotton dispute):

- On 27 September 2002, Brazil requested consultations with the US regarding prohibited and actionable subsidies provided to US producers, users and/or exporters of upland cotton, as well as legislation, regulations, statutory instruments and amendments thereto providing such subsidies (including export credits), grants, and any other assistance to the US producers, users and exporters of upland cotton.
- On 8 September 2004, the Panel Report recommended that the United States "withdraw" export credit guarantees and payments to domestic users and exporters, and "take appropriate steps to remove the adverse effects or withdraw" the mandatory price-contingent subsidy measures.

While Brazil was fighting the US through the WTO's Dispute Settlement Mechanism against a heavily subsidized cotton industry, a group of four least-developed African countries — Benin, Burkina Faso, Chad, and Mali — also known as "Cotton-4" have been the leading protagonist for the reduction of US cotton subsidies through negotiations. The four introduced a "Sectoral Initiative in Favour of Cotton", presented by Burkina Faso's President Blaise Compaoré during the Trade Negotiations Committee on 10 June 2003. In addition to concerns over subsidies, the cotton industries of some countries are criticized for employing child Labour and damaging workers' health by exposure to pesticides used in production. The Environmental Justice Foundation has campaigned against the prevalent use of forced child and adult Labour in cotton production in Uzbekistan, the world's third largest cotton exporter. The international production and trade situation has led to "fair trade" cotton clothing and footwear, joining a rapidly growing market for organic clothing, fair fashion or "ethical fashion". The fair trade system was initiated in 2005 with producers from Cameroon, Mali and Senegal.

TRADE

Cotton is bought and sold by investors and price speculators as a tradable commodity on 2 different stock exchanges in the United States of America.

- Cotton futures contracts are traded on the New York Mercantile Exchange (NYMEX) under the ticker symbol TT. They are delivered every year in March, May, July, October, and December.
- Cotton #2 futures contracts are traded on the New York Board of Trade (NYBOT) under the ticker symbol CT. They are delivered every year in March, May, July, October, and December.

CRITICAL TEMPERATURES

- Favorable travel temperature range: below 25 °C (77 °F)
- Optimum travel temperature: 21 °C (70 °F)
- Glow temperature: 205 °C (401 °F)
- Fire point: 210 °C (410 °F)
- Autoignition temperature: 407 °C (765 °F)
- Autoignition temperature (for oily cotton): 120 °C (248 °F)

A temperature range of 25 to 35 °C (77 to 95 °F) is the optimal range for mold development. At temperatures below 0 °C (32 °F), rotting of wet cotton stops. Damaged cotton is sometimes stored at these temperatures to prevent further deterioration.

BRITISH STANDARD YARN MEASURES

- 1 thread = 55 in or 140 cm
- 1 skein or rap = 80 threads (120 yd or 110 m)
- 1 hank = 7 skeins (840 yd or 770 m)
- 1 spindle = 18 hanks (15,120 yd or 13.83 km)

COTTON GENOME

A public genome sequencing effort of cotton was initiated in 2007 by a consortium of public researchers. They agreed on a strategy to sequence the genome of cultivated, tetraploid cotton. "Tetraploid" means that cultivated cotton actually has two separate genomes within its nucleus, referred to as the

A and D genomes. The sequencing consortium first agreed to sequence the D-genome relative of cultivated cotton (G. raimondii, a wild Central American cotton species) because of its small size and limited number of repetitive elements. It is nearly one-third the number of bases of tetraploid cotton (AD), and each chromosome is only present once. The A genome of G. arboreum would be sequenced next. Its genome is roughly twice the size of G. raimondii's. Part of the difference in size between the two genomes is the amplification of retrotransposons (GORGE). Once both diploid genomes are assembled, then research could begin sequencing the actual genomes of cultivated cotton varieties. This strategy is out of necessity; if one were to sequence the tetraploid genome without model diploid genomes, the euchromatic DNA sequences of the AD genomes would co-assemble and the repetitive elements of AD genomes would assembly independently into A and D sequences respectively. Then there would be no way to untangle the mess of AD sequences without comparing them to their diploid counterparts.

The public sector effort continues with the goal to create a high-quality, draft genome sequence from reads generated by all sources. The public-sector effort has generated Sanger reads of BACs, fosmids, and plasmids as well as 454 reads. These later types of reads will be instrumental in assembling an initial draft of the D genome. In 2010, two companies (Monsanto and Illumina), completed enough Illumina sequencing to cover the D genome of G. raimondii about 50x. They announced that they would donate their raw reads to the public. This public relations effort gave them some recognition for sequencing the cotton genome. Once the D genome is assembled from all of this raw material, it will undoubtedly assist in the assembly of the AD genomes of cultivated varieties of cotton, but a lot of hard work remains.

STRUCTURE AND PROPERTIES OF COTTON FIBERS

FIBRE STRUCTURE AND FORMATION

The botanical name of American Upland cotton is Gossypium Hirsutum and has been developed from cottons of Central America. Upland varieties represent approximately 97per cent of U.S. production.

Each cotton Fibre is composed of concentric layers. The cuticle layer on the Fibre itself is separable from the Fibre and consists of wax and pectin materials. The primary wall, the most peripheral layer of the Fibre, is composed of cellulosic crystalline fibrils. The secondary wall of the Fibre consists of three distinct layers. All three layers of the secondary wall include closely packed parallel fibrils with spiral winding of 25-35° and represent the majority of cellulose within the Fibre. The innermost part of cotton Fibre- the lumen- is composed of the remains of the cell contents. Before boll opening, the lumen is filled with liquid containing the cell nucleus and protoplasm. The twists and convolutions of the dried Fibre are due to the removal of this liquid. The cross section of the Fibre is bean-shaped, swelling almost round when moisture absorption takes place.

The overall contents are broken down into the following components.

RAW COTTON COMPONENTS

80-90%	Cellulose
6-8%	Water
0.5 - 1%	Waxes and fats
0 - 1.5%	Proteins
4 - 6%	Hemicelluloses and pectin's
1 - 1.8%	Ash

During scouring (treatment of the Fibre with caustic soda), natural waxes and fats in the Fibre are saponified and pectin's and other non-cellulose materials are released, so that the impurities can be removed by just rinsing away. After scouring, a bleaching solution (consisting of a stabilized oxidizing agent) interacts with the Fibre and the natural Colour is removed. Bleaching takes place at elevated temperature for a fixed period of time.

Mercerization is another process of improving sorption properties of cotton. Cotton Fibre is immersed into 18- 25per cent solution of sodium hydroxide often under tension. The Fibre obtains better luster and sorption during mercerization.

After scouring and bleaching, the Fibre is 99per cent cellulose. Cellulose is a polymer consisting of anhydroglucose units connected with 1,4 oxygen bridges in the beta position. The hydroxyl groups on the cellulose units enable hydrogen bonding between two adjacent polymer chains. The degree of polymerization of cotton is 9,000-15,000. Cellulose shows approximately 66per cent crystallinity, which can be determined by X-ray diffraction, infrared spectroscopy and density methods.

Each crystal unit consists of five chains of anhydroglucose units, parallel to the fibril axis. One chain is located at each of the corners of the cell and one runs through the center of the cell. The dimensions of the cell are a = 0.835nm, b = 1.03 nm and c = 0.79 nm. The angle between ab and BC planes is 84° for normal cellulose, *i.e.*, Cellulose I.

REPEAT UNIT OF CELLULOSE

The current consensus regarding cellulose crystallinity (X-ray diffraction) is that fibers are essentially 100per cent crystalline and that very small crystalline units imperfectly packed together cause the observed disorder.

The density method used to determine cellulose crystallinity is based on the density gradient column, where two solvents of different densities are partially mixed. Degree of Crystallinity is, then, determined from the density of the sample, while densities of crystalline and amorphous cellulose forms are known (1.505 and 1.556 respectively). Orientation of untreated cotton Fibre is poor because the crystallites are contained in the micro fibrils of the secondary wall, oriented in the steep spiral (25-30°) to the Fibre axis.

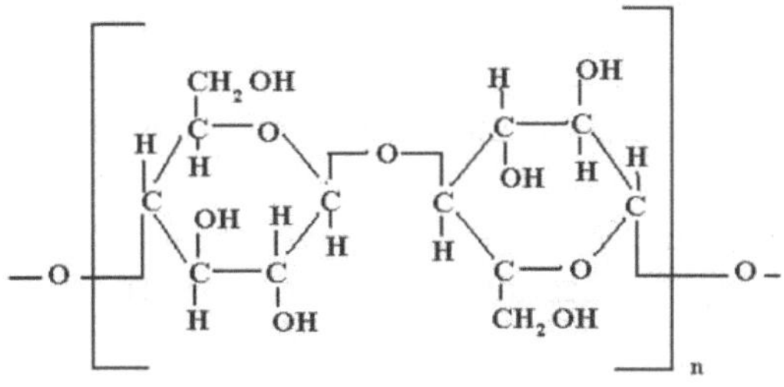

PHYSICAL PROPERTIES OF COTTON

FIBRE LENGTH

Fibre length is described as "the average length of the longer one-half of the fibers (upper half mean length)" This measure is taken by scanning a "beard " of parallel fibers through a sensing region. The beard is formed from the fibers taken from the sample, clasped in a holding clamp and combed to align the fibers. Typical lengths of Upland cottons might range from 0.79 to 1.36in.

Cottons come from the cotton plant; the longer strand types such as Pima or Sea Island produce the finest types of cotton fabrics.

LENGTH UNIFORMITY

Length uniformity or uniformity ratio is determined as " a ratio between the mean length and the upper half mean length of the fibers and is expressed as a percentage". Typical comparisons are illustrated below.

LENGTH UNIFORMITY	UNIFORMITY INDEX [%]
Very High	>85
High	83-85
Intermediate	80-82
Low	77-79
Very Low	<77

Low uniformity index shows that there might be a high content of short fibers, which lowers the quality of the future textile product.

FIBRE STRENGTH

Fibre strength is measured in grams per denier. It is determined as the force necessary to break the beard of fibers, clamped in two sets of jaws, (1/8 inch apart). Typical tensile levels are illustrated. The breaking strength of cotton is about 3.0~4.9 g/denier, and the breaking elongation is about 8~10per cent.

LENGTH UNIFORMITY	UNIFORMITY INDEX [%]
Very High	>85
High	83-85
Intermediate	80-82
Low	77-79
Very Low	<77

MICRONAIRE

Micronaire measurements reflect Fibre fineness and maturity. A constant mass (2.34 grams) of cotton fibers is compressed into a space of known volume and air permeability measurements of this compressed sample are taken. These, when converted to appropriate number, denote Micronaire values.

Cotton Range	Micronaire Reading
Premium	3.7-4.2
Base Range	4.3-4.9
Discount Range	>5.0

COLOUR

The Colour of cotton samples is determined from two parameters: degree of reflectance (Rd) and yellowness (+b). Degree of reflectance shows the brightness of the sample and yellowness depicts the degree of cotton pigmentation. A defined area located in a Nickerson-Hunter cotton colorimeter diagram represents each Colour code. The Colour of the fibers is affected by climatic conditions, impact of insects and fungi, type of soil, storage conditions etc. There is five recognized groups of Colour: white, gray, spotted, tinged, and yellow stained. As the Colour of cotton deteriorates, the process ability of the fibers decreases.

Work at the University of Tennessee has led to Colour measurement using both a spectrometer CIE-based average Colour measurement and a Colour uniformity measurement using image analysis to improve the accuracy and provide additional measurement for Colour grading. Later the investigators developed two Colour grading systems using expert system and neural networks.

TRASH

A trash measurement describes the amount of non-lint materials (such as parts of cotton plant) in the Fibre. Trash content is assessed from scanning the cotton sample surface with a video camera and calculating the percentage of the surface area occupied by trash particles. The values of trash content should be within the range from 0 to 1.6per cent. Trash content is highly correlated to leaf grade of the sample.

LEAF GRADE

Leaf grade is provided visually as the amount of cotton plant particles within the sample. There are seven leaf grades (#1-#7) and one below grade (#8).

PREPARATION

Preparation is the classer's interpretation of Fibre process ability in terms of degree of roughness or smoothness of ginned cotton.

EXTRANEOUS MATTER

Extraneous matter is all the material in the sample other than Fibre and leaf. The classer either as "light" or "heavy" determines the degree of extraneous matter.

NEPS

A nep is a small tangled Fibre knot often caused by processing. Neps can be measured by the AFIS nep tester and reported as the total number of neps per 0.5 grams of the Fibre and average size in millimeters. Nep formation reflects the mechanical processing stage, especially from the point of view of the quality and condition of the machinery used.

CHEMICAL PROPERTIES OF COTTON

Cotton swells in a high humidity environment, in water and in concentrated solutions of certain acids, salts and bases. The swelling effect is usually attributed to the sorption of highly hydrated ions. The moisture regain for cotton is about 7.1~8.5per cent and the moisture absorption is 7~8per cent.

Cotton is attacked by hot dilute or cold concentrated acid solutions. Acid hydrolysis of cellulose produces hydro-celluloses. Cold weak acids do not affect it. The fibers show excellent resistance to alkalis. There are a few other solvents that will dissolve cotton completely. One of them is a copper complex of cupramonium hydroxide and cupriethylene diamine (Schweitzer's reagent)

Cotton degradation is usually attributed to oxidation, hydrolysis or both. Oxidation of cellulose can lead to two types of so-called oxy-cellulose, depending on the environment, in which the oxidation takes place.

INSERT FORMULA OR EQUATION: OXY-CELLULOSE

Also, cotton can degrade by exposure to visible and ultraviolet light, especially in the presence of high temperatures around 250~397° C and humidity. Cotton fibers are extremely susceptible to any biological degradation (microorganisms, fungi etc.)

OPTICAL PROPERTIES OF COTTON

Cotton fibers show double refraction when observed in polarized light. Even though various effects can be observed, second order yellow and second order blue is characteristic colors of cellulosic fibers. A typical birefringence value as shown in the table of physical properties, is 0.047.

COTTON CLASSIFICATION

Cotton classification is used to determine the quality of the cotton Fibre in terms of grade, length and Micronaire. USDA classification specifically identifies the characteristics of Fibre length, length uniformity, strength, Micronaire, Colour, preparation, leaf and extraneous matter. In the past, these qualities were classified just by hand-and-eye of an experienced classer. Since 1991, all classification has been carried out with a set of up-to-date instruments, called "HVI"(High Volume Instrumentation) classification. However, measuring techniques of other qualities of cotton Fibre, such as Fibre maturity and short Fibre content, are also being developed.

COTTON IN NON-WOVENS

Cotton is the most important apparel Fibre throughout the world. It is a Fibre that was used fairly extensively during the early, developmental period of the Non-wovens business primarily because the emerging dry-laid producers came from the textile industry and had an intimate knowledge of cotton and its processing characteristics.

It was in the early part of 20th Century that a few cotton mills in the US wanted to find ways to upgrade the waste cotton fibers into saleable products. The first method used was bonding the short cotton fibers (Fibre waste) with latex and resin. These products were used mainly as industrial wipes. After World War II, products like draperies, tablecloths, napkins and wiping towels were developed. It was realized that woven fabrics have much better properties than Non-wovens; so, the approach was to claim the market where superior qualities of woven or knit fabrics were not essential but where qualities better than those of paper were needed.

As the quality requirements for Non-woven fabrics increased and particularly as the need for white, clean fabric emerged; the use of raw cotton became unacceptable and was abandoned by the industry except for a few isolated product areas. Within the last decade, bleached cotton Fibre suitable for processing on conventional Non-woven equipment has become available and has substantially increased interest in this Fibre. This is particularly true in medical and Health care applications, wiping and wiper markets, and some apparel markets.

The raw cotton consists of about 96per cent cellulose and 4per cent of waxes, pectin, and other pertinacious and plant material. These minor constituents that must be removed in the scouring and bleaching process to give the soft, clean, white, absorbent Fibre that is satisfactory for the Non-wovens industry after the application of an appropriate finishing oil. The Fibre length of cotton is important, particularly as to its process ability. Longer staple cotton (0.75 in. to 1.25 in.) is satisfactory for Non-woven production. The Fibre has excellent absorbency and feels comfortable against the skin. It has fairly good strength both wet and dry, and has moderate dimensional stability and elastic recovery. But the resilience of cotton is relatively low, unless it is cross-linked by a chemical treatment. In Non-woven applications, the purity and absorbency of bleached cotton are utilized in growing medical and Health care applications.

The spun lace process usually produces such fabrics. For similar reasons, cotton spun lace fabrics are well accepted in personal and related wipes, especially in Japan and the ASIAN region. In a sense, bleached cotton Fibre for Non-woven application is a relatively new Fibre. It is a comparatively expensive Fibre and available from only a few sources. Consequently, its use still is restricted to specialized applications. This situation is likely to change in the future as the price is further reduced and availability increased.

FIBRE PROCESSING

About 30per cent of world cotton machines harvest production. Australia, Israel and USA are the only countries where all cottons are picked by machines. Fifteen percent of world cotton production is ginned on roller gins and almost all rest of cotton is saw ginned in most countries.Cotton fibers in non-wovens are generally used in their bleached form.

A lot of research and development has taken place for the efficient production of bleached fibers. The Kier bleaching process produces most of the bleached cotton fibers. Since cotton of lesser grades is useful for non-wovens, a conventional cleaning system does not suffice. This might include a coarse wire carding, called Cotton Master Cleaners, for cleaning the cotton.

- The conventional bleaching method for cottons meant for non-wovens is a 9 step process are:
  - Fibre opening and cleaning
  - Alkali scouring application
  - Alkali reaction stage
  - Rinsing
  - Bleach application
  - Bleach reaction stage
  - Rinsing
  - Finish application
  - Drying

A continuous textile processing system and method have been disclosed recently for producing a Non-woven web containing bleached cotton fibers in a single line system which includes a supply of fibers such as a bale opening device, The final Non-woven web consisting of bleached cotton fibers may be made into highly purified and absorbent wipes, pads, and other articles for medical, industrial, or domestic use.

Finally, there is opening and bale formation.

- Cotton Incorporated patented a processing line, which promised better productivity and quality. It consists of:

a) Fibre opening and Cleaning
b) Formation of web
c) Steam purging and Alkali impregnation onto the sandwiched cotton web between 2 porous conveyors.
d) After reaction, a pressure squeezing operation.
e) Similar processes for bleaching and then finishing.

- The recent system for scouring a bleaching of cotton Fibre is the Continuous Wet Finishing Technique' patented by Lawrence Girard and Walter E Meyer and assigned to Greenville Machinery Corporation. It consists of:

a. Opening and Cleaning
b. Conversion of fibers into a bat, weighing 10-30 ounces/sq. yard, by Needle punching or Air-lay technique.
c. Scouring
d. Bleaching
e. Finishing
f. Washing
g. Drying
h. Fibre opening

Advantages of Continuous Finishing Techniques are:

a) Uniformity of scouring and bleaching
b) Uniformity of finish application
c) Shorter time in process for the materials
d) Lower water consumption and less effluent for treatment
e) The ability to provide additional chemical treatments to the cotton.

COST OF PRODUCING COTTON

The international cotton advisory committee (ICAC) undertakes a survey of the cost of the production of cotton every three years based on the data from 31 countries.

Several factors are considered, such as land rent, fertilizers, insect control, irrigation, harvesting and ginning.

The cost of seed cotton is more than $500 in USA to produce one hectare of seed cotton. The net cost of producing lint from one hectare (the value of seed and land rent were excluded from the total cost) is highest in Australia (US$1,056) followed by the USA (US$889), Pakistan (US$814), Zimbabwe (US$426) and China (US$416). It is most expensive to produce a kilogram of lint in the USA (US$1.20), Australia (US$0.75) and china (US$0.48).

WEB PROCESSING WITH COTTON

Cotton fibers are used in the manufacture of Non-wovens either alone or in a blend. The various processes for the manufacture of non-wovens are:

HYDROENTANGLEMENT

This method of bonding provides strength to the Non-wovens, comparable to woven fabric of the same basis weight. This method yields high strength without interfering with the absorbency, tensile strength and aesthetic properties of cotton.

This type of Non-wovens can be wet processed like the conventional woven textiles for bleaching, dyeing and finishing. To manufacture soft loose Non-wovens, partially entangled webs are produced by subjecting cotton webs to low water jet pressures (approx. 300-500 psi). These types of webs can be wet processed in a pad/batch state. The limitations of this process are that production has been limited to Fibre blends because of problems in recycling water and the quality of bleached cotton.

NEEDLE PUNCHING

Needle punched cotton provides highly efficient filter media based on the irregular Fibre shape and absorption properties. Increased tenacity in the wet condition can be an important advantage for cotton filters. To build strength, scrim materials can be used as in bed blankets and industrial fabrics. Needles of 36-42 gauges have been found appropriate for the production of cotton needle punched Non-wovens. For very heavy fabrics, use is made of gauge 32 and for finer fabrics 40-42 gauge needles are being used.

THERMAL BONDING

In this process cotton webs with blends of thermoplastic fibers are passed between 2 hot rollers (Calendar rollers). The thermoplastic Fibre softens/melts and bonds the web. The initial work was done with polyester as the thermoplastic Fibre. Later polypropylene was extended for the study because of economics, density and melting temperature considerations. This was mainly to study the application as a diaper lining material. Substantial work is still being done to develop this type of Non-wovens.

OTHER BONDING SYSTEMS

a. Impregnating the web with a resin or other adhesive material.

b. Stripping off of the web with adhesive, which bonds the fibers together at regular intervals.

c. Stitch bonding: cotton web is stitched like in sewing and the product performance depends on web weight, stitch/inch and type of sewing thread.

APPLICATIONS AND MANUFACTURERS OF COTTON NON-WOVENS

Cotton Non-wovens are used as swabs, puffs, wipes, filters, weddings, personal care products like in diapers and feminine hygiene products, semi-durable segments like bedding, household furnishing, pillow fillers, etc.

MANUFACTURERS OF COTTON

- BARHARDT MANUFACTURING
- BBA NON-WOVENS VERATEC
- BRANNOC FIBERS Ltd.
- COTTON INCORPORATED
- IHSAN SONS (PVT) LIMITED
- LEIGH FIBERS
- TEXTILES AND NON-WOVENS DIRECTORY

RECENT RESEARCH

- New instrumentation to measure cotton contamination.
- Cotton linters to replace the traditional 100per cent wood pulp fibers for producing absorbent cores for disposable diapers and famine pads
- New quality measurements of small sample cotton are being developed
- Cotton is being blended with kenaf fibers to improve the softness and hand
- Buckeye Technologies has developed 100per cent natural cotton for tampon manufacture
- Clustering analysis is developed for cotton trash classification
- New method to improve the dyeability of cotton with reactive dyes.

RECENT DOVELOPMENTS IN COTTON

Amazing Footage Of Seal Having Sexual Intercourse With Penguin Caught On Camera! (Buzzwok)

COLORED COTTON

Cotton Fibre is dyed with chemical dyes in order to get wide range of colors. These chemical dyes and their finishing demands large amount of water in turn when these water is disposed they cause soil and water pollution. Many dyes are of chemical origin; particularly the azure ones and these are not environment friendly. Hence many countries, including India, have prohibited use of these dyes.

The negative effects of dyeing can be reduced by naturally colored cotton. This colored cotton is developed by gene transplantation. Crossing the genes from wild cotton varieties with the cultivated white ones develops this colored cotton. The research is being conducted at The University of Agricultural Sciences (UAS), Dharwad Karnataka India, to promote the cultivation of natural colored cotton. The colors that have been developed are White, Orange, Red, Yellow, Green, Purple, Brown, Blue, And Black. These negative effects of dyeing can be avoided by extensive research and growth of colored cotton. (33).

BT COTTON

Cotton requires severe pesticide in order to combat numerous pests after some years of use of pesticide by farmers these pests develop resistance to Particular pesticide. This resistance force farmers to use more amounts of pesticides. BT Cotton is developed by transgenic technique of implanting Bacillus Thuringiens bacterial gene in to cottonseeds, which makes the cotton plant and seeds resistant to majority of pests including bollworm (A. Lepidoptera), Tobacco budworm (Heliothis virescens). Bt cotton is now one of the most widely used transgenic crops. It is currently grown throughout the United States. More than 2 million acres of Bt cotton are grown in the United States alone. Other countries include China, India, and Australia. (34) According to Dept. of Agricultural and Resource Economics, University of Arizona. *Bt* cotton planted from 1996 to 1998 is estimated to have yielded 5per cent more on average than if traditional and decreased the quantity of foliar spray.

COTTON'S FUTURE TRENDS

The world's cotton Fibre production is approximately 89 million bales. In 1997, a production forecast shows that the U.S. is the largest cotton producer (18.4 million bales), followed by China (17.5 million bales), India (12.8 million bales), Pakistan (8.0 million bales) and the former U. S. S. R. republics (7.7 million bales). Other important cotton producers are Australia, Egypt, Turkey, Brazil, Argentina, Paraguay, Greece and Mexico. The highest cotton consumption is attributed to China (21.2 million bales), India (12.9 million bales) and U.S. (11.3 million bales).

- SUPPLIES: The world production will increase a little bit. The 1998 U.S cotton crop is best described as a disaster due to cool wet spring in the west and inadequate rainfall in the southeast.

- CONSUMPTION: World cotton consumption is lagging a bit behind production. After a surge in the mid-1980s, world cotton consumption has been rather flat. But the long term potential for cotton demand remains large.

All cotton plantings for 1999 are expected to total 14.6 million acres, 9 percent above 1998, and 5 percent greater than 1997. Upland cotton is expected to total 14.2 million acres, up 9 percent from last year. Growers planted 318,200 acres of American-Pima cotton. This is a 3per cent decrease from last year's number, but 27per cent higher than the acreage of 2 years ago. Planting in Georgia started extremely slow due to a severely dry spring, but by June 1 was nearly on pace with average. Conversely, Texas experienced a near normal planting season although some replanting was necessary due to wind and hail damage.

CONCLUSION

Cotton Non-wovens can be recycled, re-used or disposed off by natural degradation conditions. Cotton is a readily renewable resource with long-term supply assurance. Extensive research works is improving bleached Fibre quality and quantity. Non-woven industries are producing various types of Non-wovens with different manufacturing techniques, for better production. Cotton share of the textile Fibre market has been steadily increasing and will continue to increase as cotton-containing items is preferred by the consumers.

5

Wool Fibres

Wool is the textile Fibre obtained from sheep and certain other animals, including cashmere from goats, mohair from goats, qiviut from muskoxen, angora from rabbits, and other types of wool from camelids. Wool has several qualities that distinguish it from hair or fur: it is crimped, it is elastic, and it grows in staples (clusters).

CHARACTERISTICS

Wool's scaling and crimp make it easier to spin the fleece by helping the individual fibers attach to each other, so they stay together. Because of the crimp, wool fabrics have greater bulk than other textiles, and they hold air, which causes the fabric to retain heat. Insulation works both ways: Bedouins and Tuaregs use wool clothes to keep heat out and protect the body.

Felting of wool occurs upon hammering or other mechanical agitation as the microscopic barbs on the surface of wool fibers hook together.

The amount of crimp corresponds to the fineness of the wool fibers. A fine wool like Merino may have up to 100 crimps per inch, while the coarser wools like karakul may have as few as one or two. In contrast, hair has little if any scale and no crimp, and little ability to bind into yarn. On sheep, the hair part of the fleece is called kemp. The relative amounts of kemp to wool vary from breed to breed and make some fleeces more desirable for spinning, felting, or carding into batts for quilts or other insulating products, including the famous tweed cloth of Scotland.

Wool fibers readily absorb moisture, but are not hollow. Wool can absorb almost one-third of its own weight in water. Wool absorbs sound like many other fabrics. It is generally a creamy white Colour, although some breeds of sheep produce natural colors, such as black, brown, silver, and random mixes.

Wool ignites at a higher temperature than cotton and some synthetic fibers. It has a lower rate of flame spread, a lower rate of heat release, a lower heat of combustion, and does not melt or drip; it forms a char which is insulating and self-extinguishing, and it contributes less to toxic gases and smoke than other flooring products when used in carpets. Wool carpets are specified for high safety environments, such as trains and aircraft. Wool is usually specified for garments for firefighters, soldiers, and others in occupations where they are exposed to the likelihood of fire.

Wool is considered by the medical profession to be hypoallergenic.

PROCESSING

SHEARING

Sheep shearing is the process by which the woolen fleece of a sheep is cut off. After shearing, the wool is separated into four main categories: fleece (which makes up the vast bulk), broken, bellies, and locks. The quality of fleeces is determined by a technique known as wool classing, whereby a qualified person called a wool classer groups wools of similar gradings together to maximize the return for the farmer or sheep owner. In Australia and New Zealand, before being auctioned, all Merino fleece wool is objectively measured for micron, yield (including the amount of vegetable matter), staple length, staple strength, and sometimes Colour and comfort factor.

SCOURING

Wool straight off a sheep, known as "greasy wool" or "wool in the grease", contains a high level of valuable lanolin, as well as dead skin, sweat residue, pesticides, and vegetable matter. Before the wool can be used for commercial purposes, it must be scoured, a process of cleaning the greasy wool. Scouring may be as simple as a bath in warm water or as complicated as an industrial process using detergent and alkali in specialized equipment. In north west England, special potash pits were constructed to produce potash used in the manufacture of a soft soap for scouring locally produced white wool.

In commercial wool, vegetable matter is often removed by chemical carbonization. In less-processed wools, vegetable matter may be removed by hand and some of the lanolin left intact through the use of gentler detergents. This semigrease wool can be worked into yarn and knitted into particularly water-resistant mittens or sweaters, such as those of the Aran Island fishermen. Lanolin removed from wool is widely used in cosmetic products such as hand creams.

QUALITY

The quality of wool is determined by its Fibre diameter, crimp, yield, Colour, and staple strength. Fibre diameter is the single most important wool characteristic determining quality and price.

Merino wool is typically 3–5 inches in length and is very fine (between 12 and 24 microns). The finest and most valuable wool comes from Merino hoggets. Wool taken from sheep produced for meat is typically more coarse, and has fibers 1.5 to 6 in (38 to 152 mm) in length. Damage or breaks in the wool can occur if the sheep is stressed while it is growing its fleece, resulting in a thin spot where the fleece is likely to break.

Wool is also separated into grades based on the measurement of the wool's diameter in microns and also its style. These grades may vary depending on the breed or purpose of the wool. For example:

- <15.5 - Ultrafine Merino
- 15.6–18.5 – Superfine Merino
- 18.6–20 – Fine Merino
- 20.1–23 – Medium Merino
- >23 - Strong Merino
- Comeback: 21–26 microns, white, 90–180 mm long
- Fine crossbred: 27–31 microns, Corriedales, etc.
- Medium crossbred: 32–35 microns
- Downs: 23–34 microns, typically lacks luster and brightness. Examples, Aussiedown, Dorset Horn, Suffolk, etc.
- Coarse crossbred: >36 microns
- Carpet wools: 35–45 microns

Any wool finer than 25 microns can be used for garments, while coarser grades are used for outerwear or rugs. The finer the wool, the softer it is, while coarser grades are more durable and less prone to pilling.

The finest Australian and New Zealand Merino wools are known as 1PP, which is the industry benchmark of excellence for Merino wool 16.9 microns and finer. This style represents the top level of fineness, character, Colour, and style as determined on the basis of a series of parameters in accordance with the original dictates of British wool as applied today by the Australian Wool Exchange (AWEX) Council. Only a few dozen of the millions of bales auctioned every year can be classified and marked 1PP.

HISTORY

Wild sheep were more hairy than woolly. Although sheep were domesticated 9 to 11 thousand years ago, archaeological evidence from statuary found at sites in Iran suggests selection for woolly sheep may have begun around 6000 BC, with the earliest woven wool garments having only been dated to two to three thousand years later. Woolly-sheep were introduced into Europe from the Near East in the early part of the 4th millennium BC. The oldest known European wool textile, ca. 1500 BC, was preserved in a Danish bog.

Prior to invention of shears—probably in the Iron Age—the wool was plucked out by hand or by bronze combs. In Roman times, wool, linen, and leather clothed the European population; the cotton of India was a curiosity of which only naturalists had heard, and silk, imported along the Silk Road from China, was an extravagant luxury. Pliny the Elder records in his *Natural History* that the reputation for producing the finest wool was enjoyed by Tarentum, where selective breeding had produced sheep with superior fleeces, but which required special care.

In medieval times, as trade connections expanded, the Champagne fairs revolved around the production of wool cloth in small centers such as Provins. The network developed by the annual fairs meant the woolens of Provins might find their way to Naples, Sicily, Cyprus, Majorca, Spain, and even Constantinople. The wool trade developed into serious business, a generator of capital. In the 13th century, the wool trade became the economic engine of the Low Countries and central Italy. By the end of the 14th century, Italy predominated, though Italian production turned to silk in the 16th century. Both industries, based on the export of English raw wool, were rivaled only by the 15th-century sheepwalks of Castile and were a significant source of income to the English crown, which in 1275 had imposed an export tax on wool called the "Great Custom". The importance of wool to the English economy can be seen in the fact that since the 14th century, the presiding officer of the House of Lords has sat on the "Woolsack", a chair stuffed with wool.

Economies of scale were instituted in the Cistercian houses, which had accumulated great tracts of land during the 12th and early 13th centuries, when land prices were low and Labour still scarce. Raw wool was baled and shipped from North Sea ports to the textile cities of Flanders, notably Ypres and Ghent, where it was dyed and worked up as cloth. At the time of the Black Death, English textile industries accounted for about 10per cent of English wool production; the English textile trade grew during the 15th century, to the point where export of wool was discouraged. Over the centuries, various British laws controlled the wool trade or required the use of wool even in burials. The smuggling of wool out of the country, known as owling, was at one time punishable by the cutting off of a hand. After the Restoration, fine English woolens began to compete with silks in the international market, partly aided by the Navigation Acts; in 1699, the English crown forbade its American colonies to trade wool with anyone but England herself.

A great deal of the value of woolen textiles was in the dyeing and finishing of the woven product. In each of the centers of the textile trade, the manufacturing process came to be subdivided into a collection of trades, overseen by an entrepreneur in a system called by the English the "putting-out" system, or "cottage industry", and the *Verlagssystem* by the Germans. In this system of producing wool cloth, until recently perpetuated in the production of Harris tweeds, the entrepreneur provides the raw materials and an advance, the remainder being paid upon delivery of the product. Written contracts bound the artisans to specified terms. Fernand Braudel traces the appearance of the system in the 13th-century economic boom, quoting a document of 1275. The system effectively bypassed the guilds' restrictions.

Before the flowering of the Renaissance, the Medici and other great banking houses of Florence had built their wealth and banking system on their textile industry based on wool, overseen by the Arte della Lana, the wool guild: wool textile interests guided Florentine policies. Francesco Datini, the "merchant of Prato", established in 1383 an *Arte della Lana* for that small Tuscan city. The sheepwalks of Castile shaped the landscape and the fortunes of the *meseta* that lies in the heart of the Iberian peninsula; in the 16th century, a unified Spain allowed export of Merino lambs only with royal permission. The German wool market – based on sheep of Spanish origin – did not overtake British wool until comparatively late.

The Industrial Revolution introduced mass production technology into wool and wool cloth manufacturing. Australia's colonial economy was based on sheep raising, and the Australian wool trade eventually overtook that of the Germans by 1845, furnishing wool for Bradford, which developed as the heart of industrialized woolens production.

Due to decreasing demand with increased use of synthetic fibers, wool production is much less than what it was in the past. The collapse in the price of wool began in late 1966 with a 40per cent drop; with occasional interruptions, the price has tended down. The result has been sharply reduced production and movement of resources into production of other commodities, in the case of sheep growers, to production of meat.

Superwash wool (or washable wool) technology first appeared in the early 1970s to produce wool that has been specially treated so it is machine washable and may be tumble-dried. This wool is produced using an acid bath that removes the "scales" from the Fibre, or by coating the Fibre with a polymer that prevents the scales from attaching to each other and causing shrinkage. This process results in a Fibre that holds longevity and durability over synthetic materials, while retaining its shape.

In December 2004, a bale of the then world's finest wool, averaging 11.8 microns, sold for AU$3,000 per kilogram at auction in Melbourne, Victoria. This fleece wool tested with an average yield of 74.5per cent, 68 mm long, and had 40 newtons per kilotex strength. The result was A$279,000 for the bale. The finest bale of wool ever auctioned was sold for a seasonal record of AU$2690 per kilo during June 2008. This bale was produced by the Hillcreston Pinehill Partnership and measured 11.6 microns, 72.1per cent yield, and had a 43 newtons per kilotex strength measurement. The bale realized $247,480 and was exported to India.

In 2007, a new wool suit was developed and sold in Japan that can be washed in the shower, and which dries off ready to wear within hours with no ironing required. The suit was developed using Australian Merino wool, and it enables woven products made from wool, such as suits, trousers, and skirts, to be cleaned using a domestic shower at home.

In December 2006, the General Assembly of the United Nations proclaimed 2009 to be the International Year of Natural Fibres, so as to raise the profile of wool and other natural fibers.

PRODUCTION

Global wool production is about 1.3 million tonnes per year, of which 60per cent goes into apparel. Australia is the leading producer of wool which is mostly from Merino sheep. New Zealand is the second-largest producer of wool, and the largest producer of crossbred wool. China is the third-largest producer of wool. Breeds such as Lincoln, Romney, Drysdale, and Elliotdale produce coarser fibers, and wool from these sheep is usually used for making carpets.

In the United States, Texas, New Mexico, and Colorado have large commercial sheep flocks and their mainstay is the Rambouillet (or French Merino). Also, a thriving home-flock contingent of small-scale farmers raise small hobby flocks of specialty sheep for the hand-spinning market. These small-scale farmers offer a wide selection of fleece. Global woolclip (total amount of wool shorn) 2004/2005

1. Australia: 25per cent of global woolclip (475 million kg greasy, 2004/2005)
2. China: 18per cent
3. United States: 17per cent
4. New Zealand: 11per cent
5. Argentina: 3per cent
6. Turkey: 2per cent
7. Iran: 2per cent
8. United Kingdom: 2per cent
9. India: 2per cent
10. Sudan: 2per cent
11. South Africa: 1per cent

Organic wool is becoming more and more popular. This wool is very limited in supply and much of it comes from New Zealand and Australia. It is becoming easier to find in clothing and other products, but these products often carry a higher price. Wool is environmentally preferable (as compared to petroleum-based nylon or polypropylene) as a material for carpets, as well, in particular when combined with a natural binding and the use of formaldehyde-free glues.

Animal rights groups have noted issues with the production of wool, such as mulesing.

MARKETING

AUSTRALIA

About 85per cent of wool sold in Australia is sold by open cry auction.

'Sale by sample' is a method in which a mechanical claw takes a sample from each bale in a line or lot of wool. These grab samples are bulked, objectively measured, and a sample of not less than 4 kg is displayed in a box for the buyer to examine. The Australian Wool Exchange conducts sales primarily in Sydney, Melbourne, Newcastle, and Fremantle. About 80 brokers and agents work throughout Australia.About 7per cent of Australian wool is sold by private treaty on farms or to local wool-handling facilities. This option gives wool growers benefit from reduced transport, warehousing, and selling costs. This method is preferred for small lots or mixed butts to make savings on reclassing and testing.

About 5per cent of Australian wool is sold over the internet on an electronic offer board. This option gives wool growers the ability to set firm price targets, reoffer passed-in wool, and offer lots to the market quickly and efficiently. This method works well for tested lots, as buyers use these results to make a purchase. About 97per cent of wool is sold without sample inspection; however, as of December 2009, 59per cent of wool listed had been passed in from auction. Growers through certain brokers can allocate their wool to a sale and at what price their wool will be reserved.

Sale by tender can achieve considerable cost savings on wool clips large enough to make it worthwhile for potential buyers to submit tenders. Some marketing firms sell wool on a consignment basis, obtaining a fixed percentage as commission.Forward selling: Some buyers offer a secure price for forward delivery of wool based on estimated measurements or the results of previous clips. Prices are quoted at current market rates and are locked in for the season. Premiums and discounts are added to cover variations in micron, yield, tensile strength, etc., which are confirmed by actual test results when available.

Another method of selling wool includes sales direct to wool mills.

OTHER COUNTRIES

The British Wool Marketing Board operates a central marketing system for UK fleece wool with the aim of achieving the best possible net returns for farmers.

Less than half of New Zealand's wool is sold at auction, while around 45per cent of farmers sell wool directly to private buyers and end-users.

United States sheep producers market wool with private or cooperative wool warehouses, but wool pools are common in many states. In some cases, wool is pooled in a local market area, but sold through a wool warehouse. Wool offered with objective measurement test results is preferred. Imported apparel wool and carpet wool goes directly to central markets, where it is handled by the large merchants and manufacturers.

YARN

Virgin wool is spun for the first time.

Shoddy or recycled wool is made by cutting or tearing apart existing wool fabric and respinning the resulting fibers. As this process makes the wool fibers shorter, the remanufactured fabric is inferior to the original. The recycled wool may be mixed with raw wool, wool noil, or another Fibre such as cotton to increase the average Fibre length. Such yarns are typically used as weft yarns with a cotton warp. This process was invented in the Heavy Woollen District of West Yorkshire and created a microeconomy in this area for many years.

Rag is a sturdy wool Fibre made into yarn and used in many rugged applications such as gloves.

Worsted' is a strong, long-staple, combed wool yarn with a hard surface.

Woolen is a soft, short-staple, carded wool yarn typically used for knitting. In traditional weaving, woolen weft yarn (for softness and warmth) is frequently combined with a worsted warp yarn for strength on the loom.

USES

In addition to clothing, wool has been used for blankets, horse rugs, saddle cloths, carpeting, felt, wool insulation and upholstery. Wool felt covers piano hammers, and it is used to absorb Odours and noise in heavy machinery and stereo speakers. Ancient Greeks lined their helmets with felt, and Roman legionnaires used breastplates made of wool felt.

Wool has also been traditionally used to cover cloth diapers. Wool Fibre exteriors are hydrophobic (repel water) and the interior of the wool Fibre is hygroscopic (attracts water); this makes a wool garment able to cover a wet diaper while inhibiting wicking, so outer garments remain dry. Wool felted and treated with lanolin is water resistant, air permeable, and slightly antibacterial, so it resists the buildup of Odour. Some modern cloth diapers use felted wool fabric for covers, and there are several modern commercial knitting patterns for wool diaper covers.Initial studies of woolen underwear have found it prevented heat and sweat rashes because it more readily absorbs the moisture than other fibers.

Merino wool has been used in baby sleep products such as swaddle baby wrap blankets and infant sleeping bags.

As an animal protein, it can be used as a soil fertilizer, being a slow-release source of nitrogen.Researchers at the Royal Melbourne Institute of Technology school of fashion and textiles have discovered a blend of wool and kevlar, the synthetic Fibre widely used in body armor, was lighter, cheaper and worked better in damp conditions than kevlar alone. Kevlar, when used alone, loses about 20per cent of its effectiveness when wet, so required an expensive waterproofing process. Wool increased friction in a vest with 28–30 layers of fabric, to provide the same level of bullet resistance as 36 layers of Kevlar alone.

CARBON FOOTPRINT

The average greenhouse gas footprint of wool in manufacturing carpets estimated at 5.48 kg CO_2 equivalent per kg, when produced in Europe. The carbon footprint of woolen products is difficult to determine and varies considerably according to factors ranging from the conditions in which sheep are bred to the processes used to manufacture the products. The carbon footprint of a 264.85g woolly jumper made from New Zealand merino wool averages about 1.667 kg CO_2 equivalent at the point of purchase by a consumer, implying a carbon footprint of around 6 kg per kg of finished woolen product.

EVENTS

A buyer of Merino wool, Ermenegildo Zegna, has offered awards for Australian wool producers. In 1963, the first Ermenegildo Zegna Perpetual Trophy was presented in Tasmania for growers of "Superfine skirted Merino fleece". In 1980, a national award, the Ermenegildo Zegna Trophy for Extrafine Wool Production, was launched. In 2004, this award became known as the Ermenegildo Zegna Unprotected Wool Trophy. In 1998, an Ermenegildo Zegna Protected Wool Trophy was launched for fleece from sheep coated for around nine months of the year.

In 2002, the Ermenegildo Zegna Vellus Aureum Trophy was launched for wool that is 13.9 microns or finer. Wool from Australia, New Zealand, Argentina, and South Africa may enter, and a winner is named from each country. In April 2008, New Zealand won the Ermenegildo Zegna Vellus Aureum Trophy for the first time with a fleece that measured 10.8 microns. This contest awards the winning fleece weight with the same weight in gold as a prize, hence the name.

In 2010, an ultrafine, 10-micron fleece, from Windradeen, near Pyramul, New South Wales, won the Ermenegildo Zegna Vellus Aureum International Trophy.

Since 2000, Loro Piana has awarded a cup for the world's finest bale of wool that produces just enough fabric for 50 tailor-made suits. The prize is awarded to an Australian or New Zealand wool grower who produces the year's finest bale.

The New England Merino Field days which display local studs, wool, and sheep are held during January, in even numbered years around the Walcha, New South Wales district. The Annual Wool Fashion Awards, which showcase the use of Merino wool by fashion designers, are hosted by the city of Armidale, New South Wales, in March each year. This event encourages young and established fashion designers to display their talents. During each May, Armidale hosts the annual New England Wool Expo to display wool fashions, handicrafts, demonstrations, shearing competitions, yard dog trials, and more.

In July, the annual Australian Sheep and Wool Show is held in Bendigo, Victoria. This is the largest sheep and wool show in the world, with goats and alpacas, as well as woolcraft competitions and displays, fleece competitions, sheepdog trials, shearing, and wool handling. The largest competition in the world for objectively measured fleeces is the Australian Fleece Competition, which is held annually at Bendigo. In 2008, 475 entries came from all states of Australia, with first and second prizes going to the Northern Tablelands, New South Wales fleeces.

WOOL FIBRES

Wool production and use dates back approximately 10,000 years in Asia Minor. People living in the Mesopotamian Plain at that time used sheep for three basic human needs: food, clothing and shelter. As spinning and weaving skills developed woollens became a greater part of people's lives. The warmth of wool clothing and the mobility of sheep allowed people to spread Civilization beyond the warm climate of the Mesopotamia. Between 3000-1000 BC the Persians, Greeks and Romans distributed sheep and wool throughout Europe. The Romans took sheep everywhere they built their Empire including the British Isles. From here the British took sheep to all their colonies.

FIBRE STRUCTURE

Wool is different to other fibres because of its chemical structure. This chemical structure influences its texture, elasticity, staple and crimp formation. Wool is a protein fibre, composed of more than 20 amino acids. These amino acids form protein polymers. Wool also contains small amounts of fat, calcium and sodium.

TYPES OF WOOL YARN

There are two types of wool yarn – woollens and worsteds.

- Woolens: Woolens is a general term describing various fabrics woven from woollen yarn that is spun from the shorter wool fibres. These shorter fibres are not combed to lie flat as in the worsted yarn. This results in soft surface textures and finishes and the weave of individual yarns does not show as clearly as in worsted fabrics.
- Worsteds: Worsted is a general term for fabrics woven from worsted yarns that contain longer fibres spun from combed wool. Worsted wool refers to tightly woven, smooth, clear finished goods in a variety of twill and other stronger weaves.

Worsteds undergo several processes:

- Scouring – Washed to remove dust, suint (sweat) and wool wax.
- Carding – Rolled with a roller that is covered with teeth tease apart the staples of wool, laying the fibres nearly parallel to form a soft rope called a 'sliver'.
- Combing – Combed to separate short from long fibres, ensuring that the long fibres are laid parallel to produce a combed sliver called a 'top'.

- Drawing – Drawing out of tops into the thickness of one, to thoroughly blend the wool and ensure evenness or regularity of the resulting 'roving'.
- Finisher drawing – Drawing to reduce the roving thickness to suit the spinning operation and further improve evenness.
- Spinning – Inserting twist into the yarn to give strength to the finished yarn.

WORSTED VS WOOLLEN FABRICS

- Worsted fabrics are often more expensive than wool spun products due to the longer raw material to resultant yarn processing route used.
- Worsted fabric is stronger and wears better than a woollen spun fabric of equivalent weave construction and fabric weight.
- Worsted fabrics are preferred for trousers, suitings, other garments and upholstery fabrics where a smooth finish is required.
- Woollen spun fabrics are used for jackets, coats, skirts, upholstery fabrics, rugs and blankets where bulk and textured finishes are desirable.

Wool has many beneficial properties which have led to its long history of use.

- Insulation. Wool insulates against heat and cold. It is comfortable in both hot and cold weather because it absorbs moisture vapour. The crimp in the wool fibres makes them stand apart from each other trapping insulating air between the fibres. Still air is one of the best insulators found in nature. In hot weather the absorption/evaporation process works to help keep the body cooler.
- Fire resistant. Wool does not have to be specially treated to become non-flammable. A fabric made entirely of wool is difficult to ignite, burns slowly, and has limited ability to sustain a flame. Wool does not melt when burned and so cannot stick to the skin and cause serious burns.
- Water repellent. Although wool can absorb moisture, it repels liquids. It is naturally hydrophobic. The scales on the outside of the fibre cause liquid to roll off the surface of the fabric. Even if wool does eventually get wet it generates heat and keeps the body warm, not cold and clammy.
- Elastic. Wool has greater elasticity than any other plant or animal fibre. Wool can be twisted, turned and stretched and will still return to its natural shape.
- Durable. The interlocking protein molecules in the fibres of wool have the power to elongate, stretch and recover, creating an extremely

robust fabric that will last. Each wool fibre is made up of millions of 'coiled springs' that stretch and give rather than break.

- Static resistant. Wool has very little tendency to collect static electricity because wool naturally absorbs moisture from the air. Wool garments are much less likely to 'spark' or cling to the body.
- Noise insulation. Wool absorbs noise and reduces noise levels.
- Dirt resistant. Wool's ability to absorb moisture and therefore its low build-up of static electricity means that wool does not attract lint and dust from the air. The crimp in the fibre and the scales on the outside of the fibre deep dirt from penetrating the fabric.
- Versatile. Different sheep breeds with their own unique fibre characteristics provide different wools for a wide range of products.
- Dye-ability. Wool is easy to dye. The scales on the surface of the wool fibre diffuse light giving less reflection and a softer colour. The proteins in the core of the fibre absorb and combine with a wide variety of dyes and allows the wool to hold its colour.
- Comfort. Wool is comfortable to wear because of its elasticity, and moisture absorbing qualities.
- Fashionable. Wool drapes well, is alive, flexible and tailors easily, making it sought after by fashion designers.

Despite the elastic properties and its fire resistance, wool garments and products are often chemically treated as follows:

- Shrink proofing
- Fire proofing
- Moth proofing.

It is important to look at how the fibre was produced, as some animal treatments can leave chemical residue in the fibre. Post shearing treatments are also a cause for concern. Organically grown fibres can still be treated with toxic chemicals for the 'proofings' mentioned above, and these chemicals can cause health problems. Because of the scales on the wool fibres wool is often itchy and can cause an allergic reaction in some people.

Australia is the world's largest producer of wool.

AUSTRALIAN WOOL PRODUCTION

Did you know...

- In 2003/2004 there were 106 million sheep shorn in Australia producing 480 million kg of greasy wool.
- It is expected that the amount of wool produced in 2004/2005 will rise by 4per cent to 500 million kg of greasy wool from 112 million sheep shorn.
- The Australian flock was composed of 84.6per cent Merino, 11.3per cent crossbred, 4.1per cent other breeds at 30 June 2003.

- The wheat-sheep zone of Australia contains around 55per cent of the Australian sheep flock, the high rainfall zone contains around 33per cent and the pastoral zone contains around 12per cent.
- At 30 June 2003, 73per cent of the total wool produced in Australia is produced by less than 40per cent of wool producing farms.

Did you know...

- Australia is the world's largest producer of wool, producing over a quarter of the world's greasy wool in 2003/2004.
- While Australia produces more wool than any other country, China has the largest sheep population. Australia has the second largest sheep population.
- Australia exports wool to 50 countries. China is the biggest purchaser of Australian wool, with exports to China valued at $1.1 billion in 2003/2004.
- In 2003/2004, wool accounted for 2.5per cent of Australia's total exports, 3.4per cent of Australia's primary industry (agricultural/ mining) exports, and 9.4per cent of Australia's agricultural exports, ranking third behind beef and wheat.
- Wool exports were valued at $2.8 billion in 2003/2004.

Did you know...

- Australian wool accounted for 48per cent of the total used in global wool apparel in 2003/2004.
- In 2003, wool accounted for 2.4per cent of total world fibre use and about 3.5per cent of world apparel fibre use (with Australian wool approximately 2.3per cent of world apparel fibre use).
- Casual leisurewear is the largest clothing sector accounting for 70per cent of the total apparel market.
- Young adults, who dominate global discretionary apparel expenditure, are influenced more by price and performance rather than the textile fibre used.
- Worsted fabrics are made with longer fibres that produce a surface that's smooth to touch. Woollens are made with shorter fibres that stand up from the surface and give the fabric a hairy touch.
- Wool insulates against heat and cold, is healthy, water repellent, fire resistant, naturally elastic, wears longer, is versatile, resists static, insulates against noise, resists dirt, is easy to sew, is comfortable and dyes beautifully.

6

Synthetic Fibre

Synthetic fibers or fibres are the result of extensive research by scientists to improve on naturally occurring animal and plant fibers. In general, synthetic fibers are created by forcing, usually through extrusion, Fibre forming materials through holes (called spinnerets) into the air and water forming a thread. Before synthetic fibers were developed, artificially manufactured fibers were made from polymers obtained from petro chemicals. These fibers are called synthetic fibers and also called artificial fibers.Some of the fibers are made from plant cellulose.They are called as "cellulose fibers"

HISTORY

EARLY EXPERIMENTS

Sir Joseph Swan invented the first synthetic Fibre in the early 1880s. His Fibre was drawn from a cellulose liquid, formed by chemically modifying the Fibre contained in tree bark.

The synthetic Fibre produced through this process was chemically similar in its potential applications to the carbon filament Swan had developed for his incandescent light bulb, but Swan soon realized the potential of the Fibre to revolutionise textile manufacturing. In 1885 he unveiled fabrics he had manufactured from his synthetic material at the International Inventions Exhibition in London.

The next step was taken by Hilaire de Chardonnet, a French engineer and industrialist, who invented the first artificial silk, which he called "Chardonnet silk". In the late 1870s, Chardonnet was working with Louis Pasteur on a remedy to the epidemic that was destroying French silk worms. Failure to clean up a spill in the darkroom resulted in Chardonnet's discovery of nitrocellulose as a potential replacement for real silk.

Realizing the value of such a discovery, Chardonnet began to develop his new product, which he displayed at the Paris Exhibition of 1889. Unfortunately, Chardonnet's material was extremely flammable, and was subsequently replaced with other, more stable materials.

COMMERCIAL PRODUCTS

The first successful process was developed in 1894 by English chemist Charles Frederick Cross, and his collaborators Edward John Bevan and Clayton Beadle. They named the Fibre "viscose", because the reaction product of carbon disulfide and cellulose in basic conditions gave a highly viscous solution of xanthate. The first commercial viscose rayon was produced by the UK company Courtaulds Fibers in 1905.

The name "rayon" was adopted in 1924, with "viscose" being used for the viscous organic liquid used to make both rayon and cellophane. A similar product known as cellulose acetate was discovered in 1865. Rayon and acetate are both artificial fibers, but not truly synthetic, being made from wood.

Nylon, the first synthetic Fibre, was developed by Wallace Carothers, an American researcher at the chemical firm DuPont in the 1930s. It soon made its debut in the United States as a replacement for silk, just in time for the introduction of rationing during World War II. Its novel use as a material for women's stockings overshadowed more practical uses, such as a replacement for the silk in parachutes and other military uses like ropes.

The first polyester Fibre was introduced by John Rex Whinfield and James Tennant Dickson, British chemists working at the Calico Printers' Association, in 1941. They produced and patented the first polyester fibre which they named Terylene, also known as Dacron, equal to or surpassing nylon in toughness and resilience. ICI and DuPont went on to produce their own versions of the fibre.

DESCRIPTION

Synthetic fibers are made from synthesized polymers or small molecules. The compounds that are used to make these fibers come from raw materials such as petroleum based chemicals or petrochemicals.

These materials are polymerized into a long, linear chemical that bond two adjacent carbon atoms. Differing chemical compounds will be used to produce different types of fibers.

Synthetic fibers account for about half of all Fibre usage, with applications in every field of Fibre and textile technology. Although many classes of Fibre based on synthetic polymers have been evaluated as potentially valuable commercial products, four of them - nylon, polyester, acrylic and polyolefin - dominate the market. These four account for approximately 98 percent by volume of synthetic Fibre production, with polyester alone accounting for around 60 per cent.

There are several methods of manufacturing synthetic fibers but the most common is the Melt-Spinning Process. It involves heating the Fibre until it begins to melt, then you must draw out the melt with tweezers as quickly as possible. The next step would be to draw the molecules by aligning them in a parallel arrangement. This brings the fibers closer together and allows them to crystallize and orient. Lastly, is Heat-Setting. This utilizes heat to permeate the shape/dimensions of the fabrics made from heat-sensitive fibers.

ADVANTAGES

A great advantage of synthetic fibers is that they are more durable than most natural fibers.

In addition, many synthetic fibers offer consumer-friendly functions such as stretching, waterproofing and stain resistance. Overtime, things like sunlight and oils from human skin cause fibers in various fabrics to break down and wear away. Natural fibers are much more sensitive to these elements than synthetic blends. This is mainly because natural products tend to be biodegradable.

In addition, natural fibers often fall prey to moth and carpet beetles, whose larvae feast on things like cotton, wool and silk. Synthetic fibers are not a good food source for these fabric-damaging insects.

As an added advantage, synthetic fibers do not break down easily when exposed to light, water, or oil. Compared to natural fibers, many synthetic fibers are more water resistant and stain resistant. Some are even specially enhanced to withstand damage from water or stains. Some fabrics are also designed to stretch in specific ways, which makes them more comfortable to wear.

In many cases, synthetic fibers are environmental fabric choices. Cotton is incredibly resource intensive, as it takes a lot of water to farm cotton. Wool is not much better, as the sheep that produce wool need water, food and a lot of grazing land in order to survive. Although synthetic Fibre production does involve some carbon emissions, the environmental footprint of many fibers is much lower in comparison to natural fibers.

Many synthetic fibers create highly attractive fabrics. Modern synthetic fabrics can look and feel as luxurious as silk or wool.

DISADVANTAGES

Most of synthetic fibers disadvantages are relate to their low melting temperature:

- Synthetic fibers burn more readily than natural.
- Prone to heat damage. Melt relatively easily.
- Prone to damage by hot washing.
- More electrostatic charge is generated by rubbing than with natural fibres.
- Not skin friendly, so uncomfortable for long wearing.
- Allergic to some persons.
- Non-biodegradable in comparison to natural fibres.

COMMON SYNTHETIC FIBERS

Common synthetic fibers include:

- Nylon (1931)
- Modacrylic (1949)
- Olefin (1949)
- Acrylic (1950)
- Polyester (1953)
- Carbon Fibre (1958)

Specialty synthetic fibers include:

- Rayon (1894) artificial silk
- Vinyon (1939)
- Saran (1941)
- Spandex (1959)
- Vinalon (1939)
- Aramids (1961) - known as Nomex, Kevlar and Twaron
- Modal (1960's)
- Dyneema/Spectra (1979)
- PBI (Polybenzimidazole Fibre) (1983)
- Sulfar (1983)
- Lyocell (1992) (artificial, not synthetic)
- PLA (2002)
- M-5 (PIPD Fibre)
- Orlon
- Zylon (PBO Fibre)
- Vectran (TLCP Fibre) made from Vectra LCP polymer
- Derclon used in manufacture of rugs

Other synthetic materials used in fibers include:

- Acrylonitrile rubber (1930)

Modern fibers that are made from older artificial materials include:

- Glass Fibre (1938) is used for:
  - industrial, automotive, and home insulation (glass wool)
  - reinforcement of composite materials (glass-reinforced plastic, glass Fibre reinforced concrete)
  - specialty papers in battery separators and filtration
- Metallic Fibre (1946) is used for:
  - adding metallic properties to clothing for the purpose of fashion (usually made with composite plastic and metal foils)
  - elimination and prevention of static charge build-up
  - conducting electricity to transmit information
  - conduction of heat

In the horticulture industry synthetics are often used in soils to help the plants grow better. Examples are:

- expanded polystyrene flakes
- urea-formaldehyde foam resin
- polyurethane foam
- phenolic resin foam

MINERAL FIBERS

MINERAL FIBERS CAN BE PARTICULARLY STRONG BECAUSE THEY ARE FORMED WITH A LOW NUMBER OF SURFACE DEFECTS, ASBESTOS IS A COMMON ONE.GLASS FIBRE

Glass Fibre (also spelled glass fibre) is a material consisting of numerous extremely fine fibers of glass.

Glassmakers throughout history have experimented with glass fibers, but mass manufacture of glass Fibre was only made possible with the invention of finer machine tooling. In 1893, Edward Drummond Libbey exhibited a dress at the World's Columbian Exposition incorporating glass fibers with the diameter and texture of silk fibers. This was first worn by the popular stage actress of the time Georgia Cayvan. Glass fibres can also occur naturally, as Pele's hair.

Glass wool, which is one product called "fiberglass" today, was invented in 1938 by Russell Games Slayter of Owens-Corning, as a material to be used as thermal building insulation. It is marketed under the trade name Fiberglas, which has become a genericized trademark. Glass Fibre when used as a thermal insulating material, is specially manufactured with a bonding agent to trap many small air cells, resulting in the characteristically air-filled low-density "glass wool" family of products.

Glass Fibre has roughly comparable mechanical properties to other fibers such as polymers and carbon Fibre. Although not as strong or as rigid as carbon Fibre, it is much cheaper and significantly less brittle when used in composites. Glass fibers are therefore used as a reinforcing agent for many polymer products; to form a very strong and relatively lightweight Fibre-reinforced polymer (FRP) composite material called glass-reinforced plastic (GRP), also popularly known as "fiberglass". This structural material product contains little air, is more dense than glass wool, and is not an especially good thermal insulator.

Fibre formation

Glass Fibre is formed when thin strands of silica-based or other formulation glass are extruded into many fibers with small diameters suitable for textile processing. The technique of heating and drawing glass into fine fibers has been known for millennia; however, the use of these fibers for textile applications is more recent. Until this time, all glass Fibre had been manufactured as staple (that is, clusters of short lengths of Fibre). The first commercial production of glass Fibre was in 1936. In 1938 Owens-Illinois Glass Company and Corning

Glass Works joined to form the Owens-Corning Fiberglas Corporation. When the two companies joined to produce and promote glass Fibre, they introduced continuous filament glass fibers. Owens-Corning is still the major glass-Fibre producer in the market today.

Composition

The most common types of glass Fibre used in fiberglass is E-glass, which is alumino-borosilicate glass with less than 1per cent w/w alkali oxides, mainly used for glass-reinforced plastics. Other types of glass used are A-glass (Alkali-lime glass with little or no boron oxide), E-CR-glass (Electrical/Chemical Resistance; alumino-lime silicate with less than 1per cent w/w alkali oxides, with high acid resistance), C-glass (alkali-lime glass with high boron oxide content, used for glass staple fibers and insulation), D-glass (borosilicate glass, named for its low Dielectric constant), R-glass (alumino silicate glass without MgO and CaO with high mechanical requirements as reinforcement), and S-glass (alumino silicate glass without CaO but with high MgO content with high tensile strength).

Naming and use

Pure silica (silicon dioxide), when cooled as fused quartz into a glass with no true melting point, can be used as a glass Fibre for fiberglass, but has the drawback that it must be worked at very high temperatures. In order to lower the necessary work temperature, other materials are introduced as "fluxing agents" (*i.e.*, components to lower the melting point). Ordinary A-glass ("A" for "alkali-lime") or soda lime glass, crushed and ready to be remelted, as so-called cullet glass, was the first type of glass used for fiberglass. E-glass ("E" because of initial electrical application), is alkali free, and was the first glass formulation used for continuous filament formation. It now makes up most of the fiberglass production in the world, and also is the single largest consumer of boron minerals globally. It is susceptible to chloride ion attack and is a poor choice for marine applications. S-glass ("S" for "stiff") is used when high tensile strength (modulus) is important, and is thus an important building and aircraft epoxy composite. The same substance is known as R-glass ("R" for "reinforcement") in Europe). C-glass ("C" for "chemical resistance") and T-glass ("T" is for "thermal insulator" — a North American variant of C-glass) are resistant to chemical attack; both are often found in insulation-grades of blown fiberglass.

Chemistry

The basis of textile-grade glass fibers is silica, SiO_2. In its pure form it exists as a polymer, $(SiO_2)_n$. It has no true melting point but softens up to 1200 °C, where it starts to degrade. At 1713 °C, most of the molecules can move about freely. If the glass is extruded and cooled quickly at this temperature, it will be unable to form an ordered structure. In the polymer it forms SiO_4 groups which are configured as a tetrahedron with the silicon atom at the center, and four oxygen atoms at the corners. These atoms then form a network bonded at the corners by sharing the oxygen atoms.

The vitreous and crystalline states of silica (glass and quartz) have similar energy levels on a molecular basis, also implying that the glassy form is extremely stable. In order to induce crystallization, it must be heated to temperatures above 1200 °C for long periods of time.

Although pure silica is a perfectly viable glass and glass Fibre, it must be worked with at very high temperatures, which is a drawback unless its specific chemical properties are needed. It is usual to introduce impurities into the glass in the form of other materials to lower its working temperature. These materials also impart various other properties to the glass that may be beneficial in different applications. The first type of glass used for Fibre was soda lime glass or A-glass ("A" for the alkali it contains). It is not very resistant to alkali. A new type, E-glass, was formed; this is an alumino-borosilicate glass that is alkali-free (<2per cent). This was the first glass formulation used for continuous filament formation. E-glass still makes up most of the glass Fibre production in the world. Its particular components may differ slightly in percentage, but must fall within a specific range.

The letter E is used because it was originally for electrical applications. S-glass (S for "stiff") is a high-strength formulation for use when tensile strength is the most important property. C-glass was developed to resist attack from chemicals, mostly acids that destroy E-glass. T-glass is a North American variant of C-glass. A-glass is an industry term for cullet glass, often bottles, made into Fibre. AR-glass is alkali-resistant glass. Most glass fibers have limited solubility in water but are very dependent on pH. Chloride ions will also attack and dissolve E-glass surfaces.

E-glass does not really melt, but softens instead, the softening point being "the temperature at which a 0.55–0.77 mm diameter Fibre 235 mm long, elongates under its own weight at 1 mm/min when suspended vertically and heated at the rate of 5 °C per minute". The strain point is reached when the glass has a viscosity of 10 poise. The annealing point, which is the temperature where the internal stresses are reduced to an acceptable commercial limit in 15 minutes, is marked by a viscosity of 10 poise.

Properties

Thermal

Glass fibers are useful thermal insulators because of their high ratio of surface area to weight. However, the increased surface area makes them much more susceptible to chemical attack. By trapping air within them, blocks of glass Fibre make good thermal insulation, with a thermal conductivity of the order of 0.05 W/(m·K).

Tensile

The strength of glass is usually tested and reported for "virgin" or pristine fibers—those that have just been manufactured. The freshest, thinnest fibers are the strongest because the thinner fibers are more ductile. The more the surface is scratched, the less the resulting tenacity. Because glass has an amorphous structure, its properties are the same along the Fibre and across the Fibre. Humidity is an important factor in the tensile strength. Moisture is easily adsorbed and can worsen microscopic cracks and surface defects, and lessen tenacity.

In contrast to carbon Fibre, glass can undergo more elongation before it breaks. There is a correlation between bending diameter of the filament and the filament diameter. The viscosity of the molten glass is very important for manufacturing success. During drawing (pulling of the glass to reduce Fibre circumference), the viscosity must be relatively low. If it is too high, the Fibre will break during drawing. However, if it is too low, the glass will form droplets rather than drawing out into Fibre.

Manufacturing processes

Melting

There are two main types of glass Fibre manufacture and two main types of glass Fibre product. First, Fibre is made either from a direct melt process or a marble remelt process. Both start with the raw materials in solid form. The materials are mixed together and melted in a furnace. Then, for the marble process, the molten material is sheared and rolled into marbles which are cooled and packaged. The marbles are taken to the Fibre manufacturing facility where they are inserted into a can and remelted. The molten glass is extruded to the bushing to be formed into Fibre. In the direct melt process, the molten glass in the furnace goes directly to the bushing for formation.

Formation

The bushing plate is the most important part of the machinery for making the Fibre. This is a small metal furnace containing nozzles for the Fibre to be formed through. It is almost always made of platinum alloyed with rhodium for durability. Platinum is used because the glass melt has a natural affinity for wetting it. When bushings were first used they were 100per cent platinum, and the glass wetted the bushing so easily that it ran under the plate after exiting the nozzle and accumulated on the underside. Also, due to its cost and the tendency to wear, the platinum was alloyed with rhodium. In the direct melt process, the bushing serves as a collector for the molten glass. It is heated slightly to keep the glass at the correct temperature for Fibre formation. In the marble melt process, the bushing acts more like a furnace as it melts more of the material.

Bushings are the major expense in Fibre glass production. The nozzle design is also critical. The number of nozzles ranges from 200 to 4000 in multiples of 200. The important part of the nozzle in continuous filament manufacture is the thickness of its walls in the exit region. It was found that inserting a counterbore here reduced wetting. Today, the nozzles are designed to have a minimum thickness at the exit. As glass flows through the nozzle, it forms a drop which is suspended from the end. As it falls, it leaves a thread attached by the meniscus to the nozzle as long as the viscosity is in the correct range for Fibre formation. The smaller the annular ring of the nozzle and the thinner the wall at exit, the faster the drop will form and fall away, and the lower its tendency to wet the vertical part of the nozzle. The surface tension of the glass is what influences the formation of the meniscus. For E-glass it should be around 400 mN/m.

The attenuation (drawing) speed is important in the nozzle design. Although slowing this speed down can make coarser Fibre, it is uneconomic to run at speeds for which the nozzles were not designed.

Continuous filament process

In the continuous filament process, after the Fibre is drawn, a size is applied. This size helps protect the Fibre as it is wound onto a bobbin. The particular size applied relates to end-use. While some sizes are processing aids, others make the Fibre have an affinity for a certain resin, if the Fibre is to be used in a composite. Size is usually added at 0.5–2.0per cent by weight. Winding then takes place at around 1000 m/min.

Staple Fibre process

For staple Fibre production, there are a number of ways to manufacture the Fibre. The glass can be blown or blasted with heat or steam after exiting the formation machine. Usually these fibers are made into some sort of mat. The most common process used is the rotary process. Here, the glass enters a rotating spinner, and due to centrifugal force is thrown out horizontally. The air jets push it down vertically, and binder is applied. Then the mat is vacuumed to a screen and the binder is cured in the oven.

Safety

Glass Fibre has increased in popularity since the discovery that asbestos causes cancer and its subsequent removal from most products. However, the safety of glass Fibre is also being called into question, as research shows that the composition of this material (asbestos and glass Fibre are both silicate fibers) can cause similar toxicity as asbestos.

1970s studies on rats found that fibrous glass of less than 3 micrometers in diameter and greater than 20 micrometers in length is a "potent carcinogen". Likewise, the International Agency for Research on Cancer found it "may reasonably be anticipated to be a carcinogen" in 1990. The American Conference of Governmental Industrial Hygienists, on the other hand, says that there is insufficient evidence, and that glass Fibre is in group A4: "Not classifiable as a human carcinogen".

The North American Insulation Manufacturers Association (NAIMA) claims that glass Fibre is fundamentally different from asbestos, since it is man-made instead of naturally-occurring. They claim that glass Fibre "dissolves in the lungs", while asbestos remains in the body for life. Although both glass Fibre and asbestos are made from silica filaments, NAIMA claims that asbestos is more dangerous because of its crystalline structure, which causes it to cleave into smaller, more dangerous pieces, citing the U.S. Department of Health and Human Services:

- Synthetic vitreous fibers [Fibre glass] differ from asbestos in two ways that may provide at least partial explanations for their lower toxicity. Because most synthetic vitreous fibers are not crystalline like asbestos, they do not split longitudinally to form thinner fibers. They also generally have markedly less biopersistence in biological tissues than asbestos fibers because they can undergo dissolution and transverse breakage.

A 1998 study using rats found that the biopersistence of synthetic fibers after one year was 0.04–10per cent, but 27per cent for amosite asbestos. Fibers that persisted longer were found to be more carcinogenic.

Glass-reinforced plastic (fiberglass)

Glass-reinforced plastic (GRP) is a composite material or Fibre-reinforced plastic made of a plastic reinforced by fine glass fibers. Like graphite-reinforced plastic, the composite material is commonly referred to as fiberglass. The glass can be in the form of a chopped strand mat (CSM) or a woven fabric.

As with many other composite materials (such as reinforced concrete), the two materials act together, each overcoming the deficits of the other. Whereas the plastic resins are strong in compressive loading and relatively weak in tensile strength, the glass fibers are very strong in tension but tend not to resist compression. By combining the two materials, GRP becomes a material that resists both compressive and tensile forces well. The two materials may be used uniformly or the glass may be specifically placed in those portions of the structure that will experience tensile loads.

Uses

Uses for regular glass Fibre include mats and fabrics for thermal insulation, electrical insulation, sound insulation, high-strength fabrics or heat- and corrosion-resistant fabrics. It is also used to reinforce various materials, such as tent poles, pole vault poles, arrows, bows and crossbows, translucent roofing panels, automobile bodies, hockey sticks, surfboards, boat hulls, and paper honeycomb. It has been used for medical purposes in casts. Glass Fibre is extensively used for making FRP tanks and vessels.

Open-weave glass Fibre grids are used to reinforce asphalt pavement. Non-woven glass Fibre/polymer blend mats are used saturated with asphalt emulsion and overlaid with asphalt, producing a waterproof, crack-resistant membrane. Use of glass-Fibre reinforced polymer rebar instead of steel rebar shows promise in areas where avoidance of steel corrosion is desired.

Role of recycling in glass Fibre manufacturing

MANUFACTURERS OF GLASS-FIBRE INSULATION CAN USE RECYCLED GLASS. RECYCLED GLASS FIBRE HAS UP TO A 40PER CENT RECYCLED GLASS.CARBON (FIBRE)

Carbon Fibre, alternatively graphite Fibre or CF, is a material consisting of fibers about 5–10 ìm in diameter and composed mostly of carbon atoms.

To produce carbon Fibre, the carbon atoms are bonded together in crystals that are more or less aligned parallel to the long axis of the Fibre as the crystal alignment gives the Fibre high strength-to-volume ratio (making it strong for its size). Several thousand carbon fibers are bundled together to form a tow, which may be used by itself or woven into a fabric.

The properties of carbon fibers, such as high stiffness, high tensile strength, low weight, high chemical resistance, high temperature tolerance and low thermal expansion, make them very popular in aerospace, civil engineering, military, and motorsports, along with other competition sports. However, they are relatively expensive when compared to similar fibers, such as glass fibers or plastic fibers.

Carbon fibers are usually combined with other materials to form a composite. When combined with a plastic resin and wound or molded it forms carbon Fibre reinforced polymer (often referred to as carbon Fibre) which has a very high strength-to-weight ratio, and is extremely rigid although somewhat brittle. However, carbon fibers are also composited with other materials, such as with graphite to form carbon-carbon composites, which have a very high heat tolerance.

In 1879, Thomas Edison baked cotton threads or bamboo slivers at high temperatures carbonizing them into an all-carbon Fibre filament used in the first incandescent light bulb to be heated by electricity. In 1880, Lewis Latimer developed a reliable carbon wire filament for the incandescent light bulb, heated by electricity. In 1958, Roger Bacon created high-performance carbon fibers at the Union Carbide Parma Technical Center, now GrafTech International Holdings, Inc., located outside of Cleveland, Ohio. Those fibers were manufactured by heating strands of rayon until they carbonized. This process proved to be inefficient, as the resulting fibers contained only about 20per cent carbon and had low strength and stiffness properties. In the early 1960s, a process was developed by Dr. Akio Shindo at Agency of Industrial Science and Technology of Japan, using polyacrylonitrile (PAN) as a raw material. This had produced a carbon Fibre that contained about 55per cent carbon.

The high potential strength of carbon Fibre was realized in 1963 in a process developed by W. Watt, L. N. Phillips, and W. Johnson at the Royal Aircraft Establishment at Farnborough, Hampshire. The process was patented by the UK Ministry of Defence, then licensed by the NRDC to three British companies: Rolls-Royce already making carbon Fibre, Morganite, and Courtaulds. Within a few years, after successful use in 1968 of a *Hyfil* carbon-fibre fan assembly in the Conways of the Vickers VC10s operated by BOAC, Rolls-Royce took advantage of the new material's properties to break into the American market with its RB-211 aero-engine with carbon-Fibre compressor blades. Unfortunately, the blades proved vulnerable to damage from bird impact. This problem and others caused Rolls-Royce such setbacks that the company was nationalized in 1971. The carbon-Fibre production plant was sold off to form "Bristol Composites".

In the late 1960s, the Japanese took the lead in manufacturing PAN-based carbon fibers. The 1970 joint technology agreement allowed Union Carbide to manufacture the Japan's Toray Industries superior product and USA to dominate the market. Morganite decided that carbon-Fibre production was peripheral to its core business, leaving Courtaulds as the only big UK manufacturer. Continuing collaboration with the staff at Farnborough proved helpful in the quest for higher quality and improvements in the speed of production as Courtaulds developed two main markets: aerospace and sports equipment. However Courtaulds's big advantage as manufacturer of the "Courtelle" precursor now became a weakness. Courtelle's low cost and ready availability were potential advantages, but the water-based inorganic process used to produce it made the product susceptible to impurities that did not affect the organic process used by other carbon-Fibre manufacturers.

Nevertheless, during the 1980s Courtaulds continued to be a major supplier of carbon Fibre for the sports-goods market, with Mitsubishi its main customer until a move to expand, including building a production plant in California, turned out badly. The investment did not generate the anticipated returns, leading to a decision to pull out of the area and Courtaulds ceased carbon-Fibre production in 1991. Ironically the one surviving UK carbon-Fibre manufacturer continued to thrive making Fibre based on Courtaulds's precursor. Inverness-based RK Carbon Fibres Ltd concentrated on producing carbon Fibre for industrial applications, removing the need to compete at the quality levels reached by overseas manufacturers.

During the 1960s, experimental work to find alternative raw materials led to the introduction of carbon fibers made from a petroleum pitch derived from oil processing. These fibers contained about 85per cent carbon and had excellent flexural strength.

Also, during this period, the Japanese Government heavily supported carbon Fibre development at home and several Japanese companies such as Toray, Nippon Carbon, Toho Rayon and Mitsubishi started their own development and production. As they subsequently advanced to become market leaders, companies in USA and Europe were encouraged to take up these activities as well, either through their own developments or contractual acquisition of carbon Fibre knowledge. These companies included Hercules, BASF and Celanese USA and Akzo in Europe.

Since the late 1970s, further types of carbon Fibre yarn entered the global market, offering higher tensile strength and higher elastic modulus. For example, T400 from Toray with a tensile strength of 4,000 MPa and M40, a modulus of 400 GPa. Intermediate carbon fibers, such as IM 600 from Toho Reyon with up to 6,000 MPa were developed. Carbon fibers from Toray, Celanese and Akzo found their way to aerospace application from secondary to primary parts first in military and later in civil aircraft as in McDonnell Douglas, Boeing and Airbus planes. By 2000 the industrial applications for highly sophisticated machine parts in middle Europe was becoming more important.

Structure and properties

Carbon Fibre is frequently supplied in the form of a continuous tow wound onto a reel. The tow is a bundle of thousands of continuous individual carbon filaments held together and protected by an organic coating, or size, such as polyethylene oxide (PEO) or polyvinyl alcohol (PVA). The tow can be conveniently unwound from the reel for use. Each carbon filament in the tow is a continuous cylinder with a diameter of 5–8 micrometers and consists almost exclusively of carbon. The earliest generation (*e.g.* T300, HTA and AS4) had diameters of 16–22 micrometers. Later fibers (*e.g.* IM6 or IM600) have diameters that are approximately 5 micrometers.

The atomic structure of carbon Fibre is similar to that of graphite, consisting of sheets of carbon atoms (graphene sheets) arranged in a regular hexagonal pattern, the difference being in the way these sheets interlock. Graphite is a crystalline material in which the sheets are stacked parallel to one another in regular fashion. The intermolecular forces between the sheets are relatively weak Van der Waals forces, giving graphite its soft and brittle characteristics.

Depending upon the precursor to make the Fibre, carbon Fibre may be turbostratic or graphitic, or have a hybrid structure with both graphitic and turbostratic parts present. In turbostratic carbon Fibre the sheets of carbon atoms are haphazardly folded, or crumpled, together. Carbon fibers derived from Polyacrylonitrile (PAN) are turbostratic, whereas carbon fibers derived from mesophase pitch are graphitic after heat treatment at temperatures exceeding 2200 °C. Turbostratic carbon fibers tend to have high tensile strength, whereas heat-treated mesophase-pitch-derived carbon fibers have high Young's modulus (*i.e.*, high stiffness or resistance to extension under load) and high thermal conductivity.

Applications

The global demand on carbon Fibre composites was valued at roughly US$10.8 billion in 2009, which declined 8–10per cent from the previous year. It is expected to reach US$13.2 billion by 2012 and to increase to US$18.6 billion by 2015 with an annual growth rate of 7per cent or more. Strongest demands come from aircraft and aerospace, wind energy, as well as from the automotive industry with optimized resin systems.

Composite materials

Carbon Fibre is most notably used to reinforce composite materials, particularly the class of materials known as carbon Fibre or graphite reinforced polymers. Non-polymer materials can also be used as the matrix for carbon fibers. Due to the formation of metal carbides and corrosion considerations, carbon has seen limited success in metal matrix composite applications. Reinforced carbon-carbon (RCC) consists of carbon Fibre-reinforced graphite, and is used structurally in high-temperature applications. The Fibre also finds use in filtration of high-temperature gases, as an electrode with high surface area and impeccable corrosion resistance, and as an anti-static component. Molding a thin layer of carbon fibers significantly improves fire resistance of polymers or thermoset composites because a dense, compact layer of carbon fibers efficiently reflects heat.

The increasing use of carbon Fibre composites is displacing aluminum from aerospace applications in Favour of other metals because of galvanic corrosion issues.

Textiles

Precursors for carbon fibers are polyacrylonitrile (PAN), rayon and pitch. Carbon Fibre filament yarns are used in several processing techniques: the direct uses are for prepregging, filament winding, pultrusion, weaving, braiding, etc. Carbon Fibre yarn is rated by the linear density (weight per unit length, *i.e.* 1 g/1000 m = 1 tex) or by number of filaments per yarn count, in thousands.

For example, 200 tex for 3,000 filaments of carbon Fibre is three times as strong as 1,000 carbon filament yarn, but is also three times as heavy. This thread can then be used to weave a carbon Fibre filament fabric or cloth. The appearance of this fabric generally depends on the linear density of the yarn and the weave chosen. Some commonly used types of weave are twill, satin and plain. Carbon filament yarns can be also knitted or braided.

Microelectrodes

Carbon fibers are used for fabrication of carbon-Fibre microelectrodes. In this application typically a single carbon Fibre with diameter of 5–7pm is sealed in a glass capillary. At the tip the capillary is either sealed with epoxy and polished to make carbon-Fibre disk microelectrode or the Fibre is cut to a length of 75–50 pm to make carbon-Fibre cylinder electrode. Carbon-Fibre microelectrodes are used either in amperometry or fast-scan cyclic voltammetry for detection of biochemical signaling.

Catalysis

PAN-based nanofibers can efficiently catalyze the first step in the making of synthetic gasoline (syngas) and other energy-rich products out of carbon dioxide. The process uses a "co-catalyst" system in three steps: (1) EMIM–CO2 complex formation; (2) adsorption of EMIM–CO2 complex on reduced carbon atoms and (3) Carbon monoxide formation.

The first step uses an ionic liquid, while graphitic carbon structures doped with other reactive atoms replaced silver to produce the final output. The carbon nanofibre catalyst exhibited negligible overpotential (0.17 V) for carbon dioxide reduction and more than an order of magnitude higher current density compared with silver under similar experimental conditions. The reduction derived from the reduced carbons rather than to electronegative nitrogen dopants. The performance came from the nanofibrillar structure and high binding energy of key intermediates to the carbon nanofibre surfaces.

Synthesis

Each carbon filament is produced from a polymer such as polyacrylonitrile (PAN), rayon, or petroleum pitch, known as a precursor. For synthetic polymers such as PAN or rayon, the precursor is first spun into filament yarns, using chemical and mechanical processes to initially align the polymer atoms in a way to enhance the final physical properties of the completed carbon Fibre. Precursor compositions and mechanical processes used during spinning filament yarns may vary among manufacturers. After drawing or spinning, the polymer filament yarns are then heated to drive off non-carbon atoms (carbonization), producing the final carbon Fibre. The carbon fibers filament yarns may be further treated to improve handling qualities, then wound on to bobbins.

A common method of manufacture involves heating the spun PAN filaments to approximately 300 °C in air, which breaks many of the hydrogen bonds and oxidizes the material. The oxidized PAN is then placed into a furnace having an inert atmosphere of a gas such as argon, and heated to approximately 2000 °C, which induces graphitization of the material, changing the molecular bond structure. When heated in the correct conditions, these chains bond side-to-side (ladder polymers), forming narrow graphene sheets which eventually merge to form a single, columnar filament. The result is usually 93–95per cent carbon. Lower-quality Fibre can be manufactured using pitch or rayon as the precursor instead of PAN. The carbon can become further enhanced, as high modulus, or high strength carbon, by heat treatment processes. Carbon heated in the range of 1500–2000 °C (carbonization) exhibits the highest tensile strength (820,000 psi, 5,650 MPa or N/mm^2), while carbon Fibre heated from 2500 to 3000 °C (graphitizing) exhibits a higher modulus of elasticity (77,000,000 psi or 531 GPa or 531 kN/mm^2).

Manufacturers

Major manufacturers of carbon fibers include Toho Tenax, Cytec Industries, Formosa Plastics, Hexcel, Mitsubishi Rayon, SGL Carbon, Toray Industries and Zoltek. Manufacturers typically make different grades of fibers for different applications. Higher modulus carbon fibers are typically more expensive.

BASALT FIBRE

Basalt Fibre is a material made from extremely fine fibers of basalt, which is composed of the minerals plagioclase, pyroxene, and olivine. It is similar to carbon Fibre and fiberglass, having better physicomechanical properties than fiberglass, but being significantly cheaper than carbon Fibre. It is used as a fireproof textile in the aerospace and automotive industries and can also be used as a composite to produce products such as camera tripods.

Basalt Fibre is made from a single material, crushed basalt, from a carefully chosen quarry source and unlike other materials such as glass Fibre, essentially no materials are added. The basalt is simply washed and then melted.

The manufacture of basalt Fibre requires the melting of the quarried basalt rock at about 1,400 °C (2,550 °F). The molten rock is then extruded through small nozzles to produce continuous filaments of basalt Fibre. There are three main manufacturing techniques, which are centrifugal-blowing, centrifugal-multiroll and die-blowing. The fibers typically have a filament diameter of between 9 and 13 μm which is far enough above the respiratory limit of 5 μm. They also have a high elastic modulus, resulting in excellent specific tenacity—three times that of steel.

The first attempts to produce basalt Fibre were made in the United States in 1923. These were further developed after World War II by researchers in the USA, Europe and the Soviet Union especially for military and aerospace applications. Since declassification in 1995 basalt fibers have been used in a wider range of civilian applications.

Uses

- Franky Morales 3 skate by USD
- Heat protection
- Friction materials
- High pressure vessels (*e.g.* tanks and gas cylinders)
- Load bearing profiles
- Windmill blades
- Lamp posts
- Ship hulls
- Car bodies
- Sports equipment
- Concrete reinforcement (*e.g.* for bridges and buildings)
- Speaker cones
- Cavity wall ties

METALLIC FIBRE

Metallic fibers are manufactured fibers composed of metal, plastic-coated metal, metal-coated plastic, or a core completely covered by metal. Gold and silver have been used since ancient times as yarns for fabric decoration. More recently, aluminum yarns, aluminized plastic yarns, and aluminized nylon yarns have replaced gold and silver. Metallic filaments can be coated with transparent films to minimize tarnishing.

Metal Fibre may also be shaved from wire (steel wool), bundle drawn from larger diameter wire, cast from molten metal, or grown around a seed (often carbon).

Gold and silver have been used since ancient times as decoration in the clothing and textiles of kings, leaders, nobility and people of status. Many of these elegant textiles can be found in museums around the world. Historically, the metallic thread was constructed by wrapping a metal strip around a Fibre core (cotton or silk), often in such a way as to reveal the Colour of the Fibre core to enhance visual quality of the decoration. Ancient textiles and clothing woven from wholly or partly gold threads is sometimes referred to as *Cloth of Gold*. They have been woven on Byzantine looms from the 7th to 9th Centuries, and after that in Sicily, Cyprus, Lucca, and Venice. Weaving also flourished in the 12th Century during the legacy of Genghis Khan when art and trade flourished under Mongol rule in China and some Middle Eastern areas. The Dobeckmum Company produced the first modern metallic Fibre in 1946. In the past, aluminum was usually the base in a metallic Fibre. More recently stainless steel has become a base as well. It is more difficult to work with but provides properties to the yarn that allows it to be used in more high tech applications.

Coated metallic filaments helps to minimize tarnishing. When suitable adhesives and films are used, they are not affected by salt water, chlorinated water in swimming pools or climatic conditions. If possible anything made with metallic fibers should be dry-cleaned, if there is no care label. Ironing can be problematic because the heat from the iron, especially at high temperatures, can melt the fibers.

Production method

There are two basic processes that are used in manufacturing metallic fibers. The most common is the *laminating process*, which seals a layer of aluminum between two layers of acetate or polyester film. These fibers are then cut into lengthwise strips for yarns and wound onto bobbins. The metal can be colored and sealed in a clear film, the adhesive can be colored, or the film can be colored before laminating. There are many different variations of Colour and effect that can be made in metallic fibers, producing a wide range of looks.

Metallic fibers can also be made by using the *metalizing process*. This process involves heating the metal until it vaporizes then depositing it at a high pressure onto the polyester film. This process produces thinner, more flexible, more durable, and more comfortable fibers.

Metal Fibre may also be shaved from wire (steel wool), bundle drawn from larger diameter wire (smallest Fibre is produced by this method), cast from molten metal, or grown around a seed (often carbon). Bundle drawn metal Fibre can be produced to sizes smaller than one micrometre in diameter.

Producers

Currently metallic fibers are manufactured primarily in Europe with only three manufacturers still producing metallic yarn in the United States. Metlon Corporation is one of the remaining manufacturers in the U.S. that stocks a wide variety of laminated and non-laminated metallic yarns.

Trademarks

The Lurex Company has manufactured metallic fibers in Europe for over fifty years. They produce a wide variety of metallic Fibre products including fibers used in apparel fabric, embroidery, braids, knitting, military regalia, trimmings, ropes, cords, and lace surface decoration. The majority of Lurex fibers have a polyamdie film covering the metal strand but polyester and viscose are also used. The fibers are also treated with a lubricant called P.W., a mineral-based oil, which helps provide ease of use.

Metlon Corporation is a trademark of Metallic Yarns in the United States and has been producing metallic yarns for over sixty years. Metlon produces their metallic yarn by wrapping single slit yarns with two ends of nylon. One end of nylon is wrapped clockwise and the other end is wrapped counterclockwise around the metallic yarn. The most commonly used nylon is either 15 denier or 20 denier, but heavier deniers are used for special purposes.

Wagner-Tech-Textil GmbH and the brand name rexess® is probably the last metallic yarn producer in Europe. All metallic yarns rexess, in many different colours, are Oeko-Tex Standart 100 certified.

Uses

The most common uses for metallic fibers is upholstery fabric and textiles such as lamé and brocade. Many people also use metallic fibers in weaving and needlepoint. Increasingly common today are metallic fibers in clothing, anything from party and evening wear to club clothing, cold weather and survival clothing, and everyday wear. Metallic yarns are woven, braided, and knit into many fashionable fabrics and trims. For additional variety, metallic yarns are twisted with other fibers such as wool, nylon, cotton, and synthetic blends to produce yarns which add novelty effects to the end cloth or trim. Stainless steel and other metal fibers are used in communication lines such as phone lines and cable television lines. Stainless steel fibers are also used in carpets. They are dispersed throughout the carpet with other fibers so they are not detected. The presence of the fibers helps to conduct electricity so that the static shock is reduced. These types of carpets are often used in computer-use areas where the chance of producing static is much greater. Other uses include tire cord, missile nose cones, work clothing such as protective suits, space suits, and cut resistant gloves for butchers and other people working near bladed or dangerous machinery.

POLYMER FIBERS

Polymer fibers are a subset of man-made fibers, which are based on synthetic chemicals (often from petrochemical sources) rather than arising from natural materials by a purely physical process. These fibers are made from:

- polyamide nylon
- PET or PBT polyester
- phenol-formaldehyde (PF)
- polyvinyl chloride Fibre (PVC) vinyon
- polyolefins (PP and PE) olefin Fibre
- acrylic polyesters, pure polyester PAN fibers are used to make carbon Fibre by roasting them in a low oxygen environment. Traditional acrylic Fibre is used more often as a synthetic replacement for wool. Carbon fibers and PF fibers are noted as two resin-based fibers that are not thermoplastic, most others can be melted.

- aromatic polyamids (aramids) such as Twaron, Kevlar and Nomex thermally degrade at high temperatures and do not melt. These fibers have strong bonding between polymer chains
- polyethylene (PE), eventually with extremely long chains/ HMPE (*e.g.* Dyneema or Spectra).
- Elastomers can even be used, *e.g.* spandex although urethane fibers are starting to replace spandex technology.
- polyurethane Fibre
- Elastolefin

Coextruded fibers have two distinct polymers forming the Fibre, usually as a core-sheath or side-by-side. Coated fibers exist such as nickel-coated to provide static elimination, silver-coated to provide anti-bacterial properties and aluminum-coated to provide RF deflection for radar chaff. Radar chaff is actually a spool of continuous glass tow that has been aluminum coated. An aircraft-mounted high speed cutter chops it up as it spews from a moving aircraft to confuse radar signals.

ACRYLIC FIBRE

Acrylic fibers are synthetic fibers made from a polymer (polyacrylonitrile) with an average molecular weight of ~100,000, about 1900 monomer units. To be called acrylic in the U.S, the polymer must contain at least 85per cent acrylonitrile monomer. Typical comonomers are vinyl acetate or methyl acrylate. DuPont created the first acrylic fibers in 1941 and trademarked them under the name Orlon. Acrylic is also called acrilan fabric. It was first developed in the mid-1940s but was not produced in large quantities until the 1950s. Strong and warm, acrylic Fibre is often used for sweaters and tracksuits and as linings for boots and gloves, as well as in furnishing fabrics and carpets. It is manufactured as a filament, then cut into short staple lengths similar to wool hairs, and spun into yarn.

Modacrylic is a modified acrylic Fibre that contains at least 35per cent and at most 85per cent acrylonitrile monomer. The comonomers vinyl chloride, vinylidene chloride or vinyl bromide used in modacrylic give the Fibre flame retardant properties. End-uses of modacrylic include faux fur, wigs, hair extensions and protective clothing.

Production

The polymer is formed by free-radical polymerization in aqueous suspension. The Fibre is produced by dissolving the polymer in a solvent such as N,N-dimethylformamide or aqueous sodium thiocyanate, metering it through a multi-hole spinnerette and coagulating the resultant filaments in an aqueous solution of the same solvent (wet spinning) or evaporating the solvent in a stream of heated inert gas (dry spinning). Washing, stretching, drying and crimping complete the processing. Acrylic fibers are produced in a range of deniers, typically from 0.9 to 15, as cut staple or as a 500,000 to 1 million filament tow. End uses include sweaters, hats, hand-knitting yarns, socks, rugs, awnings, boat covers, and upholstery; the Fibre is also used as "PAN" precursor for carbon Fibre. Production of acrylic fibers is centered in the Far East, Turkey, India, Mexico, and South America, though a number of European producers still continue to operate, including Dralon and Fisipe. US producers have ended production, though acrylic tow and staple are still spun into yarns in the USA. Former U.S. brands of acrylic were Acrilan (Monsanto), Creslan (American Cyanamid), and Orlon (DuPont). Other brand names that are still in use include Dralon (Dralon GmbH).

Textile uses

Acrylic is lightweight, soft, and warm, with a wool-like feel. It can also be made to mimic other fibers, such as cotton, when spun on short staple equipment. Some acrylic is extruded in colored or pigmented form; other is extruded in "ecru", otherwise known as "natural," "raw white," or "undyed." Pigmented Fibre has highest light-fastness. Its fibers are very resilient compared to both other synthetics and natural fibers. Some acrylic is used in clothing as a less expensive alternative to cashmere, due to the similar feeling of the materials. Some acrylic fabrics may fuzz or pill easily. Other fibers and fabrics are designed to minimize pilling. Acrylic takes Colour well, is washable, and is generally hypoallergenic. End-uses include socks, hats, gloves, scarves, sweaters, home furnishing fabrics, and awnings.

Acrylic is resistant to moths, oils, chemicals, and is very resistant to deterioration from sunlight exposure.

Acrylic is the "workhorse" hand-crafting Fibre for crafters who knit or crochet; acrylic yarn may be perceived as "cheap" because it is typically priced lower than its natural-Fibre counterparts, and because it lacks some of their properties, including softness and the ability to felt or take acid dyes. The Fibre requires heat to "kill" or set the shape of the finished garment, and it isn't as warm as alternatives like wool. Some knitters also complain that the Fibre "squeaks" when knitted, or that it is painful to knit with because of a lack of "give" or stretch in the yarn. On the other hand, it can be useful in certain items, like garments for babies, which require constant washing, because it is machine-washable and extremely Colour-fast. Acrylic can irritate the skin of people with dermatological conditions such as eczema.

ARAMID

Aramid fibers are a class of heat-resistant and strong synthetic fibers. They are used in aerospace and military applications, for ballistic rated body armor fabric and ballistic composites, in bicycle tires, and as an asbestos substitute.

The name is a portmanteau of "aromatic polyamide". They are fibers in which the chain molecules are highly oriented along the Fibre axis, so the strength of the chemical bond can be exploited.

Aromatic polyamides were first introduced in commercial applications in the early 1960s, with a meta-aramid Fibre produced by DuPont as HT-1 and then under the trade name Nomex. This Fibre, which handles similarly to normal textile apparel fibers, is characterized by its excellent resistance to heat, as it neither melts nor ignites in normal levels of oxygen. It is used extensively in the production of protective apparel, air filtration, thermal and electrical insulation as well as a substitute for asbestos. Meta-aramid is also produced in the Netherlands and Japan by Teijin under the trade name Conex, In Korea by Toray under the trade name Arawin, in China by Yantai Tayho under the trade name New Star, by SRO Group (China) under the trade name X-Fiper, and a variant of meta-aramid in France by Kermel under the trade name Kermel.

Based on earlier research by Monsanto Company and Bayer, a Fibre – para-aramid – with much higher tenacity and elastic modulus was also developed in the 1960s–1970s by DuPont and Akzo Nobel, both profiting from their knowledge of rayon, polyester and nylon processing. Much work was done by Stephanie Kwolek in 1961 while working at DuPont, and that company was the first to introduce a *para-aramid* called Kevlar in 1973. A similar Fibre called Twaron with roughly the same chemical structure was introduced by Akzo in 1978. Due to earlier patents on the production process, Akzo and DuPont engaged in a patent dispute in the 1980s. Twaron is currently owned by the Teijin company.

Para-aramids are used in many high-tech applications, such as aerospace and military applications, for "bullet-proof" body armor fabric.

The Federal Trade Commission definition for aramid Fibre is:

- A manufactured Fibre in which the Fibre-forming substance is a long-chain synthetic polyamide in which at least 85per cent of the amide linkages, (-CO-NH-) are attached directly to two aromatic rings.

Health

During the 1990s, an *in vitro* test of aramid fibers showed they exhibited "many of the same effects on epithelial cells as did asbestos, including increased radiolabeled nucleotide incorporation into DNA and induction of ODC (ornithine decarboxylase) enzyme activity", raising the possibility of carcinogenic implications. However, in 2009, it was shown that inhaled aramid fibrils are shortened and quickly cleared from the body and pose little risk.

Production

World capacity of para-aramid production is estimated at about 41,000 tonnes/year in 2002 and increases each year by 5–10per cent. In 2007 this means a total production capacity of around 55,000 tonnes/year.

Polymer preparation

Aramids are generally prepared by the reaction between an amine group and a carboxylic acid halide group. Simple AB homopolymers may look like:

nNH$_2$-Ar-COCl ! -(NH-Ar-CO)$_n$- + nHCl

The most well-known aramids (Kevlar, Twaron, Nomex, New Star and Teijinconex) are AABB polymers. Nomex, Teijinconex and New Star contain predominantly the meta-linkage and are poly-*metaphenylene isophthalamide*s (MPIA). Kevlar and Twaron are both *p*-phenylene terephthalamides (PPTA), the simplest form of the AABB para-polyaramide. PPTA is a product of *p*-phenylene diamine (PPD) and terephthaloyl dichloride (TDC or TCl). Production of PPTA relies on a co-solvent with an ionic component (calcium chloride ($CaCl_2$)) to occupy the hydrogen bonds of the amide groups, and an organic component (N-methyl pyrrolidone (NMP)) to dissolve the aromatic polymer. Prior to the invention of this process by Leo Vollbracht, who worked at the Dutch chemical firm Akzo, no practical means of dissolving the polymer was known. The use of this system led to an extended patent dispute between Akzo and DuPont.

Spinning lh3)

After production of the polymer, the aramid Fibre is produced by spinning the dissolved polymer to a solid Fibre from a liquid chemical blend. Polymer solvent for spinning PPTA is generally 100per cent anhydrous sulfuric acid (H_2SO_4).

Appearances

- Fibre
- Chopped Fibre
- Powder
- Pulp

Other types of aramids

Besides meta-aramids like Nomex, other variations belong to the aramid Fibre range. These are mainly of the copolyamide type, best known under the brand name Technora, as developed by Teijin and introduced in 1976. The manufacturing process of Technora reacts PPD and 3,4'-diaminodiphenylether (3,4'-ODA) with terephthaloyl chloride (TCl). This relatively simple process uses only one amide solvent and therefore spinning can be done directly after the polymer production.

Aramid Fibre characteristics

Aramids share a high degree of orientation with other fibers such as ultra high molecular weight polyethylene, a characteristic that dominates their properties.

General

- good resistance to abrasion
- good resistance to organic solvents
- Non-conductive
- no melting point, degradation starts from 500°C
- low flammability
- good fabric integrity at elevated temperatures
- sensitive to acids and salts
- sensitive to ultraviolet radiation
- prone to static build-up unless finished

Para-aramids

- para-aramid fibers such as Kevlar and Twaron, provide outstanding strength-to-weight properties
- high Young's modulus
- high tenacity
- low creep
- low elongation at break (~3.5per cent)
- difficult to dye – usually solution dyed

Major industrial uses

- flame-resistant clothing (example military MIL-G-181188B suits).
- heat protective clothing and helmets
- body armor, competing with PE based Fibre products such as Dyneema and Spectra
- composite materials
- asbestos replacement (*e.g.* brake linings)
- hot air filtration fabrics
- tires, newly as Sulfron (sulfur modified Twaron)
- mechanical rubber goods reinforcement
- ropes and cables
- wicks for fire dancing
- optical Fibre cable systems
- sail cloth (not necessarily racing boat sails)
- sporting goods
- drumheads
- wind instrument reeds, such as the Fibracell brand
- loudspeaker diaphragms
- boathull material
- Fibre reinforced concrete
- reinforced thermoplastic pipes
- tennis strings (*e.g.* by Ashaway and Prince tennis companies)

- hockey sticks (normally in composition with such materials as wood and carbon)
- snowboards
- jet engine enclosures

TWARON

Twaron is the brandname of Teijin Aramid for a para-aramid. It is a heat-resistant and strong synthetic fibre developed in the early 1970s by the Dutch company AKZO, division ENKA, later Akzo Industrial Fibers. The research name of the para-aramid fibre was originally Fibre X, but it was soon called Arenka. Although the Dutch para-aramid Fibre was developed only a little later than DuPont's Kevlar, introduction of Twaron as a commercial product came much later than Kevlar due to financial problems at the AKZO company in the 1970s.

This is a chronology of the development of Twaron:

- In 1960s a research Programme starts for "Fibre X."
- In 1972 the ENKA Research laboratory develops a para-aramid called Arenka.
- In 1973 Akzo decides to use sulfuric acid (H_2SO_4) as a solvent for spinning.
- In 1976 a pilot plant is built.
- In 1977 first production starts.
- In 1984 the product is renamed Twaron.
- In 1986 commercial production is started at five locations and nine plants.
- In 1987 Twaron is introduced as a commercial product.
- In 1989 the aramid business of Akzo becomes an independent Business Unit called Twaron BV.
- Since 2000 Twaron BV is owned by the Teijin Group, now called Teijin Twaron BV and based in Arnhem, The Netherlands. The main production facilities for Twaron are in Emmen and Delfzijl.
- In 2007 Teijin Twaron expands for the fourth time in six years and also changes its name into Teijin Aramid.

Production

Polymer preparation

Twaron is a *p*-phenylene terephthalamide (P*p*PTA), the simplest form of the AABB para polyaramide. P*p*PTA is a product of *p*-phenylene diamine (PPD) and terephthaloyl dichloride (TDC). To dissolve the aromatic polymer Twaron used a co-solvent of N-methyl pyrrolidone (NMP) and an ionic component (calcium chloride $CaCl_2$) to occupy the hydrogen bonds of the amide groups.

Prior to the invention of this process by Leo Vollbracht, working at the Dutch chemical firm AKZO, no practical means of dissolving the polymer was known. The use of this system by DuPont led to a patent war between AKZO and DuPont as Dupont initially used the carcinogenic HMPT (hexamethylphosphoramide). Despite heavy research DuPont now also uses the AKZO patent to use the less hazardous NMP in the Kevlar process.

Spinning

After the production of the Twaron polymer in Delfzijl, the polymer is brought to Emmen, where fibres are produced by spinning the dissolved polymer into a solid fibre from a liquid chemical blend. Polymer solvent for spinning PPTA is generally 100per cent anhydrous (water free) sulfuric acid (H_2SO_4).

The polymer is dissolved by mixing frozen sulfuric acid in powder form with the polymer in powder form and gently heating the mixture. This process, which differs from the more difficult DuPont process, was invented by Henri Lammers and patented by AKZO.

Industrial uses

Twaron is a para-aramid and is used in automotive, construction, sport, aerospace, military and industry applications, *e.g.*, "bullet-proof" body armor, fabric, and as an asbestos substitute.

- Protective gear (heat resistant/ ballistics): flame-resistant clothing, protective clothing and helmets, cut-fast or heat-hardy gloves, sporting goods, textiles, ballistic vests
- Composites: composite materials, technical paper, asbestos replacement, hot air filtration, sailcloth, speaker woofers, boat hull material, Fibre reinforced concrete, drumheads
- Automotive: brake pads, turbo hoses, V-belts and Timing belts, tires that incorporate Sulfron (sulfur modified Twaron), mechanical rubber goods reinforcement
- Linear tension: optical Fibre cables (OFC), ropes, wire ropes, cables, umbilical cables, electrical mechanical cable (EMC), reinforced thermoplastic pipes

KEVLAR

Kevlar is the registered trademark for a para-aramid synthetic Fibre, related to other aramids such as Nomex and Technora. Developed by Stephanie Kwolek at DuPont in 1965, this high-strength material was first commercially used in the early 1970s as a replacement for steel in racing tires. Typically it is spun into ropes or fabric sheets that can be used as such or as an ingredient in composite material components.

Currently, Kevlar has many applications, ranging from bicycle tires and racing sails to body armor because of its high tensile strength-to-weight ratio; by this measure it is 5 times stronger than steel. It is also used to make modern drumheads that withstand high impact. When used as a woven material, it is suitable for mooring lines and other underwater applications. A similar Fibre called Twaron with roughly the same chemical structure was developed by Akzo in the 1970s; commercial production started in 1986, and Twaron is now manufactured by Teijin.

Poly-paraphenylene terephthalamide – branded Kevlar – was invented by Polish-American chemist Stephanie Kwolek while working for DuPont, in anticipation of a gasoline shortage. In 1964, her group began searching for a new lightweight strong Fibre to use for light but strong tires. The polymers she had been working with at the time, poly-p-Phenylene-terephthalate and polybenzamide, formed liquid crystal while in solution, something unique to those polymers at the time.

The solution was "cloudy, opalescent upon being stirred, and of low viscosity" and usually was thrown away. However, Kwolek persuaded the technician, Charles Smullen, who ran the "spinneret", to test her solution, and was amazed to find that the Fibre did not break, unlike nylon. Her supervisor and her laboratory director understood the significance of her accidental discovery and a new field of polymer chemistry quickly arose. By 1971, modern Kevlar was introduced. However, Kwolek was not very involved in developing the applications of Kevlar.

Production

Kevlar is synthesized in solution from the monomers 1,4-phenylene-diamine (*para*-phenylenediamine) and terephthaloyl chloride in a condensation reaction yielding hydrochloric acid as a byproduct. The result has liquid-crystalline Behaviour, and mechanical drawing orients the polymer chains in the fiber's direction. Hexamethylphosphoramide (HMPA) was the solvent initially used for the polymerization, but for safety reasons, DuPont replaced it by a solution of *N*-methyl-pyrrolidone and calcium chloride. As this process had been patented by Akzo in the production of Twaron, a patent war ensued.

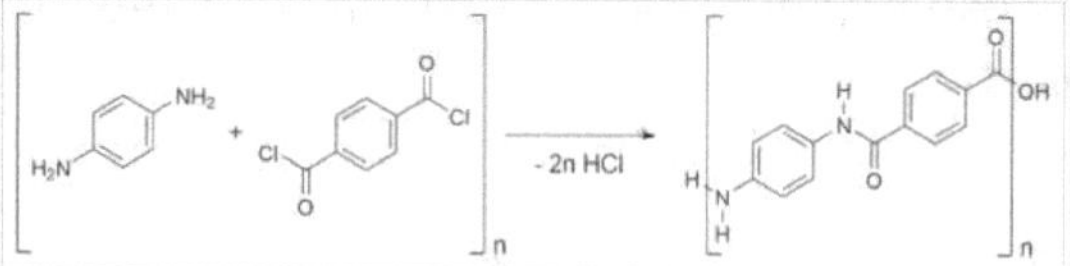

The reaction of 1,4-phenylene-diamine (*para*-phenylenediamine) with terephthaloyl chloride yielding kevlar

Kevlar (poly paraphenylene terephthalamide) production is expensive because of the difficulties arising from using concentrated sulfuric acid, needed to keep the water-insoluble polymer in solution during its synthesis and spinning.

Several grades of Kevlar are available:

1. *Kevlar K-29* – in industrial applications, such as cables, asbestos replacement, brake linings, and body/vehicle armor.
2. *Kevlar K49* – high modulus used in cable and rope products.
3. *Kevlar K100* – colored version of Kevlar
4. *Kevlar K119* – higher-elongation, flexible and more fatigue resistant
5. *Kevlar K129* – higher tenacity for ballistic applications
6. *Kevlar AP* – 15per cent higher tensile strength than K-29
7. *Kevlar XP* – lighter weight resin and KM2 plus Fibre combination
8. *Kevlar KM2* – enhanced ballistic resistance for armor applications

The ultraviolet component of sunlight degrades and decomposes Kevlar, a problem known as UV degradation, and so it is rarely used outdoors without protection against sunlight.

Structure and properties

When Kevlar is spun, the resulting Fibre has a tensile strength of about 3,620 MPa, and a relative density of 1.44. The polymer owes its high strength to the many inter-chain bonds. These inter-molecular hydrogen bonds form between the carbonyl groups and N*H* centers. Additional strength is derived from aromatic stacking interactions between adjacent strands. These interactions have a greater influence on Kevlar than the van der Waals interactions and chain length that typically influence the properties of other synthetic polymers and fibers such as Dyneema. The presence of salts and certain other impurities, especially calcium, could interfere with the strand interactions and care is taken to avoid inclusion in its production. Kevlar's structure consists of relatively rigid molecules which tend to form mostly planar sheet-like structures rather like silk protein.

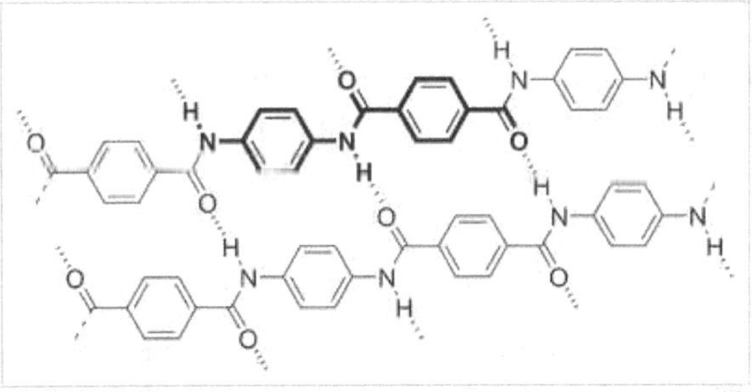

Molecular structure of Kevlar: **bold** represents a monomer unit, **dashed** lines indicate hydrogen bonds.

Thermal properties

Kevlar maintains its strength and resilience down to cryogenic temperatures ("196 °C); in fact, it is slightly stronger at low temperatures. At higher temperatures the tensile strength is immediately reduced by about 10–20per cent, and after some hours the strength progressively reduces further. For example at 160 °C (320 °F) about 10per cent reduction in strength occurs after 500 hours. At 260 °C (500 °F) 50per cent strength reduction occurs after 70 hours.

Applications

Protection

Cryogenics [sssh]

Kevlar is often used in the field of cryogenics for its low thermal conductivity and high strength relative to other materials for suspension purposes. It is most often used to suspend a paramagnetic salt enclosure from a superconducting magnet mandrel in order to minimize any heat leaks to the paramagnetic material. It is also used as a thermal standoff or structural support where low heat leaks are desired.

Armor [sssh]

Kevlar is a well-known component of personal armor such as combat helmets, ballistic face masks, and ballistic vests. The PASGT helmet and vest used by United States military forces since the 1980s both have Kevlar as a key component, as do their replacements. Other military uses include bulletproof facemasks used by sentries and spall liners used to protect the crews of armoured fighting vehicles. Even Nimitz-class aircraft carriers include Kevlar armor around vital spaces. Related civilian applications include Emergency Service's protection gear if it involves high heat (*e.g.*, tackling a fire), and Kevlar body armor such as vests for police officers, security, and SWAT.

Personal protection [sssh]

Kevlar is used to manufacture gloves, sleeves, jackets, chaps and other articles of clothing designed to protect users from cuts, abrasions and heat. Kevlar based protective gear is often considerably lighter and thinner than equivalent gear made of more traditional materials.

Sports equipment

It is used as an inner lining for some bicycle tires to prevent punctures. In table tennis, plies of Kevlar are added to custom ply blades, or paddles, in order to increase bounce and reduce weight. It is used for motorcycle safety clothing, especially in the areas featuring padding such as shoulders and elbows.

- In kyudo or Japanese archery, it may be used as an alternative to more expensive hemp for bow strings. It is one of the main materials used for paraglider suspension lines.

- In fencing it is used in the protective jackets, breeches, plastrons and the bib of the masks.
- Tennis racquets are often strung with Kevlar.
- It is even used in sails for high performance racing boats.
- It is increasingly being used in the "peto", the padded covering which protects the picadors' horses in the bullring.

Shoes [sssh]

With advancements in technology, Nike used Kevlar in shoes for the first time. It launched the Elite II Series, with enhancements to its earlier version of basketball shoes by using Kevlar in the anterior as well as the shoe laces. This was done to decrease the elasticity of the tip of the shoe in contrast to nylon used conventionally as Kevlar expanded by about 1per cent against nylon which expanded by about 30per cent. Shoes in this range included LeBron, HyperDunk and Zoom Kobe VII. However these shoes were launched at a price range much higher than average cost of basketball shoes.

It was also used as speed control patches for certain Soap Shoes models. and the laces for the adidas F50 adiZero Prime football boot.

Music

- Audio equipment: Kevlar has also been found to have useful acoustic properties for loudspeaker cones, specifically for bass and midrange drive units. Additionally, Kevlar has been used as a strength member in Fibre optic cables such as the ones used for audio data transmissions.
- Bowed string instruments: Kevlar can be used as an acoustic core on bows for string instruments. Kevlar's physical properties provide strength, flexibility, and stability for the bow's user. To date, the only manufacturer of this type of bow is CodaBow. Kevlar is also presently used as a material for tailcords (aka tailpiece adjusters), which connect the tailpiece to the endpin of bowed string instruments.
- Drumheads: Kevlar is sometimes used as a material on marching snare drums. It allows for an extremely high amount of tension, resulting in a cleaner sound. There is usually a resin poured onto the Kevlar to make the head airtight, and a nylon top layer to provide a flat striking surface. This is one of the primary types of marching snare drum heads. Remo's "Falam Slam" Patch is made with Kevlar and is used to reinforce bass drum heads where the beater strikes.
- Woodwind reeds: Kevlar is used in the woodwind reeds of Fibracell. The material of these reeds is a composite of aerospace materials designed to duplicate the way nature constructs cane reed. Very stiff but sound absorbing Kevlar fibers are suspended in a lightweight resin formulation.

Other uses

- Fire dancing: Wicks for fire dancing props are made of composite materials with Kevlar in them. Kevlar by itself does not absorb fuel very well, so it is blended with other materials such as fiberglass or cotton. Kevlar's high heat resistance allows the wicks to be reused many times.
- Frying pans: Kevlar is sometimes used as a substitute for Teflon in some non-stick frying pans.
- Rope, cable, sheath: The Fibre is used in woven rope and in cable, where the fibers are kept parallel within a polyethylene sleeve. The cables have been used in suspension bridges such as the bridge at Aberfeldy in Scotland. They have also been used to stabilize cracking concrete cooling towers by circumferential application followed by tensioning to close the cracks. Kevlar is widely used as a protective outer sheath for optical Fibre cable, as its strength protects the cable from damage and kinking. When used in this application it is commonly known by the trademarked name Parafil.
- Electricity generation: Kevlar was used by scientists at Georgia Institute of Technology as a base textile for an experiment in electricity-producing clothing. This was done by weaving zinc oxide nanowires into the fabric. If successful, the new fabric would generate about 80 milliwatts per square meter.
- Building construction: A retractable roof of over 60,000 square feet (5,575 square metres) of Kevlar was a key part of the design of Montreal's Olympic stadium for the 1976 Summer Olympics. It was spectacularly unsuccessful, as it was completed ten years late and replaced just ten years later in May 1998 after a series of problems.
- Brakes: The chopped Fibre has been used as a replacement for asbestos in brake pads. Dust produced from asbestos brakes is toxic, while aramids are a benign substitute.
- Expansion joints and hoses: Kevlar can be found as a reinforcing layer in rubber bellows expansion joints and rubber hoses, for use in high temperature applications, and for its high strength. It is also found as a braid layer used on the outside of hose assemblies, to add protection against sharp objects.
- Particle physics: A thin Kevlar window has been used by the NA48 experiment at CERN to separate a vacuum vessel from a vessel at nearly atmospheric pressure, both 192 cm in diameter. The window has provided vacuum tightness combined with reasonably small amount of material (only 0.3per cent to 0.4per cent of radiation length).
- Smartphones: The Motorola RAZR Family and the Motorola Droid Maxx have a kevlar backplate, chosen over other materials such as carbon Fibre due to its resilience and lack of interference with signal transmission.

- Marine Current Turbine and Wind turbine: The Kevlar Fibre/epoxy matrix composite materials can be used in marine current turbine (MCT) or wind turbine due to their high specific strength and light weight compared to other fibers.

Composite materials

Aramid fibers are widely used for reinforcing composite materials, often in combination with carbon Fibre and glass Fibre. The matrix for high performance composites is usually epoxy resin. Typical applications include monocoque bodies for F1 racing cars, helicopter rotor blades, tennis, table tennis, badminton and squash rackets, kayaks, cricket bats, and field hockey, ice hockey and lacrosse sticks.

TECHNORA

Technora is an aramid that is useful for a variety of applications that require high strength or chemical resistance. It is a brand name of the company Teijin. It was also used to suspend the NASA Mars rover Opportunity from its parachute during descent.

Production

Technora is produced by condensation polymerization of terephthaloyl chloride (TCl) with a mixture of p-phenylenediamine (PPD) and 3,4'-diaminodiphenylether (3,4'-ODA). The polymer is closely related to Teijin's Twaron or DuPont's Kevlar. Technora is derived from two different diamines, 3,4'-ODA and PPD, whereas Twaron is derived from PPD alone. This relatively simple process uses only one amide solvent, and therefore spinning can be done directly after the polymer production.

Major industrial uses

- Automotive and other industries:
  - Turbo hoses
  - high pressure hoses
  - Timing and V-belts
  - mechanical rubber goods reinforcement
- Linear tension
  - Optical Fibre cables (OFC)
  - ropes, wire ropes and cables
  - Umbilical cables
  - Electrical mechanical cable (EMC)
  - Windsurfing sails
  - Hangglider sails
- Drumheads

NOMEX

Nomex is a registered trademark for flame-resistant meta-aramid material developed in the early 1960s by DuPont and first marketed in 1967.

Nomex and related aramid polymers are related to nylon, but have aromatic backbones, and hence are more rigid and more durable. Nomex is the premier example of a *meta* variant of the aramids (Kevlar is a *para* aramid). Unlike Kevlar, Nomex cannot align during filament formation and has poorer strength. However, it has excellent thermal, chemical, and radiation resistance for a polymer material.

The polymer is produced by condensation reaction from the monomers m-phenylenediamine and isophthaloyl chloride.

It is sold in both Fibre and sheet forms and is used as a fabric wherever resistance from heat and flame is required. Nomex sheet is actually a calendered paper and made in a similar fashion. Nomex Type 410 paper is the original and one of the larger grade types made, mostly for electrical insulation purposes. Nomex Fibre is made in the USA and in Spain (Asturias).

Wilfred Sweeny (1926–2011), the DuPont scientist responsible for discoveries leading to Nomex, earned a DuPont Lavoisier Medal partly for this work in 2002.

The paper is used in electrical laminates such as circuit boards and transformer cores as well as fireproof honeycomb structures where it is saturated with a phenolic resin. Honeycomb structures such as these, as well as mylar-Nomex laminates are used extensively in aircraft construction. Both the firefighting and vehicle racing industries use Nomex to create clothing and equipment that can withstand intense heat.

A Nomex hood is a common piece of racing and firefighting equipment. It is placed on the head on top of a firefighter's face mask. The hood protects the portions of the head not covered by the helmet and face mask from the intense heat of the fire. Wildland firefighters wear Nomex shirts and trousers as part of their personal protective equipment during wildfire suppression activities.

Race car drivers wear driving suits constructed of Nomex and or other fire retardant materials, along with Nomex gloves, long underwear, balaclavas, socks, helmet lining and shoes to protect them in the event of a fire. The FIA and the SFI Foundation provide specifications for flame-resistant drivers clothing to be used in racing. The standards range from single layer suits that provide some protection against flash fires to much thicker multilayer SFI-15 suits required by the National Hot Rod Association that can protect a driver for up to 30 seconds against the intense heat and almost invisible flames generated by the nitromethane, ethanol and methanol fuels that are used in championship drag racing.

Military pilots and aircrew wear flight suits made of over 92 percent Nomex to protect them from the possibility of cockpit fires and other mishaps.

Recently, troops riding in ground vehicles have also begun wearing Nomex. The remaining material is typically Kevlar thread used to hold the fabric together at the seams.

Military tank drivers also typically use Nomex hoods as protection against fire and extreme cold.

In the U.S. space Programme, Nomex has been used for the Thermal Micrometeoroid Garment on the Extravehicular Mobility Unit (in conjunction with Kevlar and Gore-Tex) and ACES pressure suit, both for fire and extreme environment (water immersion to near vacuum) protection, and as thermal blankets on the payload bay doors, fuselage, and upper wing surfaces of the Space Shuttle Orbiter.

It has also been used for the airbags for the Mars Pathfinder and Mars Exploration Rover missions, the Galileo atmospheric probe, the Cassini-Huygens Titan probe, as an external covering on the AERCam Sprint, and is planned to be incorporated into NASA's upcoming Crew Exploration Vehicle.

Nomex has also been used for its acoustic qualities, the first time being used in Troy, NY, at Rensselaer Polytechnic Institute's Experimental Media and Performing Arts Center (EMPAC's) main concert hall. A ceiling canopy of Nomex reflects high and mid frequency sound, providing reverberation, while letting lower frequency sound partially pass through the canopy. According to RPI President Shirley Ann Jackson, EMPAC is the first venue in the world to use Nomex for acoustic reasons.

Nomex (like Kevlar) is also used in the production of loudspeaker drivers.

Honeycomb-structured Nomex paper is also used as a spacer between layers of lead in the ATLAS Liquid Argon Calorimeter.

The death of race car drivers in fiery crashes, notably those of Fireball Roberts at Charlotte, and Eddie Sachs and Dave MacDonald at Indianapolis, all in 1964, meant something had to be done. In early 1966 *Competition Press and Autoweek* reported: "During the past season, experimental driving suits were worn by Walt Hansgen, Masten Gregory, Marvin Panch and Group 44's Bob Tullius; these four representing a fairly good cross section in the sport. The goal was to get use-test information on the comfort and laundering characteristics of Nomex.

The Chrysler-Plymouth team at the recent Motor Trend 500 at Riverside also wore these suits."

MICROFIBER

Microfiber or microfibre is synthetic Fibre finer than one denier or decitex/thread. This is smaller than the diameter of a strand of silk, which is itself about 1/5 the diameter of a human hair. The most common types of microfibers are made from polyesters, polyamides (*e.g.*, nylon, Kevlar, Nomex, trogamide), or a conjugation of polyester, polyamide, and polypropylene (Prolen). Microfiber is used to make mats, knits, and weaves for apparel, upholstery, industrial filters, and cleaning products. The shape, size, and combinations of synthetic fibers are selected for specific characteristics, including softness, toughness, absorption, water repellency, electrodynamics, and filtering capabilities.

Production of ultra-fine fibers (finer than 0.7 denier) dates back to the late 1950s, using melt-blown spinning and flash spinning techniques. However, only fine staples of random length could be manufactured and very few applications could be found.

Experiments to produce ultra-fine fibers of a continuous filament type were made subsequently, the most promising of which were run in Japan during the 1960s by Dr. Miyoshi Okamoto, a scientist at Toray Industries. Okamoto's discoveries, together with those of Dr. Toyohiko Hikota, resulted in many industrial applications. Among these was Ultrasuede, one of the first successful synthetic microfibers, which found its way onto the market in the 1970s. Microfiber's use in the textile industry then expanded. Microfibers were first publicized in the early 1990s in Sweden and saw success as a product in Europe over the course of the decade. In 2007, Rubbermaid began a line of microfiber products for American markets, the first major company to do so.

Apparel

Clothing

Microfiber fabric is often used for athletic wear, such as cycling jerseys, because the microfiber material wicks moisture (sweat) away from the body, keeping the wearer cool and dry. Microfiber is also very elastic, making it suitable for undergarments. However, the US Marine Corps banned synthetic fabrics in forward environments due to melting and burn risk.

Microfiber is also used to make tough, very soft-to-the-touch materials for general clothing use, often used in skirts and jackets. Microfiber can be made into Ultrasuede, an animal-product-free imitation suede that is cheaper and easier to clean and sew than real suede.

Accessories

Microfiber is used to make many accessories that traditionally have been made from leather: wallets, handbags, backpacks, shoes, cell phone cases, and coin purses. Microfiber fabric is lightweight, durable, and somewhat water repellent, so it makes a good substitute.

Another advantage of fabric (compared to leather) is that fabric can be coated with various finishes or can be treated with anti-bacterial chemicals. Fabric can also be printed with various designs, embroidered with colored thread, or heat-embossed to create interesting textures.

Other Uses

Textiles for cleaning

In cleaning products, microfiber can be 100per cent polyester, or a blend of polyester and polyamide (nylon). It can be both a woven product or a Non-woven product, the latter most often used in limited use or disposable cloths. In the highest-quality fabrics for cleaning applications, the Fibre is split during the manufacturing process to produce multi-stranded fibres. A cross section of the split microfiber fabric under high magnification, would look like an asterisk. The split fibres and the size of the individual filaments working in conjunction with the spaces between them that make the cloths more effective than other fabrics for cleaning purposes.. The structure traps and retains the dirt and also absorbs liquids.

Unlike cotton, microfiber leaves no lint, the exception being some micro suede blends, where the surface is mechanically processed to produce a soft plush feel.

For microfiber to be most effective as a cleaning product, especially for water soluble soils and waxes, it should be a split microfibre. Non-split microfibre is little more than a very soft cloth. The main exception is for cloths used for facial cleansing and for the removal of skin oils, (sebum), sunscreens, and mosquito repellents from optical surfaces such as cameras, phones and eyeglasses where in higher end proprietary woven, 100per cent polyester cloths using 2 μm filaments, will absorb these types of oils without smearing.

Microfiber that is used in Non- sports related clothing, furniture, and other applications isn't split because it isn't designed to be absorbent, just soft. When buying Microfibre may not be labelled to designate whether it's split. A quick way to determine if microfibre is, is to run the cloth lightly over the palm of the hand. A split microfibre will Cling to any imperfections of the skin, which can be both heard and felt. Another way is to pour a small amount of water on a hard flat surface and try to push the water with the microfibre. If the water is pushed rather than being absorbed, it's not split microfiber..

In addition to the open spaces in the fibers created during the splitting process, microfiber is an effective cleaning tool because the fibers are positively charged. Dirt and dust are negatively charged so they are attracted to microfiber like a magnet. The microfiber holds on to the dust and dirt until it is released in the laundering process or when it is rinsed out.

Rags

Microfiber products used for consumer cleaning are generally constructed from split conjugated fibers of polyester and polyamide. Microfiber used for commercial cleaning products also includes many products constructed of 100per cent polyester microfiber. Fabrics made with microfibers are exceptionally soft and hold their shape well. When high-quality microfiber is combined with the right knitting process, it creates an extremely effective cleaning material. This material can hold up to eight times its weight in water. Microfiber products have exceptional ability to absorb oils, and are not hard enough to scratch even paintwork unless they have retained grit or hard particles from previous use.

Microfiber is widely used by car detailers to handle tasks such as removing wax from paintwork, quick detailing, cleaning interior, cleaning glass, and drying. Due to their fine fibers which leave no lint or dust, microfiber towels are used by car detailers and enthusiasts in a similar manner to a chamois leather.

Microfiber is used in many professional cleaning applications, for example in mops and cleaning cloths. Although microfiber mops cost more than non-microfiber mops, they may be more economical because they last longer and require less effort to use.

Microfiber textiles designed for cleaning clean on a microscopic scale. According to tests using microfiber materials to clean a surface leads to reducing the number of bacteria by 99per cent, whereas a conventional cleaning material reduces this number only by 33per cent. Microfiber cleaning tools also absorb fat and grease and their electrostatic properties give them a high dust-attracting power.

Microfiber cloths are used to clean photographic lenses as they absorb oily matter without being abrasive or leaving a residue, and are sold by major manufacturers such as Sinar, Nikon and Canon. Small microfiber cleaning cloths are commonly sold for cleaning computer screens and eyeglasses.

Microfiber is unsuitable for some cleaning applications as it accumulates dust, debris, and particles. Sensitive surfaces (such as all high-tech coated surfaces *e.g.* CRT, LCD and plasma screens) can easily be damaged by a microfiber cloth if it has picked up grit or other abrasive particles during use. One way to minimize the risk of damage to flat surfaces is to use a flat, non-rugged microfiber cloth, as these tend to be less prone to retaining grit.

Rags made of microfiber must only be washed in regular washing detergent, not oily, self-softening, soap-based detergents. Fabric softener must not be used. The oils in the softener and self-softening detergents will clog up the fibers and make them less effective until the oils are washed out.

Insulation

Microfiber materials such as PrimaLoft are used for thermal insulation as a replacement for down feather insulation in sleeping bags and outdoor equipment, due to their better retention of heat when damp or wet.

Basketballs

With microfiber basketballs already popular worldwide and in FIBA, the

NBA proposed the use of a microfiber ball for the 2006–07 season. The ball, which is manufactured by Spalding, does not require a "break-in" period of use as leather balls do, and has the ability to absorb water and oils, meaning that sweat from players touching the ball is better absorbed, making the ball less slippery.

Over the course of the season, the league received many complaints from players who found that the ball bounced differently from leather balls, and that it left cuts on their hands. On January 1, 2007, the league scrapped the use of all microfiber balls and returned to leather basketballs.

Other

Microfibers used in tablecloths, furniture, and car interiors are designed to repel wetting and consequently are difficult to stain. Microfiber tablecloths will bead liquors until they are removed and are sometimes advertised showing red wine on a white tablecloth that wipes clean with a paper towel. This and the ability to mimic suede economically are common selling points for microfiber upholstery fabrics (*e.g.* for couches).

Microfibers are used in towels especially those to be used at swimming pools as even a small towel dries the body quickly. They dry quickly and are less prone than cotton towels to become stale if not dried immediately. Microfiber towels need to be soaked in water and pressed before use, as they would otherwise repel water as microfiber tablecloths do.

Environmental and safety issues

Microfiber textiles tend to be flammable if manufactured from hydrocarbons (polyester) or carbohydrates (cellulose) and emit toxic gases when burning, more so if aromatic (PET, PS, ABS) or treated with halogenated flame retarders and aromatic dyes. Their polyester and nylon stock are made from petrochemicals, which are not a renewable resource and are not biodegradable. However, if made out of polypropylene, they are recyclable (Prolen).

For most cleaning applications they are designed for repeated use rather than being discarded after use. (An exception is the precise cleaning of optical components where a wet cloth is drawn once across the object and must not be used again as the debris collected and now embedded in the cloth may scratch the optical surface.) In many household cleaning applications (washing floors, furniture, etc.) microfiber cleaning fabrics can be used without detergents or cleaning solutions which would otherwise be needed.

There are environmental concerns about this product entering the oceanic food chain. However, no pesticides are used for producing synthetic fibres (in comparison to cotton). If these products are made of polypropylene yarn, the yarn is dope-dyed; *i.e.* no water is used for dyeing (as with cotton, where thousands of litres of water become contaminated).

MODACRYLIC

A modacrylic is a synthetic copolymer. Modacrylics are soft, strong, resilient, and dimensionally stable. They can be easily dyed, show good press and shape retention, and are quick to dry. They have outstanding resistance to chemicals and solvents, are not attacked by moths or mildew, and are Non-allergenic. Among their uses are in apparel linings, furlike outerwear, paint-roller covers, scatter rugs, carpets, and work clothing and as hair in wigs.

Commercial production of modacrylic Fibre began in 1949 by Union Carbide Corporation in the United States. Modacrylic and acrylic fibers are similar in composition and at one time were in the same category. In 1960 the Federal Trade Commission decided to separate the two fibers and establish a category for each.The Federal Trade Commission defines modacrylic fibers as manufactured fibers in which the Fibre-forming substance is any long-chain synthetic polymer composed of less than 85per cent, but at least 35per cent weight acrylonitrile units except when the polymer qualifies as rubber.

Production

Modacrylic fibers are modified acrylic fibers made from acrylonitriles, but larger amounts of other polymers are added to make the copolymers.

The modacrylic fibers are produced by polymerizing the components, dissolving the copolymer in acetone, pumping the solution into the column of warm air (dry-spun), and stretching while hot.

Modacrylics are creamy or white and are produced in tow and staple form. If looked at in cross section views they have an irregular shape. Modacrylic fibers are also produced in many different lengths, crimp levels, deniers and they can have various shrinkage potentials.

Current modacrylic Fibre producers include Solutia Inc. in the U.S. and Kaneka Corporation in Japan.

Properties

A modacrylic has properties that are similar to an acrylic. However, modacrylics are flame retardant and do not combust. The fibers are difficult to ignite and will self-extinguish. In addition to a modacrylic's flame retardant properties it has a relatively high durability that is comparable to wool. Modacrylic fibers have a moderate resistance to abrasion and a very low tenacity.

One of the most interesting properties of Modacrylic fabrics is the arc flash protection, where it has very good values.

Modacrylics are poor conductors of heat. The fabrics are soft, warm and resilient but are prone to pilling and matting. Modacrylics display high performance when it comes to appearance retention. The fibers are quite resilient and will not wrinkle. They also have great dimensional stability and high elastic recovery, which gives them the ability to hold their shape.

Care

Modacrylics are sensitive to loss of appearance due to improper care; therefore, it is important to know how to care for modacrylics. Modacrylics are resistant to acids, weak alkalis, and organic solvents. These fibers are also resistant to moths, mildew and sunlight. Modacrylic fabrics can be machine washed using warm water and tumble dried on a low setting. Modacrylic pieces can also be dry-cleaned, however, they should not be steamed and should only be tumbled on cold. Some fabrics may also be cleaned using the furrier method (a special Non- immersion cleaning process). The fibers are heat sensitive and will shrink at 250 °F and will stiffen at temperatures over 300 °F.

Uses

Modacrylics are used primarily in applications where environmental resistance or flame retardancy is necessary or required. Modacrylics have the ability to combine flame retardancy with a relatively low density, meaning protective gear is not uncomfortably heavy (i.e. shirts and trousers worn by electrical linemen). The combination of flame retardancy and low density is also useful in furnishings, draperies, and outdoor fabrics.

Modacrylics are used by the most technical fabric producers, to obtain comfortably and protective blends, to produce PPE.

Modacrylics are also commonly used in fake fur fabrics, toupées, wigs and fleece-type fabric. By mixing the various forms of fibers it is easy to create a realistic fur look. The fabrics can then be sheared or embossed to create a closer resemblance to fur. The heat-sensitivity of modacrylic also allows wigs and hairpiece to be curled and heat styled without damage.

Modacrylics are also used in fleece, knit-pile fabric backings, and Non-woven fabrics. Other uses of modacrylics include paint rollers, industrial fabrics, stuffed toys and filters.

NYLON

Nylon is a generic designation for a family of synthetic polymers known generically as aliphatic polyamides, first produced on February 28, 1935, by Wallace Carothers at DuPont's research facility at the DuPont Experimental Station. Nylon is one of the most commonly used polymers. Key representatives are nylon-6,6; nylon-6; nylon-6,9; nylon-6,10; nylon-6,12; nylon-11; nylon-12 and nylon-4,6.

Nylon is a thermoplastic, silky material, first used commercially in a nylon-bristled toothbrush (1938), followed more famously by women's stockings ("nylons"; 1940) after being introduced as a fabric at the 1939 New York World's Fair. Nylon is made of repeating units linked by amide bonds and is frequently referred to as *polyamide* (PA). Nylon was the first commercially successful synthetic thermoplastic polymer. There are two common ways of making nylon for Fibre applications. In one approach, molecules with an acid (-COOH) group on each end are reacted with molecules containing amine (-NH_2) groups on each end. The resulting nylon is named on the basis of the number of carbon atoms separating the two acid groups and the two amines. These are formed into monomers of intermediate molecular weight, which are then reacted to form long polymer chains.

Nylon was intended to be a synthetic replacement for silk and substituted for it in many different products after silk became scarce during World War II. It replaced silk in military applications such as parachutes and flak vests, and was used in many types of vehicle tires.

Nylon fibers are used in many applications, including clothes fabrics, bridal veils, package paper, carpets, musical strings, pipes, tents, and rope.

Solid nylon is used in hair combs and mechanical parts such as machine screws, gears and other low- to medium-stress components previously cast in metal. Engineering-grade nylon is processed by extrusion, casting, and injection molding. Type 6,6 Nylon 101 is the most common commercial grade of nylon, and Nylon 6 is the most common commercial grade of molded nylon. For use in tools such as spudgers, nylon is available in glass-filled variants which increase structural and impact strength and rigidity, and molybdenum sulfide-filled variants which increase lubricity. Its various properties also make it very useful as a material in additive manufacturing; specifically as a filament in consumer and professional grade fused deposition modeling 3D printers.

Chemistry

Nylons are condensation copolymers formed by reacting equal parts of a diamine and a dicarboxylic acid, so that amides are formed at both ends of each monomer in a process analogous to polypeptide biopolymers. Chemical elements included are carbon, hydrogen, nitrogen, and oxygen. The numerical suffix specifies the numbers of carbons donated by the monomers; the diamine first and the diacid second. The most common variant is nylon 6-6 which refers to the fact that the diamine (hexamethylene diamine, IUPAC name: hexane-1,6-diamine) and the diacid (adipic acid, IUPAC name: hexanedioic acid) each donate 6 carbons to the polymer chain. As with other regular copolymers like polyesters and polyurethanes, the "repeating unit" consists of one of each monomer, so that they alternate in the chain. Since each monomer in this copolymer has the same reactive group on both ends, the direction of the amide bond reverses between each monomer, unlike natural polyamide proteins which have overall directionality: C terminal ! N terminal. In the laboratory, nylon 6-6 can also be made using adipoyl chloride instead of adipic acid.

It is difficult to get the proportions exactly correct, and deviations can lead to chain termination at molecular weights less than a desirable 10,000 daltons (u). To overcome this problem, a crystalline, solid "nylon salt" can be formed at room temperature, using an exact 1:1 ratio of the acid and the base to neutralize each other. Heated to 285 °C (545 °F), the salt reacts to form nylon polymer. Above 20,000 daltons, it is impossible to spin the chains into yarn, so to combat this, some acetic acid is added to react with a free amine end group during polymer elongation to limit the molecular weight. In practice, and especially for 6,6, the monomers are often combined in a water solution. The water used to make the solution is evaporated under controlled conditions, and the increasing concentration of "salt" is polymerized to the final molecular weight.

DuPont patented nylon 6,6, so in order to compete, other companies (particularly the German BASF) developed the homopolymer nylon 6, or polycaprolactam — not a condensation polymer, but formed by a ring-opening polymerization (alternatively made by polymerizing aminocaproic acid). The peptide bond within the caprolactam is broken with the exposed active groups on each side being incorporated into two new bonds as the monomer becomes part of the polymer backbone. In this case, all amide bonds lie in the same direction, but the properties of nylon 6 are sometimes indistinguishable from those of nylon 6,6 — except for melt temperature and some Fibre properties in products like carpets and textiles. There is also nylon 9.

The 428 °F (220 °C) melting point of nylon 6 is lower than the 509 °F (265 °C) melting point of nylon 6,6.

Nylon 5,10, made from pentamethylene diamine and sebacic acid, was studied by Carothers even before nylon 6,6 and has superior properties, but is more expensive to make. In keeping with this naming convention, "nylon 6,12" (N-6,12) or "PA-6,12" is a copolymer of a 6C diamine and a 12C diacid. Similarly for N-5,10 N-6,11; N-10,12, etc. Other nylons include copolymerized dicarboxylic acid/diamine products that are *not* based upon the monomers listed above. For example, some aromatic nylons are polymerized with the addition of diacids like terephthalic acid (! Kevlar, Twaron) or isophthalic acid (! Nomex), more commonly associated with polyesters. There are copolymers of N-6,6/N6; copolymers of N-6,6/N-6/N-12; and others. Because of the way polyamides are formed, nylon would seem to be limited to unbranched, straight chains. But "star" branched nylon can be produced by the condensation of dicarboxylic acids with polyamines having three or more amino groups.

The general reaction is:

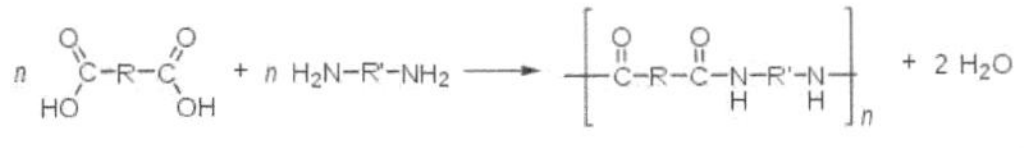

Two molecules of water are given off and the nylon is formed. Its properties are determined by the R and R' groups in the monomers. In nylon 6,6, R = 4C and R' = 6C alkanes, but one also has to include the two carboxyl carbons in the diacid to get the number it donates to the chain. In Kevlar, both R and R' are benzene rings.

Concepts of nylon production

The first approach: combining molecules with an acid (COOH) group on each end are reacted with two chemicals that contain amine (NH_2) groups on each end. This process creates nylon 6,6, made of hexamethylene diamine with six carbon atoms and adipic acid.

The second approach: a compound has an acid at one end and an amine at the other and is polymerized to form a chain with repeating units of (-NH-$[CH_2]_n$-CO-$)_x$. In other words, nylon 6 is made from a single six-carbon substance called caprolactam. In this equation, if n = 5, then nylon 6 is the assigned name (may also be referred to as polymer).

The characteristic features of nylon 6,6 include:

- Pleats and creases can be heat-set at higher temperatures
- More compact molecular structure
- Better weathering properties; better sunlight resistance
- Softer "Hand"
- Higher melting point (256 °C/492.8 °F)
- Superior colorfastness
- Excellent abrasion resistance

On the other hand, nylon 6 is easy to dye, more readily fades; it has a higher impact resistance, a more rapid moisture absorption, greater elasticity and elastic recovery.

Characteristics

- Variation of luster: nylon has the ability to be very lustrous, semilustrous or dull.
- Durability: its high tenacity fibers are used for seatbelts, tire cords, ballistic cloth and other uses.
- High elongation
- Excellent abrasion resistance
- Highly resilient (nylon fabrics are heat-set)
- Paved the way for easy-care garments
- High resistance to insects, fungi, animals, as well as molds, mildew, rot and many chemicals
- Used in carpets and nylon stockings
- Melts instead of burning
- Used in many military applications
- Good specific strength
- Transparent to infrared light ("12dB)

Bulk properties

Above their melting temperatures, T_m, thermoplastics like nylon are amorphous solids or viscous fluids in which the chains approximate random coils. Below T_m, amorphous regions alternate with regions which are lamellar crystals. The amorphous regions contribute elasticity and the crystalline regions contribute strength and rigidity. The planar amide (-CO-NH-) groups are very polar, so nylon forms multiple hydrogen bonds among adjacent strands. Because the nylon backbone is so regular and symmetrical, especially if all the amide bonds are in the *trans* configuration, nylons often have high crystallinity and make excellent fibers. The amount of crystallinity depends on the details of formation, as well as on the kind of nylon. Apparently it can never be quenched from a melt as a completely amorphous solid.

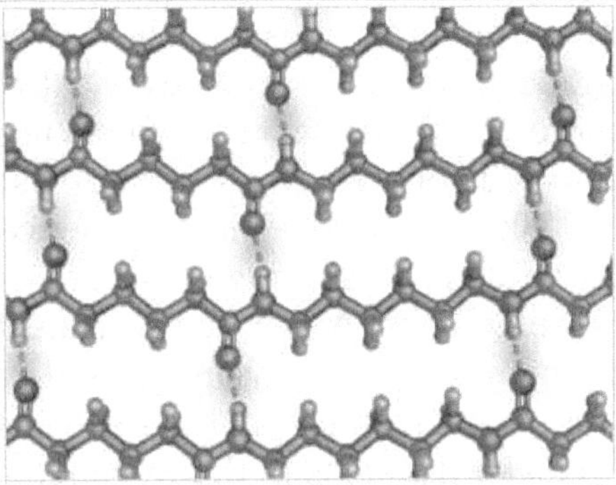

Hydrogen bonding in Nylon 6,6 (in mauve).

Nylon 6,6 can have multiple parallel strands aligned with their neighboring peptide bonds at coordinated separations of exactly 6 and 4 carbons for considerable lengths, so the carbonyl oxygens and amide hydrogens can line up to form interchain hydrogen bonds repeatedly, without interruption. Nylon 5,10 can have coordinated runs of 5 and 8 carbons. Thus parallel (but not antiparallel) strands can participate in extended, unbroken, multi-chain â-pleated sheets, a strong and tough supermolecular structure similar to that found in natural silk fibroin and the â-keratins in feathers. (Proteins have only an amino acid á-carbon separating sequential -CO-NH- groups.) Nylon 6 will form uninterrupted H-bonded sheets with mixed directionalities, but the â-sheet wrinkling is somewhat different. The three-dimensional disposition of each alkane hydrocarbon chain depends on rotations about the 109.47° tetrahedral bonds of singly bonded carbon atoms.

When extruded into fibers through pores in an industrial spinneret, the individual polymer chains tend to align because of viscous flow. If subjected to cold drawing afterwards, the fibers align further, increasing their crystallinity, and the material acquires additional tensile strength. In practice, nylon fibers are most often drawn using heated rolls at high speeds.

Block nylon tends to be less crystalline, except near the surfaces due to shearing stresses during formation. Nylon is clear and colorless, or milky, but is easily dyed. Multistranded nylon cord and rope is slippery and tends to unravel. The ends can be melted and fused with a heat source such as a flame or electrode to prevent this.When dry, polyamide is a good electrical insulator. However, polyamide is hygroscopic. The absorption of water will change some of the material's properties such as its electrical resistance. Nylon is less absorbent than wool or cotton.

Historical uses

Bill Pittendreigh, DuPont, and other individuals and corporations worked diligently during the first few months of World War II to find a way to replace Asian silk and hemp with nylon in parachutes. It was also used to make tires, tents, ropes, ponchos, and other military supplies. It was even used in the production of a high-grade paper for U.S. currency. At the outset of the war, cotton accounted for more than 80per cent of all fibers used and manufactured, and wool fibers accounted for nearly all of the rest. By August 1945, manufactured fibers had taken a market share of 25per cent, at the expense of cotton. After the war, because of shortages of both silk and nylon, nylon parachute material was sometimes repurposed to make dresses.

Some of the terpolymers based upon nylon are used every day in packaging. Nylon has been used for meat wrappings and sausage sheaths.

Nylon was used to make the stock of the Remington Nylon 66 rifle. The frame of the modern Glock pistol is made of a nylon composite.

Instrument strings

In the mid-1940s, classical guitarist Andrés Segovia mentioned the shortage of good guitar strings in the United States, particularly his favorite Pirastro catgut strings, to a number of foreign diplomats at a party, including General Lindeman of the British Embassy. A month later, the General presented Segovia with some nylon strings which he had obtained via some members of the DuPont family. Segovia found that although the strings produced a clear sound, they had a faint metallic timbre which he hoped could be eliminated.

Nylon strings were first tried on stage by Olga Coelho in New York in January, 1944.

In 1946, Segovia and string maker Albert Augustine were introduced by their mutual friend Vladimir Bobri, editor of Guitar Review. On the basis of Segovia's interest and Augustine's past experiments, they decided to pursue the development of nylon strings. DuPont, skeptical of the idea, agreed to supply the nylon if Augustine would endeavor to develop and produce the actual strings. After three years of development, Augustine demonstrated a nylon first string whose quality impressed guitarists, including Segovia, in addition to DuPont.

Wound strings, however, were more problematic. Eventually, however, after experimenting with various types of metal and smoothing and polishing techniques, Augustine was also able to produce high quality nylon wound strings.

Use in composites

Nylon can be used as the matrix material in composite materials, with reinforcing fibers like glass or carbon Fibre; such a composite has a higher density than pure nylon.

Such thermoplastic composites (25per cent to 30per cent glass Fibre) are frequently used in car components next to the engine, such as intake manifolds, where the good heat resistance of such materials makes them feasible competitors to metals.

Hydrolysis and degradation

All nylons are susceptible to hydrolysis, especially by strong acids, a reaction essentially the reverse of the synthetic reaction shown above. The molecular weight of nylon products so attacked drops fast, and cracks form quickly at the affected zones.

Lower members of the nylons (such as nylon 6) are affected more than higher members such as nylon 12. This means that nylon parts cannot be used in contact with sulfuric acid for example, such as the electrolyte used in lead–acid batteries. When being molded, nylon must be dried to prevent hydrolysis in the molding machine barrel since water at high temperatures can also degrade the polymer. The reaction is of the type:

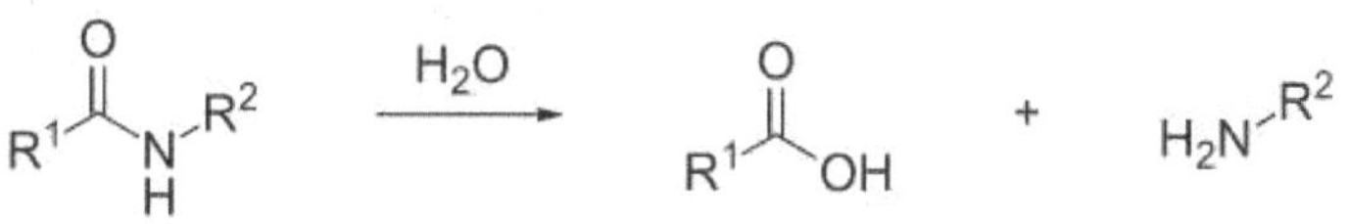

Environmental impact, incineration and recycling

Berners-Lee reckons the average greenhouse gas footprint of nylon in manufacturing carpets at 5.43kg CO2 equivalent per kilo, when produced in Europe. This gives it almost the same carbon footprint as wool, but with greater durability and therefore a lower overall carbon footprint.

Various nylons break down in fire and form hazardous smoke, and toxic fumes or ash, typically containing hydrogen cyanide. Incinerating nylons to recover the high energy used to create them is usually expensive, so most nylons reach the garbage dumps, decaying very slowly. Some recycling is done on nylon, usually creating pellets for reuse in the industry.

Etymology

In 1940, John W. Eckelberry of DuPont stated that the letters "nyl" were arbitrary and the "on" was copied from the suffixes of other fibers such as cotton and rayon. A later publication by DuPont (*Context*, vol. 7, no. 2, 1978) explained that the name was originally intended to be "No-Run" ("run" meaning "unravel"), but was modified to avoid making such an unjustified claim. Since the products was not really run-proof, the vowels were swapped to produce "nuron", which was changed to "nilon" "to make it sound less like a nerve tonic". For clarity in pronunciation, the "i" was changed to "y".

Alternative but apocryphal explanations for the naming are:

- it is a combination of the names of the two cities New York and London, NY-Lon;
- it is an acronym from WWII, a challenging slogan by the allies to their enemy at that time: Now You Lazy Old Nippon, Nippon being an alternative name for Japan, which is the country known for the traditional culture of making silk from silkworms.

OLEFIN FIBRE

Olefin Fibre is a synthetic Fibre made from a polyolefin, such as polypropylene or polyethylene. It is used in wallpaper, ropes, and vehicle interiors.

Olefin's advantages are its strength, colourfastness and comfort, its resistance to staining, mildew, abrasion, sunlight and its good bulk and cover.

Italy began production of olefin fibers in 1957. The chemist Giulio Natta successfully formulated olefin suitable for more textile applications. Both Natta and Karl Ziegler were later awarded the Nobel Prize for their work on transition metal catalysis of olefins to Fibre, also known as Ziegler–Natta catalysis. U.S. production of olefin fibers began in 1960. Olefin fibers account for 16per cent of all manufactured fibers.

Major Fibre properties

Olefin fibers have great bulk and cover while having low specific gravity. This means "warmth without the weight." The fibers have low moisture absorption, but they can wick moisture and dry quickly. Olefin is abrasion, stain, sunlight, fire, and chemical resistant. It does not dye well, but has the advantage of being colorfast. Since Olefin has a low melting point, textiles can be thermally bonded. The fibers have the lowest static of all manufactured fibers and a medium luster. One of the most important properties of olefin is its strength. It keeps its strength in wet or dry conditions and is very resilient. The Fibre can be produced for strength of different properties.

Production method

The Federal Trade Commission's official definition of olefin Fibre is "A manufactured Fibre in which the Fibre forming substance is any long-chain synthetic polymer composed of at least 85per cent by weight of ethylene, propylene, or other olefin units"

Polymerization of propylene and ethylene gases, controlled with special catalysts, creates olefin fibers. Dye is added directly to the polymer before melt spinning is applied. Additives, polymer variations and different process conditions can create a range of characteristics. High pressure production, which uses ten tons per square inch, creates a film for molded materials. Low pressure production uses a low temperature with a catalyst and hydrocarbon solvent. This process is less expensive and produces a polyethylene polymer more for textile use.

The polymer is then melted, spun, by a spinneret, into water, or air cooled. The Fibre is drawn out to six times the spun length. Gel spinning is a new method in which a gel form of polyethylene polymers is used.

Physical and chemical structure

Physical

Olefin fibers can be multi- or monofilament and staple, tow, or film yarns. The fibers are colorless and round in cross section. This cross section can be modified for different end uses. The physical characteristics are a waxy feel and colorless.

Chemical

There are two types of polymers that can be used in olefin fibers. The first, polyethylene, is a simple linear structure with repeating units. These fibers are used mainly for ropes, twines and utility fabrics.

The second type, polypropylene, is a three-dimensional structure with a backbone of carbon atoms. Methyl groups protrude from this backbone. Stereoselective polymerization orders these methyl groups to the same spatial placement. This creates a crystalline polypropylene polymer. The fibers made with these polymers can be used in apparel, furnishing and industrial products.

Manufacturers

The first commercial producer of an olefin Fibre in the United States was Hercules, Inc. (FiberVisions). Other U.S. olefin Fibre producers include Asota; American Fibers and Yarns Co; American Synthetic Fibre, LLC; Colour-Fi; FiberVisions; Foss Manufacturing Co., LLC; Drake Extrusion; Filament Fibre Technology, Inc.; TenCate Geosynthetics; Universal Fibre Systems LLC.

Trademarks according to fabric use

Producer – Allied-Signal

- A.C.E. – Tire cord, furniture webbing

Producer – DuPont

- CoolMax – Warm-weather and action wear
- Hollofil, Quallofil – Fiberfill and insulating fibers
- Sontara – Spunlaced Non-woven fabrics

- Thermostat – Cold-weather wear
- Thermoloft - Fiberfill and insulating fibers
- Tyvek - Used for house wraps to postal envelopes to clothing

Producer – Trevira

- ESP – Apparel and furnishings
- Celwet – Non-wovens
- Comfort Fibre – Staple Fibre for apparel uses
- Floor Guardian - Gym Floor Carpet Protection System
- Loftguard – Staple Fibre for industrial uses
- Polar Guard
- Lambda – Filament yarns with spun-yarn characteristics
- Serene
- Superba
- Trevira HT – Marine and military uses; ropes, cordages
- Trevira ProEarth – Recycled-content geotextiles
- Trevira XPS - Carpeting
- BTU – Cold-weather apparel

Producer - 3M

- Thinsulate - Cold-weather action wear

Uses

- Apparel: Sports and active wear, socks, thermal underwear; lining fabrics.
- Home furnishing: Olefin can be used by itself or in blends for indoor and outdoor carpets, carpet tiles, and carpet backing. The Fibre can also be used in upholstery, draperies, wall coverings, slipcovers, and floor coverings.
- Automotive: Olefin can be used for interior fabrics, sun visors, arm rests, door and side panels, trunks, parcel shelfs, and resin replacement as binder fibers.
- Industrial: In an industrial setting, olefin creates carpets; ropes, geotextiles that are in contact with the soil, filter fabrics, bagging, concrete reinforcement, and heat-sealable paper (*e.g.* tea- and coffee-bags).

Care procedures

Dry cleaning Olefin is not recommended, because many dry-cleaning solvents can swell the fibers. As Olefin dries quickly, line drying and low tumble drying with little or no heat is the recommended method of drying. Since Olefin is not absorbent, waterborne stains do not present a problem. However, oily stains are difficult to remove. Most such stains can be removed with lukewarm water and detergent, but bleach can also be used. Olefin Fibre has a low melting point (around 225 to 335 °F, depending on the polymer's grade) so items should be ironed at a very low temperature, if at all. Items such as outdoor carpets and other fabrics can be hosed off. Olefin is easy to recycle.

POLYESTER

Polyester is a category of polymers which contain the ester functional group in their main chain. Although there are many polyesters, the term "polyester" as a specific material most commonly refers to polyethylene terephthalate (PET). Polyesters include naturally occurring chemicals, such as in the cutin of plant cuticles, as well as synthetics through step-growth polymerization such as polybutyrate. Natural polyesters and a few synthetic ones are biodegradable, but most synthetic polyesters are not.

Depending on the chemical structure, polyester can be a thermoplastic or thermoset, there are also polyester resins cured by hardeners; however, the most common polyesters are thermoplastics.

Fabrics woven or knitted from polyester thread or yarn are used extensively in apparel and home furnishings, from shirts and pants to jackets and hats, bed sheets, blankets, upholstered furniture and computer mouse mats. Industrial polyester fibers, yarns and ropes are used in tyre reinforcements, fabrics for conveyor belts, safety belts, coated fabrics and plastic reinforcements with high-energy absorption.

Polyester Fibre is used as cushioning and insulating material in pillows, comforters and upholstery padding. Polyesters are also used to make bottles, films, tarpaulin, canoes, liquid crystal displays, holograms, filters, dielectric film for capacitors, film insulation for wire and insulating tapes. Polyesters are widely used as a finish on high-quality wood products such as guitars, pianos and vehicle/yacht interiors. Thixotropic properties of spray-applicable polyesters make them ideal for use on open-grain timbers, as they can quickly fill wood grain, with a high-build film thickness per coat. Cured polyesters can be sanded and polished to a high-gloss, durable finish.

While synthetic clothing in general is perceived by many as having a less natural feel compared to fabrics woven from natural fibers (such as cotton and wool), polyester fabrics can provide specific advantages over natural fabrics, such as improved wrinkle resistance, durability and high Colour retention. As a result, polyester fibers are sometimes spun together with natural fibers to produce a cloth with blended properties. Synthetic fibers also can create materials with superior water, wind and environmental resistance compared to plant-derived fibers, and are sometimes renamed so as to suggest their similarity or even superiority to natural fibers (for example, China silk, which is a term in the textiles industry for a 100per cent polyester Fibre woven to resemble the sheet and durability of insect-derived silk).

Liquid crystalline polyesters are among the first industrially used liquid crystal polymers. They are used for their mechanical properties and heat-resistance. These traits are also important in their application as an abradable seal in jet engines.

Types

Polyesters as thermoplastics may change shape after the application of heat. While combustible at high temperatures, polyesters tend to shrink away from flames and self-extinguish upon ignition. Polyester fibers have high tenacity and E-modulus as well as low water absorption and minimal shrinkage in comparison with other industrial fibers.

Unsaturated polyesters (UPR) are thermosetting resins. They are used as casting materials, fiberglass laminating resins and non-metallic auto-body fillers. Fiberglass-reinforced unsaturated polyesters find wide application in bodies of yachts and as body parts of cars.

According to the composition of their main chain, polyesters can be:

	Number of repeating units	Examples of polyesters	Examples of manufacturing methods
Aliphatic	Homopolymer	Polyglycolide or Polyglycolic acid (PGA)	Polycondensation of glycolic acid
		Polylactic acid (PLA)	Ring-opening polymerization of lactide
		Polycaprolactone (PCL)	Ring-opening polymerization of caprolactone
		Polyhydroxyalkanoate (PHA)	
		Polyhydroxybutyrate (PHB)	
	Copolymer	Polyethylene adipate (PEA)	
		Polybutylene succinate (PBS)	Polycondensation of succinic acid with 1,4-butanediol
		Poly(3-hydroxybutyrate-co-3-hydroxyvalerate) (PHBV)	Copolymerization of 3-hydroxybutanoic acid and 3-hydroxypentanoic acid, butyrolactone and valerolactone (oligomeric aluminoxane as a catalyst)
Semi-aromatic	Copolymer	Polyethylene terephthalate (PET)	Polycondensation of terephthalic acid with ethylene glycol
		Polybutylene terephthalate (PBT)	Polycondensation of terephthalic acid with 1,4-butanediol
		Polytrimethylene terephthalate (PTT)	Polycondensation of terephthalic acid with 1,3-propanediol
		Polyethylene naphthalate (PEN)	Polycondensation of at least one naphthalene dicarboxylic acid with ethylene glycol
Aromatic	Copolymer	Vectran	Polycondensation of 4-hydroxybenzoic acid and 6-hydroxynaphthalene-2-carboxylic acid

Increasing the aromatic parts of polyesters increases their glass transition temperature, melting temperature, thermal stability, chemical stability...

Polyesters can also be telechelic oligomers like the polycaprolactone diol (PCL) and the polyethylene adipate diol (PEA). They are then used as prepolymers.

Industry

Basics

Polyester is a synthetic polymer made of purified terephthalic acid (PTA) or its dimethyl ester dimethyl terephthalate (DMT) and monoethylene glycol (MEG). With 18per cent market share of all plastic materials produced, it ranges third after polyethylene (33.5per cent) and polypropylene (19.5per cent).

The main raw materials are described as follows:

- Purified terephthalic acid – PTA – CAS-No.: 100-21-0

Synonym: 1,4 benzenedicarboxylic acid,

Sum formula; $C_6H_4(COOH)_2$, mol weight: 166.13

- Dimethylterephthalate – DMT – CAS-No: 120-61-6

Synonym: 1,4 benzenedicarboxylic acid dimethyl ester

Sum formula $C_6H_4(COOCH_3)_2$, mol weight: 194.19

- Mono Ethylene Glycol – MEG – CAS No.: 107-21-1

Synonym: 1,2 ethanediol

Sum formula: $C_2H_6O_2$, mol weight: 62,07

To make a polymer of high molecular weight a catalyst is needed. The most common catalyst is antimony trioxide (or antimony tri acetate):

Antimony trioxide – ATO – CAS-No.: 1309-64-4 Molecular weight: 291.51 Sum formula: Sb_2O_3In 2008, about 10,000 tonnes Sb_2O_3 were used to produce around 49 million tonnes polyethylene terephthalate.

Polyester is described as follows:

Polyethylene Terephthalate CAS-No.: 25038-59-9 Synonym/abbreviations: polyester, PET, PES Sum Formula: H-$[C_{10}H_8O_4]$-n=60–120 OH, molelcular unit weight: 192.17

There are several reasons for the importance of Polyester:

- The relatively easy accessible raw materials PTA or DMT and MEG
- The very well understood and described simple chemical process of polyester synthesis
- The low toxicity level of all raw materials and side products during polyester production and processing
- The possibility to produce PET in a closed loop at low emissions to the environment
- The outstanding mechanical and chemical properties of polyester
- The recyclability
- The wide variety of intermediate and final products made of polyester.

In table the estimated world polyester production is shown. Main applications are textile polyester, bottle polyester resin, film polyester mainly for packaging and specialty polyesters for engineering plastics. According to this table, the world's total polyester production might exceed 50 million tons per annum before the year 2010.

Table. World polyester production

Market size per year

Product type	2002 Million tonnes/year	2008 Million tonnes/year
Textile-PET	20	39

Resin, bottle/A-PET	9	16
Film-PET	1.2	1.5
Special polyester	1	2.5
Total	31.2	49

Polyester processing

After the first stage of polymer production in the melt phase, the product stream divides into two different application areas which are mainly textile applications and packaging applications. In table, the main applications of textile and packaging of polyester is listed.

Table. Textile and packaging polyester application list

Polyester-based polymer (melt or pellet)	
Textile	Packaging
Staple fiber (PSF)	Bottles for CSD, Water, Beer, Juice, Detergents
Filaments POY, DTY, FDY	A-PET Film
Technical yarn and tire cord	Thermoforming
Non-woven and spunbond	BO-PET Biaxial oriented Film
Mono-filament	Strapping

Abbreviations: PSF = Polyester Staple Fibre; POY = Partially Oriented Yarn; DTY = Draw Textured Yarn; FDY = Fully Drawn Yarn; CSD = Carbonated Soft Drink; A-PET = Amorphous Polyester Film; BO-PET = Biaxial Oriented Polyester Film;

A comparable small market segment (much less than 1 million tonnes/ year) of polyester is used to produce engineering plastics and masterbatch.

In order to produce the polyester melt with a high efficiency, high-output processing steps like staple Fibre (50–300 tonnes/day per spinning line) or POY/ FDY (up to 600 tonnes/day split into about 10 spinning machines) are meanwhile more and more vertically integrated direct processes. This means the polymer melt is directly converted into the textile fibers or filaments without the common step of pelletizing.

We are talking about full vertical integration when polyester is produced at one site starting from crude oil or distillation products in the chain oil ! benzene ! PX ! PTA ! PET melt ! Fibre/filament or bottle-grade resin. Such integrated processes are meanwhile established in more or less interrupted processes at one production site. Eastman Chemicals were the first to introduce the idea of closing the chain from PX to PET resin with their so-called INTEGREX process. The capacity of such vertically integrated production sites is >1000 tonnes/day and can easily reach 2500 tonnes/day.

Besides the above-mentioned large processing units to produce staple Fibre or yarns, there are ten thousands of small and very small processing plants, so that one can estimate that polyester is processed and recycled in more than 10 000 plants around the globe. This is without counting all the companies involved in the supply industry, beginning with engineering and processing machines and ending with special additives, stabilizers and colors. This is a gigantic industry complex and it is still growing by 4–8per cent per year, depending on the world region.

Synthesis

Synthesis of polyesters is generally achieved by a polycondensation reaction. The general equation for the reaction of a diol with a diacid is:

$$(n+1)\ R(OH)_2 + n\ R'(COOH)_2\ !\ HO[ROOCR'COO]_nROH + 2n\ H_2O$$

Azeotrope esterification

In this classical method, an alcohol and a carboxylic acid react to form a carboxylic ester. To assemble a polymer, the water formed by the reaction must be continually removed by azeotrope distillation.

Alcoholic transesterification

Transesterification: An alcohol-terminated oligomer and an ester-terminated oligomer condense to form an ester linkage, with loss of an alcohol. R and R' are the two oligomer chains, R" is a sacrificial unit such as a methyl group (methanol is the byproduct of the esterification reaction).

Acylation (HCl method)

The acid begins as an acid chloride, and thus the polycondensation proceeds with emission of hydrochloric acid (HCl) instead of water. This method can be carried out in solution or as an enamel.

- Silyl method: In this variant of the HCl method, the carboxylic acid chloride is converted with the trimethyl silyl ether of the alcohol component and production of trimethyl silyl chloride is obtained

Ring-opening polymerization

Aliphatic polyesters can be assembled from lactones under very mild conditions, catalyzed anionically, cationically or metallorganically. A number of catalytic methods for the copolymerization of epoxides with cyclic anhydrides have also recently been shown to provide a wide array of functionalized polyesters, both saturated and unsaturated.

Biodegradation

The futuro house was made of fibreglass-reinforced polyester plastic; polyester-polyurethane, and poly(methylmethacrylate) one of them was found to be degrading by Cyanobacteria and Archaea

POLYETHYLENE

Polyethylene (abbreviated PE) or polythene (IUPAC name polyethene or poly(methylene)) is the most common plastic. The annual global production is approximately 80 million tonnes. Its primary use is in packaging (plastic bag, plastic films, geomembranes, containers including bottles, etc.). Many kinds of polyethylene are known, with most having the chemical formula $(C_2H_4)_nH_2$.

Thus PE is usually a mixture of similar organic compounds that differ in terms of the value of n.

Properties

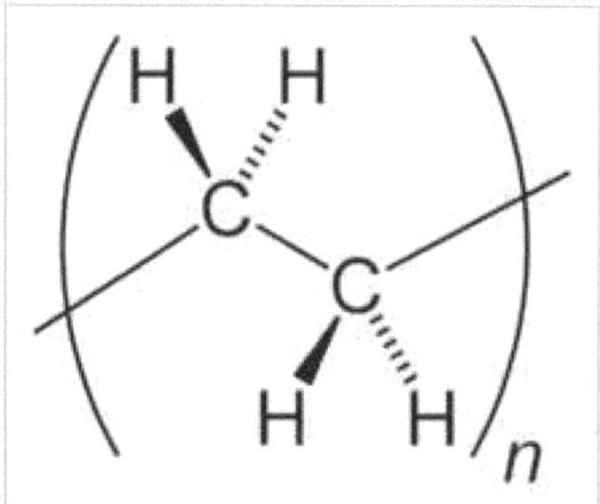

The repeating unit of **polyethylene** in the most stable staggered conformation.

Physical properties

Polyethylene is a thermoplastic polymer consisting of long hydrocarbon chains. Depending on the crystallinity and molecular weight, a melting point and glass transition may or may not be observable. The temperature at which these occur varies strongly with the type of polyethylene. For common commercial grades of medium- and high-density polyethylene the melting point is typically in the range 120 to 180 °C (248 to 356 °F). The melting point for average, commercial, low-density polyethylene is typically 105 to 115 °C (221 to 239 °F).

Chemical properties

Most LDPE, MDPE and HDPE grades have excellent chemical resistance, meaning that it is not attacked by strong acids or strong bases. It is also resistant to gentle oxidants and reducing agents. Polyethylene burns slowly with a blue flame having a yellow tip and gives off an odour of paraffin. The material continues burning on removal of the flame source and produces a drip. Crystalline samples do not dissolve at room temperature. Polyethylene (other than cross-linked polyethylene) usually can be dissolved at elevated temperatures in aromatic hydrocarbons such as toluene or xylene, or in chlorinated solvents such as trichloroethane or trichlorobenzene.

Process

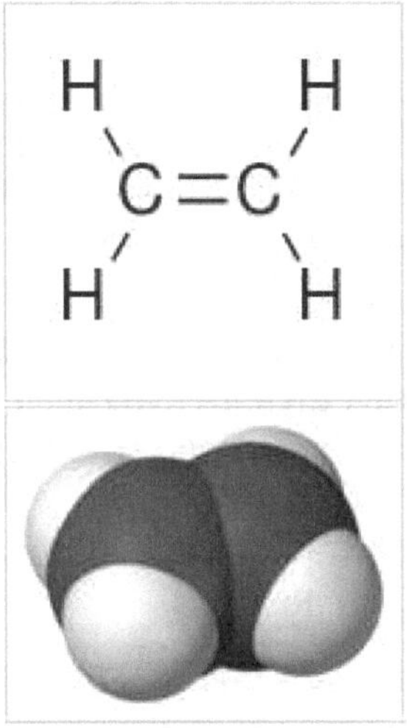

Ethylene (ethene)

Monomer

The ingredient or monomer is ethylene (IUPAC name ethene), a gaseous hydrocarbon with the formula C_2H_4, which can be viewed as a pair of methylene groups (=CH 2) connected to each other. Because the compound is highly reactive, the ethylene must be of high purity. Typical specifications are <5 ppm for water, oxygen, as well as other alkenes. Acceptable contaminants include N_2, ethane (common precursor to ethylene), and methane. Ethylene is usually produced from petrochemical sources, but also is generated by dehydration of ethanol.

Polymerization

Ethylene is a rather stable molecule that polymerizes only upon contact with catalysts. The conversion is highly exothermic. Coordination polymerization is the most pervasive technology, which means that metal chlorides or metal oxides are used.

The most common catalysts consist of titanium(III) chloride, the so-called Ziegler-Natta catalysts. Another common catalyst is the Phillips catalyst, prepared by depositing chromium(VI) oxide on silica. Ethylene can be produced through radical polymerization, but this route has only limited utility and typically requires high pressure apparatus.

Classification

Polyethylene is classified into several different categories based mostly on its density and branching. Its mechanical properties depend significantly on variables such as the extent and type of branching, the crystal structure and the molecular weight. With regard to sold volumes, the most important polyethylene grades are HDPE, LLDPE and LDPE.

- Ultra-high-molecular-weight polyethylene (UHMWPE)
- Ultra-low-molecular-weight polyethylene (ULMWPE or PE-WAX)
- High-molecular-weight polyethylene (HMWPE)
- High-density polyethylene (HDPE)
- High-density cross-linked polyethylene (HDXLPE)
- Cross-linked polyethylene (PEX or XLPE)
- Medium-density polyethylene (MDPE)
- Linear low-density polyethylene (LLDPE)
- Low-density polyethylene (LDPE)
- Very-low-density polyethylene (VLDPE)
- Chlorinated polyethylene (CPE)

Ultra-high-molecular-weight polyethylene (UHMWPE)

UHMWPE is polyethylene with a molecular weight numbering in the millions, usually between 3.1 and 5.67 million. The high molecular weight makes it a very tough material, but results in less efficient packing of the chains into the crystal structure as evidenced by densities of less than high density polyethylene (for example, 0.930–0.935 g/cm).

UHMWPE can be made through any catalyst technology, although Ziegler catalysts are most common. Because of its outstanding toughness and its cut, wear and excellent chemical resistance, UHMWPE is used in a diverse range of applications.

These include can and bottle handling machine parts, moving parts on weaving machines, bearings, gears, artificial joints, edge protection on ice rinks and butchers' chopping boards. It is commonly used for the construction of articular portions of implants used for hip and knee replacements. As Fibre, it competes with aramid in bulletproof vests.

High-density polyethylene (HDPE)

HDPE is defined by a density of greater or equal to 0.941 g/cm. HDPE has a low degree of branching and thus low intermolecular forces and tensile strength. HDPE can be produced by chromium/silica catalysts, Ziegler-Natta catalysts or metallocene catalysts. The lack of branching is ensured by an appropriate choice of catalyst (for example, chromium catalysts or Ziegler-Natta catalysts) and reaction conditions. HDPE is used in products and packaging such as milk jugs, detergent bottles, butter tubs, garbage containers and water pipes. One third of all toys are manufactured from HDPE. In 2007 the global HDPE consumption reached a volume of more than 30 million tons.

Cross-linked polyethylene (PEX or XLPE)

PEX is a medium- to high-density polyethylene containing cross-link bonds introduced into the polymer structure, changing the thermoplastic into a thermoset. The high-temperature properties of the polymer are improved, its flow is reduced and its chemical resistance is enhanced. PEX is used in some potable-water plumbing systems because tubes made of the material can be expanded to fit over a metal nipple and it will slowly return to its original shape, forming a permanent, water-tight, connection.

Medium-density polyethylene (MDPE)

MDPE is defined by a density range of 0.926–0.940 g/cm. MDPE can be produced by chromium/silica catalysts, Ziegler-Natta catalysts or metallocene catalysts. MDPE has good shock and drop resistance properties. It also is less notch-sensitive than HDPE, stress cracking resistance is better than HDPE. MDPE is typically used in gas pipes and fittings, sacks, shrink film, packaging film, carrier bags and screw closures.

Linear low-density polyethylene (LLDPE)

LLDPE is defined by a density range of 0.915–0.925 g/cm. LLDPE is a substantially linear polymer with significant numbers of short branches, commonly made by copolymerization of ethylene with short-chain alpha-olefins (for example, 1-butene, 1-hexene and 1-octene). LLDPE has higher tensile strength than LDPE, it exhibits higher impact and puncture resistance than LDPE.

Lower thickness (gauge) films can be blown, compared with LDPE, with better environmental stress cracking resistance but is not as easy to process. LLDPE is used in packaging, particularly film for bags and sheets. Lower thickness may be used compared to LDPE.

Cable covering, toys, lids, buckets, containers and pipe. While other applications are available, LLDPE is used predominantly in film applications due to its toughness, flexibility and relative transparency. Product examples range from agricultural films, saran wrap, and bubble wrap, to multilayer and composite films. In 2009 the world LLDPE market reached a volume of almost US$24 billion ($17 billion).

Low-density polyethylene (LDPE)

LDPE is defined by a density range of 0.910–0.940 g/cm. LDPE has a high degree of short and long chain branching, which means that the chains do not pack into the crystal structure as well. It has, therefore, less strong intermolecular forces as the instantaneous-dipole induced-dipole attraction is less. This results in a lower tensile strength and increased ductility. LDPE is created by free radical polymerization. The high degree of branching with long chains gives molten LDPE unique and desirable flow properties. LDPE is used for both rigid containers and plastic film applications such as plastic bags and film wrap. In 2009 the global LDPE market had a volume of about US$22.2 billion (□15.9 billion).

Very-low-density polyethylene (VLDPE)

VLDPE is defined by a density range of 0.880–0.915 g/cm. VLDPE is a substantially linear polymer with high levels of short-chain branches, commonly made by copolymerization of ethylene with short-chain alpha-olefins (for example, 1-butene, 1-hexene and 1-octene). VLDPE is most commonly produced using metallocene catalysts due to the greater co-monomer incorporation exhibited by these catalysts. VLDPEs are used for hose and tubing, ice and frozen food bags, food packaging and stretch wrap as well as impact modifiers when blended with other polymers.

Recently much research activity has focused on the nature and distribution of long chain branches in polyethylene. In HDPE a relatively small number of these branches, perhaps 1 in 100 or 1,000 branches per backbone carbon, can significantly affect the rheological properties of the polymer.

Copolymers

In addition to copolymerization with alpha-olefins, ethylene can also be copolymerized with a wide range of other monomers and ionic composition that creates ionized free radicals. Common examples include vinyl acetate (the resulting product is ethylene-vinyl acetate copolymer, or EVA, widely used in athletic-shoe sole foams) and a variety of acrylates. Applications of acrylic copolymer include packaging and sporting goods, and superplasticizer, used for cement production.

History

Polyethylene was first synthesized by the German chemist Hans von Pechmann who prepared it by accident in 1898 while investigating diazomethane. When his colleagues Eugen Bamberger and Friedrich Tschirner characterized the white, waxy substance that he had created, they recognized that it contained long -CH_2- chains and termed it *polymethylene*.

The first industrially practical polyethylene synthesis (diazomethane is a notoriously unstable substance that is generally avoided in industrial application) was discovered in 1933 by Eric Fawcett and Reginald Gibson, again by accident, at the Imperial Chemical Industries (ICI) works in Northwich, England. Upon applying extremely high pressure (several hundred atmospheres) to a mixture of ethylene and benzaldehyde they again produced a white, waxy, material. Because the reaction had been initiated by trace oxygen contamination in their apparatus, the experiment was, at first, difficult to reproduce. It was not until 1935 that another ICI chemist, Michael Perrin, developed this accident into a reproducible high-pressure synthesis for polyethylene that became the basis for industrial LDPE production beginning in 1939.

Because polyethylene was found to have very low-loss properties at very high frequency radio waves, commercial distribution in Britain was suspended on the outbreak of World War II, secrecy imposed and the new process was used to produce insulation for UHF and SHF coaxial cables of radar sets.

During World War II, further research was done on the ICI process and in 1944 Bakelite Corporation at Sabine, Texas and Du Pont at Charleston, West Virginia, began large scale commercial production under license from ICI.

The breakthrough landmark in the commercial production of polyethylene began with the development of catalyst that promote the polymerization at mild temperatures and pressures.

The first of these was a chromium trioxide–based catalyst discovered in 1951 by Robert Banks and J. Paul Hogan at Phillips Petroleum. In 1953 the German chemist Karl Ziegler developed a catalytic system based on titanium halides and organoaluminium compounds that worked at even milder conditions than the Phillips catalyst.

The Phillips catalyst is less expensive and easier to work with, however, and both methods are heavily used industrially. By the end of the 1950s both the Phillips- and Ziegler-type catalysts were being used for HDPE production. In the 1970s, the Ziegler system was improved by the incorporation of magnesium chloride.

Catalytic systems based on soluble catalysts, the metallocenes, were reported in 1976 by Walter Kaminsky and Hansjörg Sinn. The Ziegler- and metallocene-based catalysts families have proven to be very flexible at copolymerizing ethylene with other olefins and have become the basis for the wide range of polyethylene resins available today, including very low density polyethylene and linear low-density polyethylene. Such resins, in the form of UHMWPE fibers, have (as of 2005) begun to replace aramids in many high-strength applications.

Environmental issues

Although ethylene can be produced from renewables, polyethylene is mainly made from petroleum or natural gas.

Biodegrading plastics

One of the main problems of polyethelyne is that without special treatment it is not readily biodegradable, and thus accumulates. In Japan getting rid of plastics in an environmentally friendly way was the major problem discussed until the Fukushima disaster in 2011.

It was listed as a $90 billion market for solutions. Since 2008 Japan has rapidly increased the recycling of plastics, but still has a large rate of plastic wrapping which goes to waste.

In May 2008, Daniel Burd, a 16-year-old Canadian, won the Canada-Wide Science Fair in Ottawa after discovering that *Pseudomonas fluorescens*, with the help of *Sphingomonas*, can degrade over 40per cent of the weight of plastic bags in less than three months.

The thermophilic bacterium Brevibacillus borstelensis (strain 707) was isolated from a soil sample and found to use low-density polyethylene as a sole carbon source; when incubated together at 50 degrees Celsius. Biodegredation increased with time exposed to ultraviolet radiation.

In 2010 a Japanese researcher Akinori Ito released the prototype of a machine which creates oil from Polyethylene using a small, self-contained vapor distillation process.

Acinetobacter sp. 351 can degrade lower molecular weight PE oligomers. When PE is subjected to thermo and photo-oxidization, products including alkanes, alkenes, ketones, aldehydes, alcohols, carboxylic acid, keto-acids, dicarboxylic acids, lactones and esters are released.

Bio-derived polyethylene

Braskem and Toyota Tsusho Corporation started Joint marketing activities for producing polyethylene from sugar cane. Braskem will build a new facility at their existing industrial unit in Triunfo, ₹ Brazil with an annual production capacity of 200,000 short tons (180,000,000 kg), and will produce high-density polyethylene (HDPE) and low-density polyethylene (LDPE) from bioethanol derived from sugarcane.

Polyethylene can also be made from other feedstocks, including wheat grain and sugar beet. Retrieved from cane sugar, i.e. plant biomass renewable feedstock; Brazil is the first country to develop the product.

These developments are using renewable resources rather than fossil fuel, although the issue of plastic source is currently negligible in the wake of plastic waste and in particular polyethylene waste as shown above.

Joining

Commonly used methods for joining polyethylene parts together include:

- Hot gas welding
- Fastening
- Infrared welding
- Laser welding
- Ultrasonic welding
- Heat sealing
- Heat fusion

Adhesives and solvents are rarely used because polyethylene is Non-polar and has a high resistance to solvents. Pressure-sensitive adhesives (PSA) are feasible if the surface is flame treated or corona treated. Commonly used adhesives include:

- Dispersion of solvent-type PSAs
- Polyurethane contact adhesives
- Two-part polyurethane or epoxy adhesives
- Vinyl acetate copolymer hot melt adhesives

Nomenclature and general description of the process

The name polyethylene comes from the ingredient and not the resulting chemical compound, which contains no double bonds. The scientific name *polyethene* is systematically derived from the scientific name of the monomer. The alkene monomer converts to a long, sometimes *very* long, alkane in the polymerization process. In certain circumstances it is useful to use a structure-based nomenclature; in such cases IUPAC recommends poly(methylene) (poly(methanediyl) is a non-preferred alternative). The difference in names between the two systems is due to the *opening up* of the monomer's double bond upon polymerization. The name is abbreviated to PE. In a similar manner polypropylene and polystyrene are shortened to PP and PS, respectively. In the United Kingdom the polymer is commonly called polythene, from the ICI trade name, although this is not recognized scientifically.

ULTRA-HIGH-MOLECULAR-WEIGHT POLYETHYLENE

Ultra-high-molecular-weight polyethylene (UHMWPE, UHMW) is a subset of the thermoplastic polyethylene. Also known as high-modulus polyethylene, (HMPE), or high-performance polyethylene (HPPE), it has extremely long chains, with a molecular mass usually between 2 and 6 million u. The longer chain serves to transfer load more effectively to the polymer backbone by strengthening intermolecular interactions. This results in a very tough material, with the highest impact strength of any thermoplastic presently made.

UHMWPE is odorless, tasteless, and Non-toxic. It is highly resistant to corrosive chemicals except oxidizing acids; has extremely low moisture absorption and a very low coefficient of friction; is self-lubricating; and is highly resistant to abrasion, in some forms being 15 times more resistant to abrasion than carbon steel. Its coefficient of friction is significantly lower than that of nylon and acetal, and is comparable to that of polytetrafluoroethylene (PTFE, Teflon), but UHMWPE has better abrasion resistance than PTFE.

Development

Polymerisation of UHMWPE was commercialised in the 1950s by Ruhrchemie AG, which changed names over the years. Today UHMWPE powder materials, which may be directly molded into a product's final shape, are produced by Ticona, Braskem, and Mitsui. Processed UHMWPE is available commercially either as fibers or in consolidated form, such as sheets or rods. Because of its resistance to wear and impact, UHMWPE continues to find increasing industrial applications, including the automotive and bottling sectors. Since the 1960s, UHMWPE has also been the material of choice for total joint arthroplasty in orthopedic and spine implants.

UHMWPE fibers, commercialized in the late 1970s by the Dutch chemical company DSM, are widely used in ballistic protection, Defence applications, and increasingly in medical devices.

Structure and properties

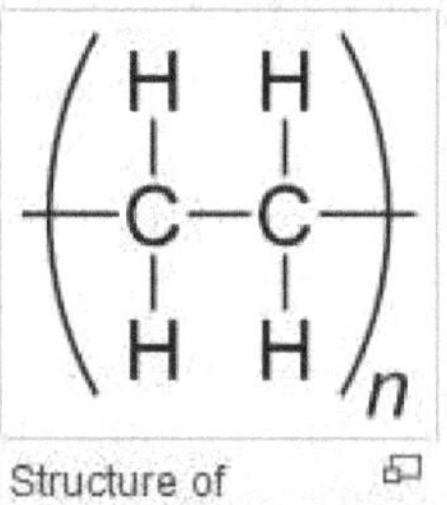

Structure of UHMWPE, with n greater than 100,000

UHMWPE is a type of polyolefin. It is made up of extremely long chains of polyethylene, which all align in the same direction. It derives its strength largely from the length of each individual molecule (chain). Van der Waals bonds between the molecules are relatively weak for each atom of overlap between the molecules, but because the molecules are very long, large overlaps can exist, adding up to the ability to carry larger shear forces from molecule to molecule. Each chain is bonded to the others with so many Van der Waals bonds that the whole of the inter-molecule strength is high. In this way, large tensile loads are not limited as much by the comparative weakness of each Van der Waals bond. When formed to fibers, the polymer chains can attain a parallel orientation greater than 95per cent and a level of crystallinity from 39per cent to 75per cent. In contrast, Kevlar derives its strength from strong bonding between relatively short molecules. The weak bonding between olefin molecules allows local thermal excitations to disrupt the crystalline order of a given chain piece-by-piece, giving it much poorer heat resistance than other high-strength fibers. Its melting point is around 130 to 136 °C (266 to 277 °F), and, according to DSM, it is not advisable to use UHMWPE fibers at temperatures exceeding 80 to 100 °C (176 to 212 °F) for long periods of time. It becomes brittle at temperatures below "150 °C ("240 °F).

The simple structure of the molecule also gives rise to surface and chemical properties that are rare in high-performance polymers. For example, the polar groups in most polymers easily bond to water. Because olefins have no such groups, UHMWPE does not absorb water readily, nor wet easily, which makes bonding it to other polymers difficult. For the same reasons, skin does not interact with it strongly, making the UHMWPE Fibre surface feel slippery. In a similar manner, aromatic polymers are often susceptible to aromatic solvents due to aromatic stacking interactions, an effect aliphatic polymers like UHMWPE are immune to. Since UHMWPE does not contain chemical groups (such as esters, amides or hydroxylic groups) that are susceptible to attack from aggressive agents, it is very resistant to water, moisture, most chemicals, UV radiation, and micro-organisms.

Under tensile load, UHMWPE will deform continually as long as the stress is present—an effect called *creep*.

When UHMWPE is annealed, the material is heated to 135 °C to 138 °C in an oven or a liquid bath of silicone oil or glycerine. The material is then cooled down at a rate of 5 °C/h to 65 °C or less. Finally, the material is wrapped in an insulating blanket for 24 hours to bring to room temperature.

Production

UHMWPE is synthesized from monomer of ethylene, which are bonded together to form the base polyethylene product. These *ultra-high-molecular-weight polyethylene* molecules are several orders of magnitude longer than those of familiar high-density polyethylene (HDPE) due to a synthesis process based on metallocene catalysts, resulting in UHMWPE molecules typically having 100,000 to 250,000 monomer units per molecule each compared to HDPE's 700 to 1,800 monomers.

UHMWPE is processed variously by compression molding, ram extrusion, gel spinning, and sintering. Several European companies began compression molding UHMW in the early 1960s. Gel-spinning arrived much later and was intended for different applications.

In gel spinning a precisely heated gel of UHMWPE is extruded through a spinneret. The extrudate is drawn through the air and then cooled in a water bath. The end-result is a Fibre with a high degree of molecular orientation, and therefore exceptional tensile strength. Gel spinning depends on isolating individual chain molecules in the solvent so that intermolecular entanglements are minimal. Entanglements make chain orientation more difficult, and lower the strength of the final product.

Applications

Fibre

Dyneema and Spectra are lightweight high-strength oriented-strand gel spun through a spinneret. They have yield strengths as high as 2.4 GPa (350,000 psi) and specific gravity as low as 0.97 (for Dyneema SK75). High-strength steels have comparable yield strengths, and low-carbon steels have yield strengths much lower (around 0.5 GPa). Since steel has a specific gravity of roughly 7.8, this gives strength-to-weight ratios for these materials in a range from 8 to 15 times higher than steel. Strength-to-weight ratios for Dyneema are about 40per cent higher than for aramid. Dyneema was invented by Albert Pennings in 1963 but made commercially available by DSM in 1990.

UHMWPE fibers are used in armor, in particular, personal armor and on occasion as vehicle armor, cut-resistant gloves, bow strings, climbing equipment, fishing line, spear lines for spearguns, high-performance sails, suspension lines on sport parachutes and paragliders, rigging in yachting, kites, and kites lines for kites sports. Spectra is also used as a high-end wakeboard line.

For personal armor, the fibers are, in general, aligned and bonded into sheets, which are then layered at various angles to give the resulting composite material strength in all directions. Recently developed additions to the US Military's Interceptor body armor, designed to offer arm and leg protection, are said to utilize a form of Spectra or Dyneema fabric. Dyneema provides puncture resistance to protective clothing in the sport of fencing.

Spun UHMWPE fibers excel as fishing line, as they have less stretch, are more abrasion-resistant, and are thinner than traditional monofilament line.

In climbing, cord and webbing made of combinations of UHMWPE and nylon yarn have gained popularity for their low weight and bulk, though, unlike their nylon counterparts, they exhibit very low elasticity, making them unsuitable for limiting forces in a fall. Also, low elasticity translates to low toughness. The fiber's very high lubricity leads to poor knot-holding ability, and has led to the recommendation to use the triple fisherman's knot rather than the traditional double fisherman's knot in 6mm UHMWPE *core* cord to avoid a particular failure mechanism of the double fisherman's, where first the sheath fails at the knot, then the core slips through.

Owing to its low density, ships' hawsers and cables can be made from the fibre, and float on sea water. "Spectra Wires" as they are called in the towboat community are commonly used for face wires as a lighter alternative to steel wires. It is used in skis and snowboards, often in combination with carbon Fibre, reinforcing the fiberglass composite material, adding stiffness and improving its flex characteristics. The UHMWPE is often used as the base layer, which contacts the snow, and includes abrasives to absorb and retain wax.

It is also used in lifting applications for manufacturing low weight, and heavy duty lifting slings. Due to its extreme abrasion resistance it is also used as an excellent corner protection for synthetic lifting slings.

High-performance lines (such as backstays) for sailing and parasailing are made of UHMWPE, due to their low stretch, high strength, and low weight.

Dyneema was used for the 30-kilometre space tether in the ESA/Russian Young Engineers' Satellite 2 of September, 2007.

Sheet

UHMWPE sheet has been used as synthetic ice in ice rinks where ambient temperatures or energy costs make it impractical to create and maintain normal ice. The material's resistance to cutting and abrasion make it highly suitable for this application.

UHMWPE sheet is also cut into small blocks to be used as a brake pad material for mountain bike trials rim brakes. The material's combination of flexibility and abrasion resistance allows it to run on a roughened rim surface to lock the braked wheels extremely securely when the brake is pulled, while still wearing at a slower rate than more common bicycle brake pad materials.

Medical

UHMWPE has over 40 years of clinical history as a successful biomaterial for use in hip, knee, and (since the 1980s), for spine implants. An online repository of information and review articles related to medical grade UHMWPE, known as the UHMWPE Lexicon, was started online in 2000.

Joint replacement components have historically been made from "GUR" resins. These powder materials are produced by Ticona, typically converted into semi-forms by companies such as Quadrant and Orthoplastics, and then machined into implant components and sterilised by device manufacturers.

UHMWPE was first used clinically in 1962 by Sir John Charnley and emerged as the dominant bearing material for total hip and knee replacements in the 1970s. Details about the "discovery" of UHMWPE for orthopedic applications by Charnley and his engineering associate Harry Craven are available Throughout its history, there were unsuccessful attempts to modify UHMWPE to improve its clinical performance until the development of highly crosslinked UHMWPE in the late 1990s.

One unsuccessful attempt to modify UHMWPE was by blending the powder with carbon fibers. This reinforced UHMWPE was released clinically as "Poly

Two" by Zimmer in the 1970s. The carbon fibers had poor compatibility with the UHMWPE matrix and its clinical performance was inferior to virgin UHMWPE.

A second attempt to modify UHMWPE was by high-pressure recrystallisation. This recrystallised UHMWPE was released clinically as "Hylamer" by DePuy in the late 1980s. When gamma irradiated in air, this material exhibited susceptibility to oxidation, resulting in inferior clinical performance related to virgin UHMWPE. Today, the poor clinical history of Hylamer is largely attributed to its sterilisation method, and there has been a resurgence of interest in studying this material (at least among certain research circles). Hylamer fell out of Favour in the United States in the late 1990s with the development of highly crosslinked UHMWPE materials, however negative clinical reports from Europe about Hylamer continue to surface in the literature.

Highly crosslinked UHMWPE materials were clinically introduced starting in 1998 and have rapidly become the standard of care for total hip replacements, at least in the United States.

These new materials are crosslinked with gamma or electron beam radiation (50–105 kGy) and then thermally processed to improve their oxidation resistance. Five-year clinical data, from several centers, are now available demonstrating their superiority relative to conventional UHMWPE for total hip replacement. Clinical studies are still underway to investigate the performance of highly crosslinked UHMWPE for knee replacement.

Since 2007, manufacturers started incorporating antioxidants into UHMWPE for hip and knee arthroplasty bearing surfaces. Vitamin E (a-tocopherol) is the most common antioxidant used in radiation-crosslinked UHMWPE for medical applications. The antioxidant helps quench free radicals that are introduced during the irradiation process, imparting improved oxidation resistance to the UHMWPE without the need for thermal treatment. Several companies have been selling antioxidant-stabilized joint replacement technologies since 2007, using both synthetic Vitamin E as well as hindered phenol-based antioxidants.

Another important medical advancement for UHMWPE in the past decade has been the increase in use of fibers for sutures. Medical-grade fibers for surgical applications are produced by DSM under the "Dyneema Purity" trade name.

Manufacturing

UHMWPE is used in the manufacture of PVC (vinyl) windows and doors, as it can stand up to the heat required to soften the PVC-based materials and is used as a form/chamber filler for the various PVC shape profiles in order for those materials to be 'bent' or shaped around a template.

UHMWPE is also used in the manufacture of Hydraulic Seals and Bearings. It is best suited for Medium mechanical duties in water, Oil Hydraulics, pneumatics, and unlubricated applications. It has a good abrasion resistance but is better suited to soft mating surfaces.

Wire/Cable

HALAR cathodic protection cable is made of a fluoropolymer insulation that exhibits superior chemical resistance. Made with dual insulation, HALAR wire can be used in any situation where chlorine and hydrogen gases are present. HALAR cable is made with a primary layer, which is most resistant to chlorine, sulfuric acid and hydrochloric acid. Following the primary layer is a high molecular weight polyethylene (HMWPE) insulation, which provides pliable strength and allows considerable abuse during installation. The HMWPE jacketing provides mechanical protection as well.

SPANDEX

Spandex, Lycra or elastane is a synthetic Fibre known for its exceptional elasticity. It is stronger and more durable than natural rubber. It is a polyester-polyurethane copolymer that was invented in 1958 by chemist Joseph Shivers at DuPont's Benger Laboratory in Waynesboro, Virginia. When introduced in 1962, it revolutionized many areas of the clothing industry.

The name "spandex" is an anagram of the word "expands". It is the preferred name in North America; in continental Europe it is referred to by variants of "elastane", *i.e. elasthanne* (France), *elastan* (Germany), *elastano* (Spain), *elastam* (Italy) and *Elasthaan* (Netherlands), and is known in the UK, Ireland, Portugal, Brazil, Argentina, Australia, New Zealand and Israel primarily as Lycra. Brand names for spandex include Lycra (made by Koch subsidiary Invista, previously a part of DuPont), Elaspan (also Invista), Acepora (Taekwang), Creora (Hyosung), INVIYA (Indorama Corporation), ROICA and Dorlastan (Asahi Kasei), Linel (Fillattice), and ESPA (Toyobo).

Spandex fibers production

Spandex fibers are produced in four different ways: melt extrusion, reaction spinning, solution dry spinning, and solution wet spinning. All of these methods include the initial step of reacting monomers to produce a prepolymer. Once the prepolymer is formed, it is reacted further in various ways and drawn out to make the fibers. The solution dry spinning method is used to produce over 94.5per cent of the world's spandex fibers.

Solution dry spinning

- Step 1: The first step is to produce the prepolymer. This is done by mixing a macroglycol with a diisocyanate monomer. The two compounds are mixed in a reaction vessel to produce a prepolymer. A typical ratio of glycol to diisocyanate is 1:2.

- Step 2: The prepolymer is further reacted with an equal amount of diamine. This reaction is known as *chain extension reaction.* The resulting solution is diluted with a solvent (DMAc) to produce the spinning solution. The solvent helps make the solution thinner and more easily handled, and then it can be pumped into the fibre production cell.
- Step 3: The spinning solution is pumped into a cylindrical spinning cell where it is cured and converted into fibres. In this cell, the polymer solution is forced through a metal plate called a spinneret. This causes the solution to be aligned in strands of liquid polymer. As the strands pass through the cell, they are heated in the presence of a nitrogen and solvent gas. This process causes the liquid polymer to react chemically and form solid strands.
- Step 4: As the fibres exit the cell, an amount of solid strands are bundled together to produce the desired thickness. Each fibre of spandex is made up of many smaller individual fibres that adhere to one another due to the natural stickiness of their surface.
- Step 5: The resulting fibres are then treated with a finishing agent which can be magnesium stearate or another polymer. This treatment prevents the fibres' sticking together and aids in textile manufacture. The fibres are then transferred through a series of rollers onto a spool.

Major spandex fibre uses

Because of its elasticity and strength (stretching up to five times its length), spandex has been incorporated into a wide range of garments, especially in skin-tight garments.

A benefit of spandex is its significant strength and elasticity and its ability to return to the original shape after stretching and faster drying than ordinary fabrics.

The types of garments which incorporate spandex include:

- Apparel and clothing articles where stretch is desired, generally for comfort and fit, such as:
  - activewear
  - athletic, aerobic, and exercise apparel
  - belts
  - bra straps and side panels
  - competitive swimwear
  - cycling jerseys and shorts
  - dance belts worn by male ballet dancers and others

  - gloves
  - hosiery
  - leggings
  - netball bodysuits
  - orthopaedic braces
  - rowing unisuits
  - cross country race suits
  - ski pants
  - skinny jeans
  - slacks
  - miniskirts
  - socks and tights
  - swimsuits/bathing suits
  - underwear
  - wetsuits
  - zentai
- Compression garments such as:
  - foundation garments
  - motion capture suits
- Shaped garments such as:
  - bra cups
  - support hose
  - surgical hose
  - superhero outfits
  - women's volleyball shorts
  - wrestling singlets
- Home furnishings, such as microbead pillows

For clothing, spandex is usually mixed with cotton or polyester, and accounts for a small percentage of the final fabric, which therefore retains most of the look and feel of the other fibers.

In North America it is rare in men's clothing, but prevalent in women's. An estimated 80per cent of clothing sold in the United States contained spandex in 2010.

VINYLON

Vinylon, also known as Vinalon, is a synthetic Fibre produced from polyvinyl alcohol, using anthracite and limestone as raw materials. Vinalon was first developed in 1939 by Ichiro Sakurada, Ri Sung Gi, and H. Kawakami. Production of this Fibre was delayed for World War II.

The fibre was largely ignored in Korea until Ri defected to North Korea in 1950. Trial production began in 1954 and in 1961 the massive February 8 Vinylon Complex was built in Hamhung. Its success and widespread usage in North

Korea is often pointed to as an example of the success of the *juche* philosophy. While Hamhung remains a major production centre for vinylon; in 1998, a vinylon factory was opened up in South Pyongan.

In early 2010, Kim Jong-il himself attended a mass rally at the February 8 Vinalon Complex in Hamhung to celebrate its reopening after 16 years of inactivity.

While Kim was often seen at political rallies or military parades, this was the first documented time he has ever attended an industrial mass rally, said an anonymous South Korean security official.

Vinylon, also known as "Juche fibre" in the DPRK, has become the national fibre of North Korea and is used for the majority of textiles, outstripping fibre such as cotton or nylon, which are produced only in small amounts in North Korea. Other than clothing, Vinalon is also used for shoes, ropes, and quilt wadding.

Textile artist Toshiko MacAdam used Vinylon in her early works, as it was more economical than nylon.

Vinylon is resistant to heat and chemicals but has numerous disadvantages: it is stiff, uncomfortable, shiny, prone to shrinking, and difficult to dye.

VINYON

Vinyon is a synthetic Fibre made from polyvinyl chloride. In some countries other than the United States, vinyon fibers are referred to as polyvinyl chloride fibers. It can bind non-woven fibers and fabrics.

It was invented in 1939. It has the same health problems associated with chlorinated polymers. In the past, Vinyon was used a substitute for plant-based filters in tea bags.

Vinyon Fibre characteristics:

- doesn't flame, but softens at low temperatures(55 C)
- high resistance to chemicals
- Moisture absorption is less than 0.5per cent and moisture regained is less than 0.1per cent
- crease resistant and elastic

Major vinyon Fibre uses:

- industrial applications as a bonding agent for non-woven fabrics and products

Production

The U.S. Federal Trade Commission definition for vinyon Fibre is "A manufactured Fibre in which the Fibre-forming substance is any long chain synthetic polymer composed of at least 85 percent by weight of vinyl chloride units (—CH2—CHCl—)."First U.S. commercial vinyon Fibre production: 1939, FMC Corporation, Fibre Division (formerly American Viscose Corporation).

ZYLON

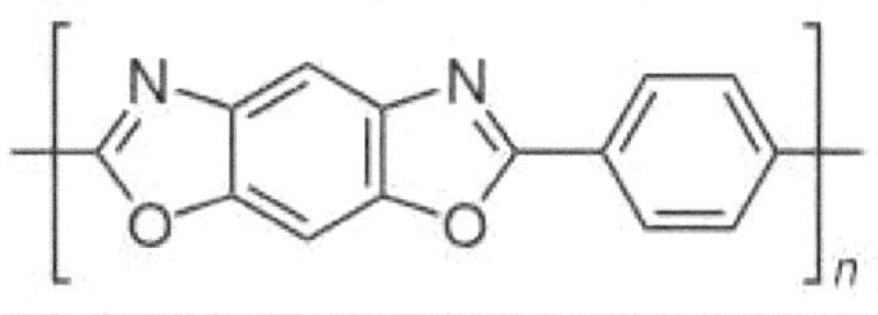

The structure of Zylon

Zylon (IUPAC name: poly(*p*-phenylene-2,6-benzobisoxazole)) is a trademarked name for a range of thermoset liquid crystalline polyoxazole. This synthetic polymer material was invented and developed by SRI International in the 1980s, and is currently manufactured by the Toyobo Corporation. In generic usage, the Fibre is referred to as PBO.

Zylon has 5.8 GPa of tensile strength, which is 1.6 times that of Kevlar. Like Kevlar, Zylon is used in a number of applications that require very high strength with excellent thermal stability. Tennis racquets, table tennis blades, various medical applications, and some of the Martian rovers are some of the better known instances. More common usage is in drive belts for snow mobiles manufactured by MBL USA Corporation.

Usage

Body armor

Zylon gained wide use in U.S. police officers body armor protection in 1998 with its introduction by Second Chance Body Armor, Inc. But protective vests constructed with Zylon became controversial in late 2003 when Oceanside, CA Police Officer Tony Zeppetella's and Forest Hills, PA Police Officer Ed Limbacher's vests failed, leaving Zeppetella mortally wounded and Limbacher seriously injured. Some studies subsequently reported that the Zylon vests may degrade rapidly, leaving wearers with significantly less protection than expected. Second Chance eventually recalled all of its zylon-containing vests, which led to its subsequent bankruptcy. In early 2005, Armor Holdings, Inc. first recalled its existing Zylon-based products, and decreased the rated lifespan warranty of new vests from 60 months to 30 months. In August 2005, AHI decided to discontinue manufacturing all of its Zylon-containing vests. This was largely based on the actions of the U.S. government's National Institute of Justice, which decertified Zylon for use in its approved models of ballistic vests for law enforcement.

Space elevator research

A competition was held in the Wirefly X Prize Cup in Las Cruces, New

Mexico, US, on October 20–21, 2006. A team from the University of British Columbia entered into the Tether Challenge, using a construction made from Zylon fibers. The house tether used by Spaceward, that the other teams would have to beat in strength by 50per cent in the 2007, was made of Zylon.

High altitude balloon science

Zylon is used by NASA in long-duration, high altitude data collection. Braided Zylon strands maintain the structure of polyethylene superpressure balloons. Zylon is the material of choice due to its low weight, high tensile strength, and thermal properties.

Motorsport

As has been the case since 2001, Zylon tethers are used in Formula One to affix the wheel to the chassis thus preventing the wheel from ejecting in to a crowded area in the event of an accident which causes the wheel to become airborne. Starting in the 2007 season, the driver's cockpit must now be clad in special anti-penetration panels made of Zylon.

In 2011, a Zylon strip was introduced to reinforce the top of the racing helmet visor and provide an overlap between the visor and helmet for additional protection after Felipe Massa's 2009 injury. The Indy Racing League began using Zylon in 2008.

Standing rigging

On modern racing yachts Zylon is used for parts of the standing rigging. It is used as shrouds and stays. The PBO (Polybenzoxazole) Fibre is degraded by UV and visible light, seawater and chafing, and is therefore protected by a synthetic melted-on jacket.

It is claimed to be 65per cent lighter than traditional rigging at 110–130per cent of the price of rodrigging. Based on laboratory tests, superior durability is also claimed. Zylon has a high Young's modulus of 270GPa, meaning that it is very stiff.

Conductive textile

Zylon has been incorporated as the base Fibre for AmberStrand-Z series conductive fibers made by Syscom Advanced Materials, Inc. Currently offered in three different tow or strand sizes, the Zylon Fibre is plated with nickel, copper, silver, or gold. The conductive Fibre is used for electronic textiles, EMI shielding in woven or knit sheets, or as a braid over wires, and for signal transmission or current conduction. This conductive Fibre combines the advantages of Zylon (strength, high temperature, durability, light weight, etc.) with the electrical properties of various metals. The conductive yarns can be sewn, braided, knit, or simply insulated like a bare wire.

Structural Rehabilitation

PBO is the strongest and stiffest of commercially available fibers used to repair and strengthen concrete and masonry structures in externally bonded composite systems. In this system, the fibers are produced in the form of a fabric mesh, and a cementitious mortar matrix bonds this to an existing structure.

7

Abrasion of Textile Surfaces

INTRODUCTION

Wear in textile materials is one of a limited number of fault factors in which an object loses its usefulness and the economic implication can be of enormous value to the industry.

The terms wear and abrasion are sometimes confused. Wear is a very general term covering the loss of material by virtually any means. As wear usually occurs by rubbing together of two surfaces, abrasion is often used as a general term to mean wear. The resistance of textile materials to abrasion as measured on a testing machine in the laboratory is generally only one of several factors contributing to wear performance or durability as experienced in the actual use of the material. While "abrasion resistance" (often stated in terms of the number of cycles on a specified machine, using a specified technique to produce a specified degree or amount of abrasion) and "durability" (defined as the ability to withstand deterioration or wearing out in use, including the effects of abrasion) are frequently related, the relationship varies with different end uses, and different factors may be necessary in any calculation of predicted durability from specific abrasion data.

Abrasion is the physical destruction of fibres, yarns, and fabrics, resulting from the rubbing of a textile surface over another surface. Textile materials can be unserviceable because of several different factors and one of the most important causes is abrasion. Abrasion occurs during wearing, using, cleaning or washing process and this may distort the fabric, cause fibres or yarns to be pulled out or remove fibre ends from the surface. Abrasion ultimately results in the loss of performance characteristics, such as strength, but it also affects the appearance of the fabric. The main factors that reduce service life of the garment are heavily dependent on its end use. But especially certain parts of apparel, such as collar, cuffs and pockets, are subjected to serious wear in use. Abrasion is a serious problem for home textiles like as carpets and upholstery fabrics, socks and technical textiles as well. Yarn abrasion is another important subject that should be considered during processing.

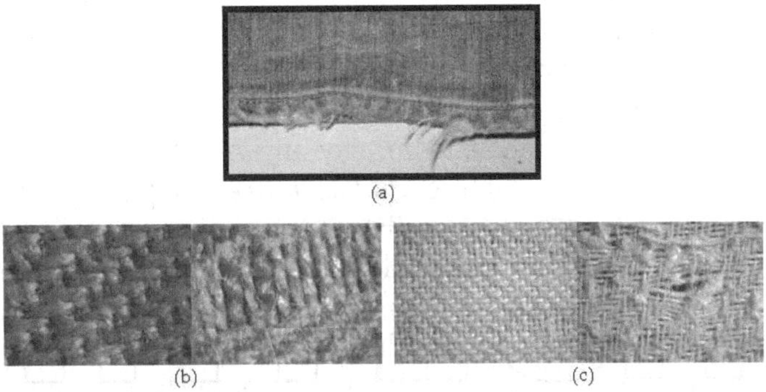

Fig. Abraded textile products (a) edge of pants, (b)(c) surface appearances of fabrics before (left) and after (right) the abrasion test.

In this chapter, detailed information about the abrasion and abrasion resistance of the textile materials are discussed. In the first part, the abrasion and wear mechanism are explained. In the second part, abrasion of the fabrics (factors affecting abrasion such as fibre, yarn and fabric properties, parameters affecting the test results, testing and evaluation methods), yarn abrasion (yarn on yarn and yarn external abrasion), abrasion characteristics of socks and technical textile fabrics are analyzed. Studies on the mentioned subjects are given as well.

ABRASION MECHANISM OF TEXTILES

Abrasive wear in textiles is caused by different conditions mainly given below:

- Friction between textile materials, such as rubbing of a jacket or coat lining on a shirt, pants pockets against pants fabric etc.
- Friction between the textile materials to the external object, such as rubbing of trousers to the seat, friction of the yarn to the needle etc.
- Friction between the fibres and dust, or grit, in a fabric that results in cutting of the fibres. This is an extremely slow process, it may be observed on flags hanging out or swimwear because of the unremoved sand.
- Friction between the fabric components. Flexing, stretching, and bending of the fibres during the usage causes fibre slippage, friction to each other and breakage.

The study of the processes of wear is part of the discipline of tribology and the mechanism of wear is very complex. Under normal mechanical and practical procedures, the wear-rate normally changes through three different stages: primary stage or early run-in period, where surfaces adapt to each other and the wear-rate might vary between high and low; secondary stage or mid-age process, where a steady rate of wearing is in motion. Most of the components operational life is comprised in this stage. Tertiary stage or old-age period is where the components are subjected to rapid failure due to a high rate of wearing. Some commonly referred to wear mechanisms include: Adhesive wear, abrasive wear, surface fatigue, fretting wear, erosive wear. Adhesive, abrasive wear and surface fatigue mechanism play an important role in the abrasion mechanism of the yarns and fabrics. *Adhesive wear*, occurs between surfaces during frictional contact and generally refers to unwanted displacement and attachment of wear debris and material compounds from one surface to another. The adhesive wear and material transfer due to direct contact and plastic deformation are the main issues in adhesive wear.

The asperities or microscopic high points or surface roughness found on each surface, define the severity on how fragments of oxides are pulled off and adds to the other surface. This is partly due to strong adhesive forces between atoms, but also due to accumulation of energy in the plastic zone between the asperities during relative motion. *Abrasive wear,* occurs when a hard rough surface slides across a softer surface.

ASTM (American Society for Testing and Materials) defines it as the loss of material due to hard particles or hard protuberances that are forced against and move along a solid surface. Abrasive wear is commonly classified according to the type of contact and the contact environment. The type of contact determines the mode of abrasive wear. The two modes of abrasive wear are known as two-body and three-body abrasive wear. Two-body wear occurs when the grits or hard particles remove material from the opposite surface. The common analogy is that of material being removed or displaced by a cutting or plowing operation.

Three-body wear occurs when the particles are not constrained, and are free to roll and slide down a surface. *Fatigue wear* of a material is caused by a cycling loading during friction. Fatigue occurs if the applied load is higher than the fatigue strength of the material. Fatigue cracks start at the material surface and spread to the subsurface regions. The cracks may connect to each other resulting in separation and delamination of the material pieces. One of the types of fatigue wear is fretting wear caused by cycling sliding of two surfaces across each other with small amplitude.

The friction force produces alternating compression-tension stresses, which result in surface fatigue. In terms of wear mechanism in textiles, abrasion first modifies the fabric surface and then affects the internal structure of the fabric, damaging it. Good abrasion resistance depends more on a high energy of rupture than on high tenacity at break. Abrasion is not influenced so much by the energy absorbed in the first deforming process, as by the work absorbed during repeated deformation. This work is manifested in the elastic energy or the recoverable portion of the total energy. Thus, to prevent abrasion damage, the material must be capable of absorbing energy and releasing that energy upon the removal of load. Energies in tension, shear, compression, and bending are all important for the evaluation of surface abrasion; however, these energies are unknown, and therefore elastic energies in tension permit at least a quantitative interpretation of abrasive damage in fibres and fabrics. Fibres in use are subject to a variety of different forces, which are repeated many times. The gradual breakdown of the internal cohesion of the individual fibres or by a gradual breakdown of the forces of structural cohesion between the fibres results fabric failure. The relative occurrence of these two phenomena depends to a great extent upon the fabric geometry, but there are limitless factors involved (*e.g.*, construction of yarns and weaves) depending on the individual Behaviour of different fibres.

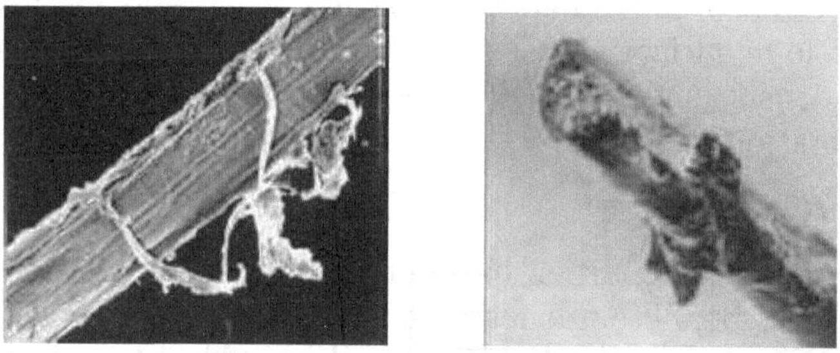

(a) (b)

Fig. Abrasion of fibres over rotating pin: (a) nylon, (b) wool

Fig. Fibre rupture occurred as abraded against standard worsted fabirc.

During the course of abrasion in textiles, fibre to fibre cohesion plays an important role, usually influenced by yarn twist or close fibre packing. Abrasion Behaviour indicates that fibre cohesion is strong in the fabric system, and it causes the shear of the fibres themselves. Frictional forces developed in the yarn due to the motion of the abrasion test are dissipated largely in the fibres by the development of tensile and shear stresses; repetition of such stresses results in fibre fatigue, which causes the loss of fibre mechanical properties, leading to rupture. Fibres in the crowns are broken down in succession, and this causes a reduction in fibre cohesion and yarn strength. In lateral abrasion cycles, frictional forces are able to displace fibre from their normal position, and these fibres ruptures through bending and flexing. In addition, it is also possible that some cracking is initiated by abrasion and then propagated by bending action. Although the mechanism of abrasion is similar, because of the differences in the measurement methods, abrasion resistance of textile materials can be studied in two parts such as fabric abrasion and yarn abrasion

ABRASION OF FABRICS

FACTORS AFFECTING ABRASION RESISTANCE OF FABRICS

Abrasion resistance of the textile materials is very complex phenomenon and affected by many factors, mainly classified as follows: Fibre, yarn, fabric properties and finishing processes. Some of these parameters affect fabric surface whereas some of them has an influence on internal structure of the fabrics. For example fibre characteristics like wool ratio and fineness play a significant role in surface abrasion, while yarn and fabric characteristics like yarn linear density and interlacing coefficient are significantly related with structural abrasion.

Fibre properties

The mechanical properties and dimensions of the fibres are important for abrasion. Fibre type, fibre fineness and fibre length are the main parameters that affect abrasion. Fibres with high elongation, elastic recovery and work of rupture have a good ability to withstand repeated distortion; hence a good degree of abrasion resistance is achieved. Nylon is generally considered to have the best abrasion resistance, followed by polyester, polypropylene. Blending either nylon or polyester with wool and cotton is found to increase abrasion resistance at the expense of other properties. Higher wool rate increase the mass loss. Acrylic and modacrylic have a lower resistance than these fibres while wool, cotton and high modulus viscose have a moderate abrasion resistance. Viscose and acetates are found to have the lowest degree of resistance to abrasion. However, synthetic fibres are produced in many different versions so that the abrasion resistance of particular variant may not conform to the general ranking of fibres. The removal of the fibres from yarn structure is one of the reasons of the abrasion. Therefore factors that affect the cohesion of yarns will influence the abrasion resistance of fabrics as well. Longer fibres incorporated into a fabric confer better abrasion resistance than short fibres because it is harder to liberate them from the fabric structure. For the same reason filament yarns are more abrasion resistant than staple yarns made from the same fibre. The using of finer fibres in the production of yarns causes increment in the number of the fibre in cross section with higher cohesion which results better abrasion resistance. So abrasion retention is better for fabrics with finer fibres.

Yarn properties

Yarn structure, count, twist and hairiness are the main properties which affect abrasion of the textile fabrics. Increasing linear density at constant fabric mass per unit area increases the abrasion resistance of the fabrics. As yarn got thinner, abrasion resistance values of knitted fabrics decrease and breaking occurs in lower cycles. Twist is another parameter affecting abrasion. There is an optimum amount of twist in a yarn to give the best abrasion resistance. At low-twist, fibres can easily be removed from the yarn so that it is gradually reduced in diameter. At high twist levels the fibres are held more tightly but the yarn is stiffer so it is unable to distort under pressure when being abraded. Yarn hairiness has a negative effect in terms of mass loss during abrasion. An increase in yarn hairiness, due to the higher level of protruding fibres from yarn surface, reduces fabric abrasion resistance.

The production method of yarn has also an influence on the abrasion resistance, such that carded fabric gives lower resistance than that of combed fabric. Even yarn structure, using long fibre and lower yarn hairiness are the reasons of that result. Knitted fabrics from ring spun yarns have better abrasion resistance than knitted fabrics from OE spun yarns. Ring spun yarns are hairier but more compactly structured than OE yarns, this well aligned compact structure doesn't promote easy fibre wear off. Compact yarn fabrics have higher abrasion resistance values compared to the ring yarn fabrics with the same fabric construction. Since the fibres of compact yarns are held more tightly within the yarn structure and higher participation of the fibres into the yarn structure exists, compact yarns have a denser and closer structure compared to the ring yarns.

The compact yarn has lower hairiness, high tensile resistance as a result of that fibre movements causing limited abrasion. Fabrics woven from compact yarns have also lower weight loss compared to those woven from ring yarns. Sirospun is a modified ring spinning process that two rovings per spindle are fed to the drafting system within specially developed condensers separately and drafted simultaneously. Fabrics knitted from sirospun yarns show better abrasion resistance than ring, air-jet and OE yarns because of the better evenness, hairiness, regular and tightly structure. However as considering the results of fabrics produced with two ply yarns, fabrics from sirospun yarns wear faster than two fold ring spun yarn. Another factor that affects the abrasion is the number of yarn plies. As the number of ply threads per yarn increases, the thickness and the mass per unit area increases and it causes an improvement in abrasion characteristics of the fabric.

Fabric properties

Fabric construction, thickness, weight, the number of yarn and interlacing per unit area are the fabric properties affecting abrasion. Weave type has a significant effect on abrasion resistance of the fabrics. Woven fabric properties will differ by changing the weave pattern which is evaluated not only as an appearance property, but also as a very important structure parameter. If one set of yarns is predominantly on the surface then this set will wear most; this effect can be used to protect the load bearing yarns preferentially. Long yarn floats and a low number of interlacings cause the continuous contact area of one yarn strand to expand and this facilitates the yarn to lose its form more easily by providing easier movement as a result of the rubbing motion. So long floats in a weave such as sateen structures are more exposed and abrade faster, usually cause breaking of the yarns and increasing the mass loss. In this way, holding the fibres in the yarn structure becomes harder and the removal of fibre becomes easier.

But the fabrics that have lower floats such as flat plain weave fabrics have better abrasion resistance than other weaves because the yarns are more tightly locked in structure and the wear is spread more evenly over all of the yarns in the fabric. Like as woven structure, knitting structure has also an important effect on abrasion characteristics of knitted fabrics. Average abrasion resistance values of interlock knitted fabrics are higher than rib and single jersey fabrics. The reason of that is more stabile, thicker and voluminous structure of the interlock fabrics. Course length for the knitted fabrics is so important that the weight loss percent after abrasion increases with increasing course length. Open, slack knitted fabric structure is abraded more than denser fabrics. The fabric mass per square meter and fabric thickness that are the main structural properties of fabrics have an effect on abrasion resistance. Higher values of these factors ensure higher abrasion resistance.

The other parameter that affects the abrasion is thread density of the fabric. The more threads per unit area in a fabric are the less force to each individual thread is, therefore the fabrics with a tight structure have higher abrasion resistance than those with a loose structure. However as the threads become jammed together they are the unable to deflect under load and thus absorb the distortion.

The literature contains papers dealing with the abrasion resistance of the specific type of the fabrics. One of them characterizes certain properties of flocked fabrics produced from different fibre type by measuring the abrasion resistance. Abrasion resistance of the flocked fabrics is related to the flock fibre length and density of flock fibre ends.

The flocked fabrics with low flock fibre density and high flock fibre length show more resistance to abrasion in comparison with the flocked fabrics which have high flock fibre density and short flock fibre length. The wet rubbing resistance of the flocked fabrics is less than the dry rubbing resistance. Another study is about the performance of upholstery fabrics woven with chenille yarns. Chenille yarn material, yarn twist, and pile length have a significant effect on the abrasion resistance of the chenille yarns and fabrics. Twist levels and pile lengths affect yarn cohesion. There is an improvement in abrasion resistance of the fabrics with increasing twist, pile length, and the use of natural fibres as pile materials, which may be due to increasing frictional Behaviour between the pile and lock yarns.

Finishing process

Finishing treatments, the types and concentration of the chemicals used in the treatment processes are also the parameters affecting the abrasion characteristics of the fabrics. Grey fabrics have lower abrasion resistance compared to dyed fabrics with the same construction. During the dyeing operation, fibres on the fabric surface will cling to it, hence the fabric will achieve a closer state, and the movement of fibres within the yarn will be limited. Laundering process affects the abrasion resistance.

The abrasion resistance of both undyed and dyed fabrics is negatively influenced by the laundering treatment. The degree of damage in fibres within the fuzz entanglements tends to increase with an increased number of launderings, and that the damage varies from small cracks and fractures to slight flaking depending on the fabric and yarn. Another process that is important for fabric abrasion is bleaching and enzymatic process.

The fabrics applied bleaching and enzymatic processes have higher abrasion resistance with regard to grey knitted fabrics. However as enzymatic treatment is applied to the dyed fabrics the abrasion tendency become worse compared to non-enzymatic dyed fabric Nano-silicone softener treatment causes decrease in abrasion resistance of the fabrics. The mass loss ratios of the samples with nano-silicone softeners are higher than mass loss ratios of the samples without nano-silicone softener. It is the probable result of fibre mobility inside the fabric which is increased by nano-silicone softener. Silicone softeners provide better wrinkle recovery, tear strength, and abrasion resistance than the cationic softener for 100per cent cotton woven fabric.

The laser fading process is acknowledged as a very strong alternative compared to the conventional physical and chemical processes used for aged-worn look on denim fabrics. Even with the lower pulse time of laser beams, abrasion resistance significantly decreases after fading process and with the higher pulse times, the decrease in abrasion resistance values is much more apparent.

METHODS FOR TESTING ABRASION RESISTANCE OF FABRICS

Most abrasion tests depend on applying energy to the fabrics and measuring their response to it. The manner of transferring the energy from machine to the fabric is different for different machines, but the basic principles are the same. There are three types of abrasion in terms of occurrence; flat, edge and flex abrasion. Therefore different abrasion test methods have been described by the abrasion type, the test head movement or testing device setup. The differences among the procedures include the type of equipment, abradant (the material that rubs against the specimen), material used (including woven, Non-woven, and knit apparel fabrics, household fabrics, industrial fabrics, and floor coverings) and assessment method.

In all of the test methods, the tested specimen is rubbed in a particular manner against an abradant which may be a fabric, or a emery sheet for either a certain amount of time for a certain number of strokes or cycles. ASTM and ISO define several methods to quantify abrasion resistance of textile materials and introduce methods for the evaluation of abraded fabrics. However, there is not a linear relationship between successive measurements using any of these methods and progressive amounts of abrasion.

Flat abrasion

Flat abrasion occurs when flat objects is rubbed to a flat material. Flat resistance tends to be good for most materials because the force of rubbing is distributed over a wide area.

However for many products flat abrasion resistance is assumed to occur when the curve is gradual or the bend is shallow such as what occurs in a shirt or jacket as it bends across the back of the wearer or on the seat of an upholstered chair.

Martindale Tester, Taber Abraser, Uniform Abrasion Tester are the instruments working on the flat abrasion mechanism. In Martindale abrasion resistance tester, circular specimens are abraded under known pressure against a standard fabric.

Abrasion resistance is measured by subjecting the specimen to rubbing motion in the form of a geometric figure, denoted Lissojous motion, that is, a straight line, which becomes a gradually widening elipse, until it forms another straight line in the opposite direction and traces the same figure again under known conditions of pressure and abrasive action. The advantage of the Martindale abrasion test is that the fabric sample gets abrasion in all directions. Stress develops along the Fibre from the force acting transverse to the Fibre axis as a result of surface friction; the magnitude of surface friction developed is directly related to the harshness of standard worsted fabric abradant.

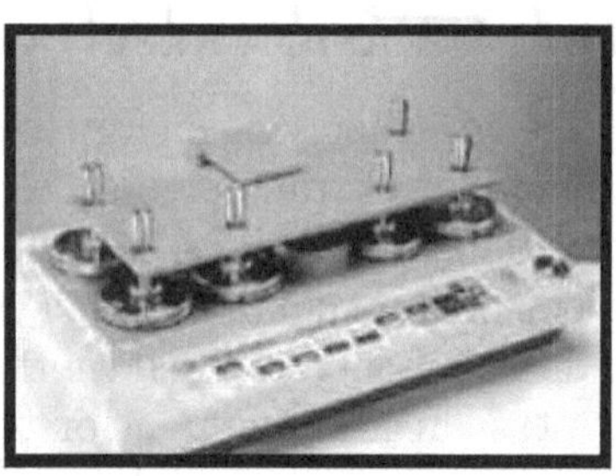

Fig. Martindale Abrasion Tester

Fig. Lissojous motion

Specimens are circular of either 38 mm or 140 mm in diameter. Polyurethane foam disk is placed under the specimen for fabric having a mass/ unit area less than 500 g/m2. The small test specimen is sitting on the large abradant and then cycled backwards and forwards. If assessment of appearance change needs to be carried out, then larger test pieces are required. The roles are reversed and the abradant is placed in the holder with the specimen as the base platform. A force of either 9 or 12 kPa (for upholstery and technical fabrics) is applied to the top of the specimen to hold it against the abradant. The standard abradant should be replaced at the start of each test and after 50 000 cycles if the test is to be continued beyond this number. The rotary platform double head method can be used for most fabrics.

The specimen is abraded using rotary rubbing action under controlled conditions of pressure and abrasive action, and subjected to multidirectional abrasion using a rotary rubbing action. Abrasive heads of rubber based compound simulate mild and harsh abrasion. The test specimen, mounted on a turntable platform, turns on a vertical axis, against the sliding rotation of two abrading wheels. The fabric is subjected to the wear action by two abrasive wheels pressing onto a rotating sample. The wheels are arranged at diametrically opposite sides of the sample so that they are rotated in the opposite direction by the rotation of the sample.

One abrading wheel rubs the specimen outward Towards the periphery and the other, inward Towards the center. The resulting abrasion marks form a pattern of crossed arcs over an area of approximately 30 cm2. Load adjustment for varying the load of the abrader wheels on the specimen is possible. In ASTM D 4158, the uniform abrasion testing machine is used. In the apparatus, a specimen is mounted in a holder and abraded uniformly in all directions in the plane and about every point of the surface of the specimen. The Uniform Abrasion Tester, consists of the abradant mounted at the lower end of a shaft, weights placed on the upper end of the shaft to produce constant pressure between abradant and specimen throughout the test, lever and cam for raising and lowering the abradant, shaft, and weights. Essentially, the surface of the abradant lies in a plane parallel to the surface supporting the specimen and presses upon the specimen. The abradant and specimen rotate in the same direction at very nearly but not quite the same angular velocity on Non-coaxial axes which are parallel to within 0.0025 mm. The small difference in speed is to permit each part of the specimen to come in contact with a different part of the abradant at each rotation. Each rotation is equivalent to one cycle.

Flex abrasion

Flex abrasion is the most common abrasion type to which a textile product is subjected. In flex abrasion the material is bent or flexed during rubbing. Flex abrasion occurs in apparel, furnishings and industrial products. Very little of the surface of most products is completely flat during usage, therefore flex abrasion tests may reflect the usage conditions more. Oscillatory cylinder, inflated diaphgram abrasion tester, flex abrasion testing machine are the instruments working on flex abrasion mechanism. In ASTM D 4157 oscillatory cylinder method, abrasion resistance is measured by subjecting the specimen to unidirectional rubbing action under known condition of pressure, tension, and abrasive action. During the test fabric is pulled tight in a frame and held stationary. Oscillating Cylinder Section, equipped with edge clamps to permit mounting of a sheet of abrasive material over its surface, capable of oscillating through an arc. Individual test specimens cut from the warp and weft directions are then rubbed back and forth using an ACT approved #10 cotton duck fabric as the abradant.

The procedure often is used for upholstery, leather and other materials. In inflated diaphgram method, a specimen is abraded by rubbing either unidirectionally or multidirectionally against an abradant having specified surface characteristics. A specimen is held in a fixed position and supported by an inflated rubber diaphragm which is held under constant pressure. The abradant is mounted upon a plate, which is rigidly supported by a double-lever assembly to provide for free movement in a direction perpendicular to the plane of the reciprocating specimen clamp. The abradant plate assembly should be well balanced to maintain a vertical pressure equivalent to a mass of 0 to 2.2 kg by means of dead weights. In ASTM D 3885, abrasion resistance is measured by subjecting the specimen to unidirectional reciprocal folding and rubbing over a specific bar under specified conditions of pressure, tension, and abrasive action by Flex Abrasion Testing Machine.

This test method is useful for pretreating material for subsequent testing for strength or barrier performance. The pressure and tension used is varied, depending on the mass and nature of the material and the end-use application. Testing machine consist of Balanced Head and Flex Block Assembly that has two parallel, smooth plates. The balanced head is rigidly supported by a double lever assembly to provide free movement in a direction perpendicular to the plate of the flex block. A positioning device is provided to position the flexing bar and yoke assembly while loading the specimen such that the edge of the flexing bar is parallel to the fold of the specimen during the test.

Edge and Flex abrasion

In edge abrasion the material is folded back on itself while it is being abraded. In products, pleats, folds, cording, cuffs, collars, hems, pocket flaps are most subject to edge abrasion. Most products show damage along edges before damage elsewhere in the product becomes apparent because the force is concentrated on a small portion of the material. So the edge abrasion is a harsh measure of fabric's resistance to wear. The accelerator abrasion tester has an action that is quite different from most other abrasion testers. In the test a free fabric specimen is driven by a rotor inside a circular chamber lined with an abrasive cloth. The specimen suffers abrasion by rubbing against itself as well as the liner. In AATCC 93 method, each specimen is subjected to flexing, rubbing, shock, compression and other mechanical forces while the specimen is tumbling. Abrasion is produced thorough the specimen by rubbing of yarn against yarn, fibre against fibre, surface again surface and surface against abradant

FACTORS AFFECTING ABRASION TEST RESULTS

During testing, the resistance to abrasion is also greatly affected by the conditions of the tests, such as the nature of the abradant, variable motion of the abradant over the area of specimen, the tension of the specimen, the pressure between the specimen and the abradant, and the condition of the specimen. The type of abrasion like flat, flex or edge abrasion or a combination of more than one affect the test results as expected. In many fabrics the abrasion resistance in the warp direction differs from that of the weft direction. Ideally the rubbing motion used by an abrasion machine should be such as to eliminate directional effects. In jacquard and textured fabrics, the test sample should include both the textured and different parts, sensitive to abrasion. Abradants can consist of anything that will cause wear.

A number of different abradants have been used in abrasion tests including standard fabrics, abrasive paper or stones (aluminum oxide or silicon carbide), steel plates and metal 'knives'. The nature of abradants and the type of action control the severity of the test. For the test to correspond with actual wear in use it is desirable that the abrasive should be similar to that encountered in service. The abradant itself wear during the test so they should be replaced after a certain usage time. Abrasion tests are all subject to variation due to changes in the abradant during specific tests. The abradant must accordingly be discarded at frequent intervals or checked periodically against a standard. With disposable abradants, the abradant is used only once or discarded after limited use. With permanent abradants that use hardened metal or equivalent surfaces, it is assumed that the abradant will not change appreciably in a specific series of tests. Similar abradants used in different laboratories will not change at the same rate, due to differences in usage. Permanent abradants may also change due to pick up of finishing or other material from test fabrics and must accordingly be cleaned at frequent intervals. The measurement of the relative amount of abrasion may also be affected by the method of evaluation and may be influenced by the Judgement of the operator.

The pressure between the abradant and the sample and the test speed affects the severity and rate at which abrasion occurs. It has been shown that using different pressures can seriously alter the ranking of a set of fabrics when using a particular abradant. Excessive pressure and testing speed can result in shorter testing time but however, the results can not simulate the normal usage conditions. The tension of the mounted specimen is also important, and it should be same for all samples. Tension during the test is provided by backing foam or inflated diaphragm.

EVALUATION METHODS OF ABRASION TESTS

There are different options for assessment of the tested fabrics as given below.

- First one is finding endpoint which counts the number of cycles until the fabric ruptures, two or more yarns have broken or a hole appears. The end point is different according to fabric type; for woven structure abrading is continued until two threads are broken while one broken yarn causing a hole for knitted fabric. Removing of the fabric fusses completely, and occurring a hole in 0.5 mm diameter is the end points for fused and for Non-woven fabrics respectively.
- Second option is assessment of the abraded fabric for a set time or number of cycles in terms of some aspect such as loss of mass, loss of strength, change in thickness or other relevant property. The loss of any property before and after abrasion is reported as loss in the relevant property or as a percentage calculated by the formula:

Loss percentage in relevant property *100, where: A = relevant property before abrasion, and B = after abrasion

According to ISO 12947-3 standard, the weight loss in certain cycles is determined based on the end breaking cycle, then at the every intervals the abraded sample is taken from the device, weighted in the sensitivity of 1 mg, percentage of the weight loss is calculated and the graph is plotted.

- The third method is the evaluation of any visual changes that occurred as a result of abrasion test. The end point is reached when there is a change in shade or appearance that is sufficient to cause a customer to complain or the visual appearance after the predetermined abrasion cycles. Visual change considers the effect that a specified number of cycles have on the lustre, colour, surface nap or pile, pilling, matting or any chance resulting from abrasion. Average number of rubs required to reach a gray scale rating of three or lower is recorded while assessing colour related changes.

None of the above assessment methods produces results that show a linear or direct comparison with one another.

YARN ABRASION

The issue of yarn abrasion has been defined both as the attrition of yarn upon itself or another surface and the resistance of yarn to suffer damage when abraded upon a surface. A yarn which is being knitted or woven into a fabric or wound onto a package runs around many guides during the process. As a yarn passes over a metallic surface, either it gets abraded or it abrades the metallic surface, which is an undesirable effect causing wear of machine parts. Yarn abrasion resistance is an important factor affecting weavability. The yarn-to-yarn friction between warp and weft yarns at every crossover is important in determining the shear property and consequently the formability of woven fabrics. The factors, which affect the yarn abrasion, can be detailed as; yarn friction, yarn type, twist multiple, hairiness, contaminants and applied lubricants. Average surface length which is highly correlated with the abrasion resistance of yarns is also introduced as a new yarn structural parameter. The length of fibres exposed on the yarn surface depends on their interactions with surrounding fibres on the yarn.

The yam abrasion resistance was found inversely proportional to the average surface length. For synthetic yarns, it is indicated that, abrasion Behaviour of the yarns is associated with the fundamental properties such as the polymer and fibre molecular weight, undrawn fibre orientation and crystalline structure, drawn fibre properties, and elevated temperature post-treatment. The coefficient of friction is a significant characteristic of textile yarns, because it defines the amount of resistive shear force a yarn will exert and have exerted upon it during fabric forming or preparation processes. As the friction between the yarn and the abradant increases, the abrasion resistance decreases. Yarn type is another important parameter for the abrasion resistance. Generally, ring-spun yarns have higher snagging tendency and lower abrasion than rotor yarns. Air-jet spun yarns are less resistant to abrasion than ring-spun yarns, because the former are bound by outer wrapper fibres only and have no internal twist.

The abrasion resistance of the sirospun yarn is better than that of the two-plied and the single yarns. Resistance to abrasion increases as the twist multiplier increases. At low twist levels, yarns can be easily splitted therefore yarns with high twist levels are abraded less than yarns with low twist levels.

The abrasion resistance of the two-plied yarn depends on both single-yarn twist and ply twist. Single-yarn twist and ply twist have a more influential effect on the yarn-to-yarn and yarn-to-pin abrasion resistances respectively of cotton two-ply yarns. It is known that during weaving, the yarns experience the abrasive action of the moving loom parts such as heddle eyes, reeds, whip rolls, and picking elements. Sizing is applied to the warp yarns to improve their abrasion resistance and strength and to reduce friction. The yarn abrasion resistance was found higher in sized yarn with a higher twist level. The size formulations used for spun yarns also contain other ingredients such as lubricants and binders. The lubricants help to reduce the friction and abrasion between the adjacent yarns and between yarns and heddles, dropwires, shuttles, rapiers, or projectiles.

MEASUREMENT METHODS OF TESTING YARN ABRASION

There are various measurement methods to measure abrasion properties of yarns, including yarn on yarn abrasion and yarn external abrasion, which comprises different test procedures.

Yarn on yarn abrasion

The yarn-on-yarn abrasion test is described in ASTM D 6611, "Standard Test Method for Wet and Dry Yarn-on-Yarn Abrasion Resistance". A length of yarn is inter-wrapped in contact with itself between three pulleys that are positioned in a defined geometry to produce a specific intersection angle. A weight is attached to one end of the yarn to apply a prescribed tension. The other end is drawn back and forth through a defined stroke at a defined speed until the yarn fails due to abrasion upon itself within the inter-wrapped region. The yarn abrasion test can be conducted in either the dry state or the wet state.

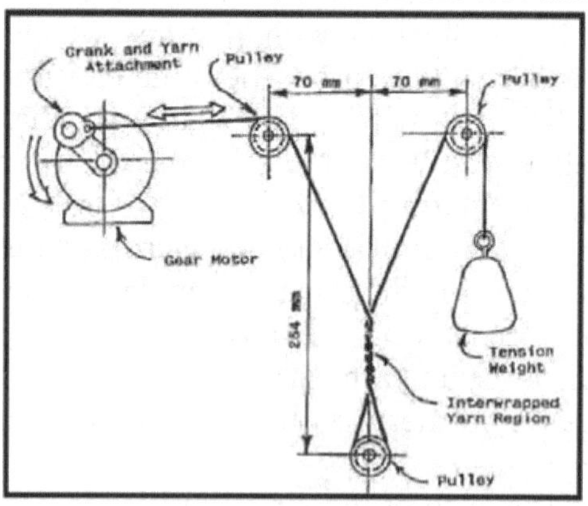

Fig. Yarn-on-yarn abrasion setup (ASTM D 6611)

Zweigle Staff-Tester G 556, measures accurately abrasion and breakage tendency before further processing. The yarn runs with constant tension via the hysteresis brake into the tester. The thread is guided over two easy-going rollers in such a way, that the running in section coils around the running-out section over an angle of approximately 180 degrees. At this coiling point there is intense thread-to-thread friction. Various metal or ceramic elements can be placed into the running path of the thread. A funnel below the coiling point collects the parts rubbed-off from the yarn or avivage. The built-in blast exhausts the dust into a small, easily removable filter. By using an accurate balance the weight of the yarn and/or avivage rubbings can be determined.

Fig. Staff tester

Lint generation test can be conducted by Lawson-Hemphill's CTT instrument to measure yarn to yarn abrasion. The instrument is a dynamic yarn transport system with the ability to maintain a specific yarn tension and let the yarn run at any selected speed up to 360 m/min. It simulates the production conditions in the testing laboratory and therefore it gives idea about the production performance.

The CTT Lint Generation Test determines how much lint a yarn will generate under certain manufacturing conditions, including: yarn-to-yarn thread paths, yarn-to-metal thread path, yarn-to-ceramic and yarn-to-needles, sinkers and reeds. During the lint test, the yarn is wrapped around itself. As the yarn is moving, the generated lint will be collected on a piece of paper under the vacuum sealed enclosure. The amount of lint that is generated is expressed as mg/1 km. Therefore, abrasion properties of different yarns can be compared based on their weight loss.

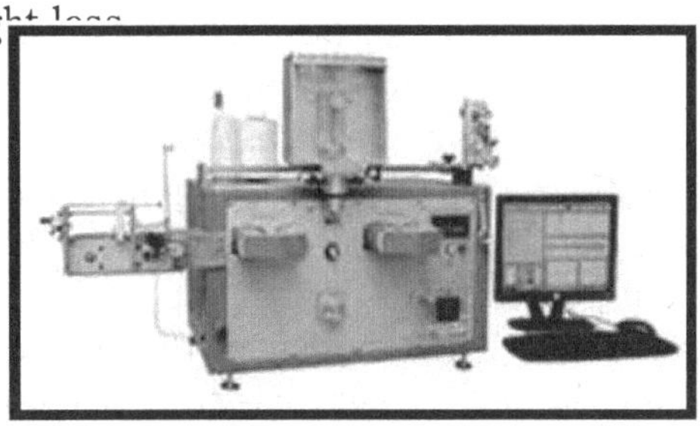

Fig. CTT Instrument with Lint Generation Test Module

Yarn external abrasion

External abrasion resistance is important in many applications and there are different measurement methods to evaluate yarn abrasion resistance. Oscillating stress tester incorporates the abrasion element consisting of three rows of case-hardened steel pegs.

Yarns hanging from the jaws just touch the pegs in the top and bottom rows, and the middle pegs can be traversed sideways to deflect the yarns around by a desired amount. In a test, the yarn hanging from the jaws passes to one side of the corresponding top and bottom pegs and around the opposite sides of the displaced middle peg, and thus abrade the yarn when reciprocated. Owen and Locke also modified the instrument to apply abrasion by the rotating steel shaft, as shown in Figure. The shaft was rotated by a belt driven from a small electromotor to provide abrasion to yarns at the point of contact.

(a) (b)

Fig. Oscillating stresses abrasion tester

Using these modified equipment, Owen and Locke subjected yarns to a fixed number of abrasion cycles and measured the breaking strength of such abraded yarns. The results were expressed in terms of percent deterioration in the breaking load, calculated as follows:

$$\text{deterioration}(\%) = \frac{MSYBA - MSYAA}{MSYBA} \times 100$$

where, *MSYBA* = mean strength of yarn before abrasion and *MSYAA* = mean strength of yarn after abrasion.

As a yarn passes over a metallic surface, either it gets abraded or it abrades the metallic surface, which is an undesirable effect causing wear of machine parts. Therefore, the abrasiveness of the yarns is another abrasion characteristic of the yarns and can be tested by Lawson-Hemphill CTT tester. In this test, the wires used in testing are secured at one end, stretched over a pulley and secured at the other end to a lever.

When the wire is cut, the weight hanging on the lever enables the activation of auto stop mechanism. Suitable weight has to be used on the lever to give support to the wire to keep it straight when the wire is passing over it without stretching it unnecessarily. The yarn is run over a tensioned standardized copper wire and the abrasion factor is measured as the length of yarn it takes to cut the wire or, the surface destruction is measured with a microscope, after running of the yarn in a fixed length.

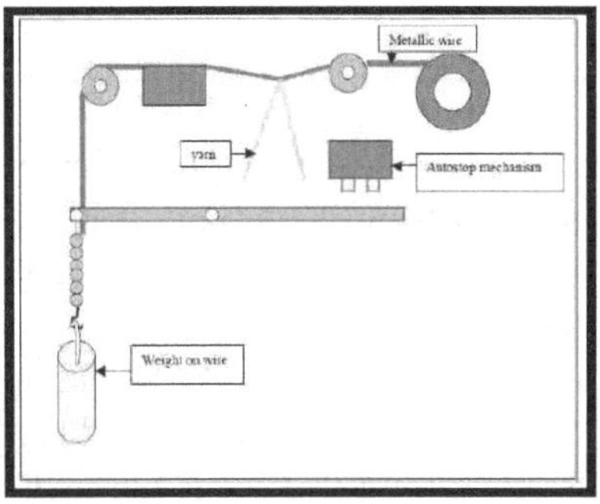

Fig. Diagram to show the CTT set up of yarn and metallic wire

Another instrument is Shirley yarn abrasion tester, which consists of two reciprocating bars: one is made of hardened steel and the other is covered with the standard abradant used in the Martindale fabric abrasion tester. Eight yarn specimens are tested simultaneously.

Yarns are threaded from the fixed holders and clipped onto the flexible holders where sensors are attached. T

he initial tension exerted on each yam is 0.5N. When a yam breaks, the flexible holder falls, a signal is sent to the control unit, and the number of rubs for that particular yarn is recorded.

The effect of yarn count and material type on the abrasiveness property is also significant for fancy yarns. Wool fancy yarns are more abrasive than the acrylic yarns and thicker yarns are more abrasive than the thinner yarns as well.

ABRASION RESISTANCE OF SOCKS

Socks, which are a necessary item of clothing, need to be comfortable, affordable and retain their quality throughout their life. The most significant problem is abrasion which can greatly reduce the materials life. Abrasion usually occurs on the heel, sole and toes of the socks.

The span life of the socks is shorter than other textile materials because of higher abrasion within the shoes, slippers or even the ground during usage. The first stage of abrasion unrevals of the loose fibres from the fabric surface, eventually breakdown of fibres and a hole occur.

If the sock consists of synthetic fibres with natural fibres, during rubbing action natural fibres, which give the desirable properties of the sock, move away, only synthetic fibres remain. This situation stated as thinning and gives the sock undesirable appearance and decreases the overall fabric thickness.

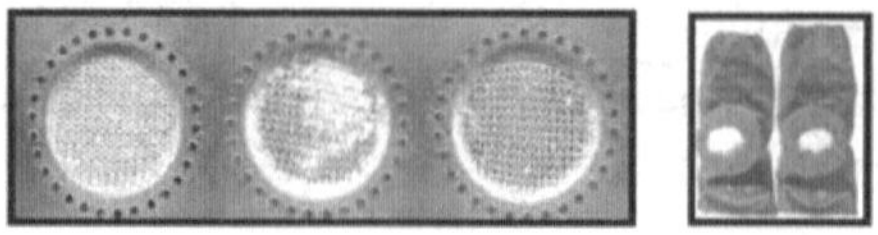

(a) (b)

FIg. (a) The thinning appearance of a Co/PA sock at the heel during abrasion resistance tests, (b) abraded socks,

The Sock Testing Consortium accepted three methods for abrasion resistance of socks. These are CSI Stoll method, ILE SCR method and in present widely used Martindale method. A modified specimen holder for the Martindale abrasion tester, which stretches the knitted material, is used for socks' abrasion. The holder takes a standard size 38mm diameter sample which is held to size by a pinned ring. A flattened rubber ball is pushed through the sample as the holder is tightened thus stretching it, test is carried out as fabric tests. The sample is inspected at suitable intervals until a hole appears or the material develops an unacceptable level of thinning.

Fig. Setting of sock kit on Martindale apparatus.

There is not too much research related with sock abrasion. In one of the research abrasion resistance of 7 different types of socks consisting of different rates of Co-PA were searched with both laboratory tests and usage tests. Pilling test device was used for abrasion resistance and abrasion time of the sample assessed. It was found that the results from the laboratory and the usage tests were different. In another study the abrasion resistance of terry socks was investigated. PA yarns for ground, wool and wool blends also cotton and cotton blends for pile were used. It was found that the result of the wool and wool blends are evenly matched to each other and abrasion resistance of the cotton socks was better than wool socks. They used different yarns for piles to increase the abrasion resistance of the socks and found that the yarns spun with wool - PES blends gave the best results. The results of the PAC–wool yarns and PAC-linen yarns followed it respectively. Increased yarn twist, adding PA and folded PAC yarns increases abrasion resistance of the socks are the other results of them. The effects of the yarn parameters and some finishing process on the abrasion resistance of socks were researched in detail. It was found that the abrasion resistance value of socks can be increased by a number of measures; usage of thicker yarns, adding PA and elastic yarns to the structure, increasing the PES ratio in Co-PES yarns. The resistance of wool socks is higher than acrylic ones and the wool ratio in woolacrylic samples has a positive influence on abrasion resistance. The abrasion resistance increases with mercerization process and decreases with the use of silicone softeners.

ABRASION RESISTANCE IN TECHNICAL TEXTILES

Abrasion resistance is one of the most important properties in special technical textile products. In order to measure the abrasion resistance of these products such as protective clothes, military fabrics, gloves, laminated fabrics, multi layered structures, special measurement procedures were developed. Car seat is one of the most important parts of the interior of a car and abrasion resistance of the seat fabric is an important parameter for the usage. It is composed of metal structure, filling and seat cover including exterior fabric, foam and support material.The foam has influence on the abrasion resistance of car seats upholstery; increased height and weight of the foam, causes less weight and thickness loss. Non-usage of foam reduced the abrasion resistance significantly. Fabric type has an important effect on abrasion resistance of car seat covers. The warp knit double bar raschel was found more resistant to abrasion than flat woven, circular knitted flat and warp knit flat fabrics. An abrasion test for geo-textiles uses a reciprocal back-and-forth rubbing motion of a sandpaper abradant against the geo-textile. The instrument used in this test is a Stoll Flex abrader modified to allow the fabric to be mounted on a stationary platform and the abrading medium on a reciprocating platform.

The abrader can be loaded to provide a constant pressure and has an adjustable speed drive to allow the abrasive action to be controlled as required. In day-to-day applications, abrasion resistance is one of the most important properties that determine the duration of the useful life of the Non-woven articles. In particular, in many industrial filtration applications, filter media can undergo severe abrasive loading. Abrasive damage can manifest itself in two ways. Firstly, the filtration area can be damaged due to repeat contact with hard and sharp particulate materials in the fluid, and can result in pinhole damage over time. This in turn weakens the body of filter media and reduces its retention efficiency, resulting in rapid solids loss. In such cases, the process would have to be stopped and the cloths either changed or occasionally repaired with patches. A second source of abrasive damage is due to attrition between the cloth and the filter machinery. The abrasion resistance of the thermally bonded Non-woven articles is significantly dependent on not only the choice of fibre, but also the construction of the fabric, which is further influenced by both the web forming and thermal bonding processes. Abrasion has a potential mode of failure for either latex or non-latex medical glove materials. Resistance to abrasion is necessary to maintain barrier integrity during routine tasks such as twisting a capped needle onto a syringe or turning a knurled knob on a piece of equipment. An alternative method for measuring durability of both latex and non-latex medical glove materials, utilizing abrasion resistance testing, has been developed.

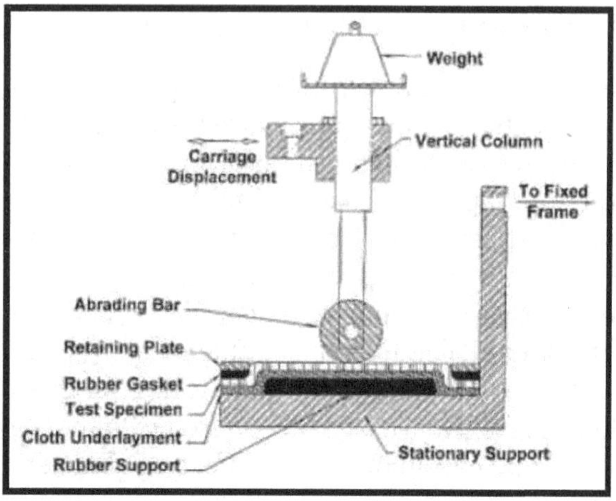

Fig. Glove abrasion resistance apparatus

This device converts the rotary motion of an adjustable-speed, DC gear motor into oscillating linear displacement of a carriage that serves as a translating support for a vertical loading column. The abrader is linearly stroked against a test specimen at a rate that is preset by adjustment of a motor speed control. Along with displaying the preset number of cycles for a particular test sequence, the system control unit also indicates real-time values of test cycles and cycle rate. Established test machines such as the Martindale, revolving drum and Taber abrader were considered for the evaluation of "Protective clothing for motorcycle riders", but they are not capable of testing the multitudinous single and combination materials and constructions present in motorcyclists' protective clothing. The Darmstadt tester and the Cambridge tester are used for this measurement. The Darmstadt machine consists of a `doughnut' of concrete, in the centre of which is situated an electric motor. In testing, the motor is run up to a specified speed at which point the specimen holders are released and freefall the short distance down the axis of the shaft into contact with the surface of the concrete. The specimens continue to spin freely around the drive shaft and in contact with the concrete until coming to a halt. The mass of test specimens both before and after testing is recorded, and the difference established. In Cambridge tester, the test specimen is mounted on a holder which is attached to the free end of a horizontal, rigid pendulum which pivots at the opposite end to the specimen holder. In testing, the pendulum is released and falls onto the moving belt. A fine copper wire of 0.14mm diameter, located across the outer face of the specimen, is cut upon contact with the moving belt and this starts an electronic timer. A second wire is exposed and cut when the specimen is abraded through, which stops the timer and records the time taken to perforate the specimen. The more prolonged the period between contact and perforation, the better-performing the material. Para-aramid laminated constructions for motorcyclists' clothing was found better than polyester laminated construction, layered construction textile, air textured nylon and leather in terms of the results of impact abrasion resistance.

CONCLUSION

In this chapter, abrasion which affects serviceability of textile materials was explained in detail. The abrasion mechanism of textiles is a complex phenomenon and associated with the properties of fibers, yarns, fabric structure and applied treatments. Abrasion in textiles such as fabrics, yarns, socks and technical textiles can be measured by different methods. Due to the technological improvements and growing demands on the properties of textile materials, it seems the development of new test techniques and equipments will continue on this issue.

8

Fibre Properties

CELLULOSIC FIBERS

Cellulose is a fibrous material of plant Origin and the basis of all natural and man-made cellulosic fibers. The natural cellulosic fibers include cotton, flax, hemp, jute, and ramie.

The major man-made cellulosic Fibre is rayon, a Fibre produced by regeneration of dissolved forms of cellulose. Cellulose is a polymeric sugar made up of repeating 1,4-8-anhydroglucose units connected to each other by 8-ether linkages.

The number of repeating units in cellulosic fibers can vary from less than 1000 to as many as 18,000, depending on the Fibre source. Cellulose is,a hemiacetal and hydrolyzes in dilute acid solutions to form glucose, a simple sugar.

The long 1inear chains of cellulose permit the hydroxyl functional groups on each anhydroglucose unit to interact with hydroxyl groups on adjacent chains through hydrogen bonding and van der Waal s forces. These strong intermolecular forces between chains, coupled with the high linearity of the cellulose molecule, account for the crystalline nature of cellulosic fibers.

It is believed that a gradual transition from alternating areas of greater molecular alignment or crystallinity to more disordered or amorphous areas occurs in cellulose. The number, size, and arrangement of crystalline regions within celluloses determine the ultimate properties of a particular Fibre.

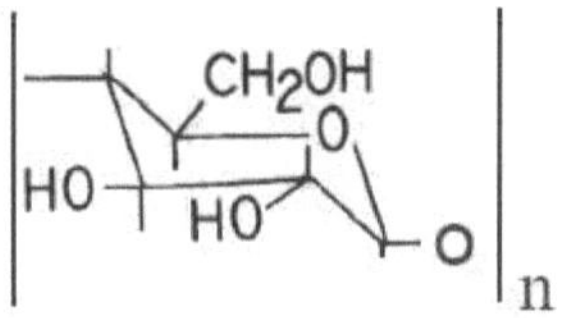

CELLULOSE

The predominant reactive groups within cellulose are the primary and secondary hydroxyl functional groups. Each repeating anhydroglucose unit contains one primary and two secondary hydroxyl functional groups which are capable of undergoing characteristic chemical reactions of hydroxyl groups. The primary hydroxyls are more accessible and reactive than secondary hydroxyls; nevertheless, both types enter into many of the chemical reactions characteristic of cellulose.

COTTON

Cotton is the most important of the natural cellulosic fibers. It still accounts for about 50per cent of the total Fibre production of the world, although man-made fibers have made significant inroads into cotton's share during the last three decades. Cotton fibers grow in the seed hair pod of cotton plants grown and cultivated in warm climates.

Structural Properties

Cotton is very nearly pure cellulose. As many as 10,000 repeating anhydroglucose units are found in the polymeric cellulosic chains of cotton. Studies have shown that all of the hydroxyl hydrogens in cotton are hydrogen bonded. These hydrogen bonds will hold several adjacent cellulose chains in close alignment to form crystalline areas called microfibrils. These microfibrils in turn align themselves with each other to form larger crystalline units called fibrils, which are visible under the electron microscope. In cotton the fibrils are laid down in a spiral fashion within the Fibre. Modern Fibre theory suggests that each cellulose molecule is present within two or more crystalline regions of cellulose will be held together. Between the crystalline regions in cotton, amorphous unordered regions are found. Voids, spaces, and irregularities in structure will occur in these amorphous areas, whereas the cellulose chains in crystalline regions will be tightly packed. Penetration of dyestuffs and chemicals occurs more readily in these amorphous regions. Approximately 70per cent of the cotton Fibre is crystalline. Individual cotton fibers are ribbonlike structures of somewhat irregular diameter with periodic twists or convolutions along the length of the Fibre. These characteristic convolutions, as well as the cross-sectional shape of cotton are caused by collapse of the mature Fibre on drying.

Fig. Cotton

Three basic areas exist within the cross section of a cotton Fibre. The primary outer wall or cuticle of cotton is a protective tough shell for the Fibre, while the secondary wall beneath the outer shell makes up the bulk of the Fibre. The fibrils within the secondary wall are packed alongside each other aligned as spiral s running the length of the Fibre. The lumen in the center of the Fibre is a narrow canal-like structure running the length of the Fibre. The lumen carries nutrients to the Fibre during growth, but on maturity the Fibre dries and the lumen collapses.

Physical Properties

The high crystallinity and associative forces between chains in cotton result in a moderately strong Fibre having a tenacity of 2-5 g/d. The hydrophilic nature of cotton and the effect of absorbed water on the hydrogen bonding within cotton cause the tensile strength of cotton to change significantly with changes in moisture content. As a result, wet cotton is about 20per cent stronger than dry cotton. Cotton breaks at elongations of less than 10per cent, and the elastic recovery of cotton is only 75per cent after only 2per cent elongation. Cotton is a relatively stiff Fibre; however, wetting of the Fibre with water plasticizes the cellulose structure, and the cotton becomes more pliable and soft. The resiliency of dry and wet cotton is poor, and many finishes have been developed to improve the wrinkle recovery characteristics of cotton. Cotton is one of the more dense fibers and has a specific gravity of 1.54. The hydroxyl groups of cotton possess great affinity for water, and the moisture regain of cotton is 7per cent-9per cent under standard conditions. At 100per cent relative humidity, cotton has 25per cent-30per cent moisture absorbency. The heat conductivity of cotton is high, and cotton fabrics feel cool to the touch. Cotton has excellent heat characteristics, and its physical properties are unchanged by heating at 120°C for moderate periods. The electrical resistivity of cotton is low at moderate relative humidities, and the Fibre has low static electricity buildup characteristics. Cotton is not dissolved by common organic solvents. Cotton is swollen slightly by water because of its hydrophilic nature, but it is soluble only in solvents capable of breaking down the associative forces within the crystalline areas of cotton. Aqueous cuprammonium hydroxide and cupriethylenediamine are such solvents.

Chemical Properties

Cotton is hydrolyzed by hot dilute or cold concentrated acids to form hydrocellulose but is not affected by dilute acids near room temperature. Cotton has excellent resistance to alkalies. Concentrated alkali solutions swell cotton, but the Fibre is not damaged. The swelling of cotton by concentrated sodium hydroxide solution is used to chemically finish cotton by a technique called mercerization. In aqueous alkali, the cotton swells to form a more circular cross section and at the same time loses convolutions. If the cotton is held fast during swelling to prevent shrinkage, the cellulose fibers deform to give a Fibre of smoother surface. After washing to remove alkali, followed by drying, the cotton Fibre retain s a more cylindrical shape and circular cross section. Although little chemical difference exists between mercerized and unmercerized cotton, mercerization does give a more reactive Fibre with a higher regain and better dyeability.

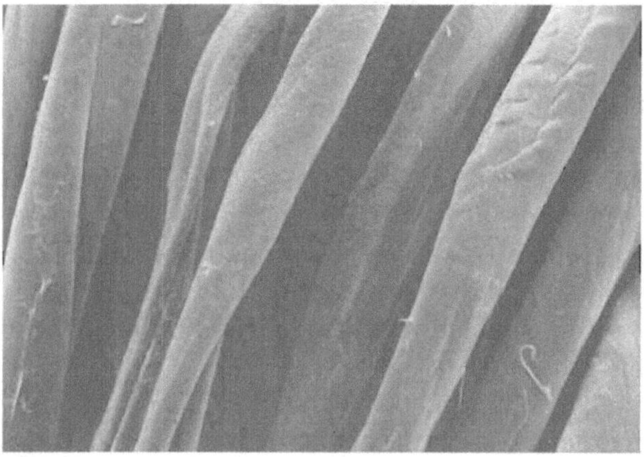

Fig. Mercerized Cotton

Dilute solutions of oxidizing and reducing agents have little effect on cotton; however, appreciable attack by concentrated solutions of hydrogen peroxide, sodium chlorite, and sodium hypochlorite is found. Most insects do not attack cotton; however, silverfish will attack cotton in the presence of starch. A major problem with cotton results from fungi and bacteria being able to grow on cotton. Mildews feed on hot moist cotton fibers, causing rotting and weakening of the fibers. Characteristic Odour and pigment staining of the cotton occurs when mildews attack.

Additives capable of protecting cotton are available and commercially applied to cotton fabrics used outdoors. These materials are often metal salts of organic compounds which are capable of inhibiting growth of mildews and similar organisms. Cotton is only slowly attacked by sunlight, since cellulose lacks for the most part groups which absorb ultraviolet light between 300 and 400 nm. Over long periods sunlight degrades cotton, causing it to lose strength and to turn yellow. Certain vat dyes tend to accelerate the rate of cotton photo degradation through a sensitization reaction called "photo tendering." Although cotton has excellent heat resistance, degradation due to oxidation becomes noticeable when cotton is heated in the air at I50°C for long periods. Spontaneous ignition and burning of cotton occurs at 390°C. At low humidities in the absence of heat and light, cotton will not deteriorate over long periods of storage.

End-Use Properties

The properties of cotton Fibre are such that it serves as nature's utility Fibre. Although cotton has some properties which are undesirable from the viewpoint of the consumer, the superior properties of cotton, coupled with its low cost, nevertheless make it a valuable Fibre in many applications. Different species of cotton produce fibers of various average lengths. In the United States lengths of cotton staples are designated as follows:

- Extra long staple: 1 3/8 - 2 inches (3.2 - 4.7 em)
- Long staple: 1 1/8 - 1 3/8 inches (2. 7 - 3. 2 em)
- Medium staple: 1 - 1 1/8 inches (2.5 - 2.7 em)
- Short staple: 7/8 - 1 inch (2.0 - 2.5 em)

The two major types of American cotton are American-Upland and American-Egyptian. Cotton has excellent hand, and the drivability of cotton fabrics is quite acceptable. Fabrics of cotton are of sat is factory appearance and have a low luster unless mercerized or resin finished. The superior absorbency of cotton, coupled with its ability to desorb moisture, makes it a very comfortable Fibre to wear.

This absorbency permits cotton to be used in applications where moisture absorption is important, such as in sheets and towels. Mildewing of cotton under hot moist conditions and its slowness of drying are undesirable properties associated with its high affinity for water. The cotton Fibre has sufficient strength in the dry and wet states to make it suitab1e for most consumer textile applications. The increased strength of cotton on wetting adds to its long useful life. Cotton wears wel1 without undue abrasion, and pills do not tend to form as it wears. Cotton's low resiliency and poor recovery from deformation means that it wrink1es easily in both the dry and wet states and exhibits inferior crease retention.

Starching of cotton improves these properties, but the effect is only temporary, and it is necessary to renew this finish after each laundering. The resistance of cotton to common household chemicals, sunlight, and heat makes it durable in most textile applications. Cotton can be dyed successfully by a wide variety of dyes, and the colorfastness of properly dyed cotton is satisfactory. Fabrics of cotton are maintained with a moderate degree of care. Cotton fabrics launder readily, and its wet strength and alkali resistance mean that cotton is resistant to repeated washings.

Stresses which occur during the spinning and weaving process will cause cotton fabrics to undergo relaxation shrinkage during initial launderings. Relaxation shrinkage can be controlled through resin stabilization or through the well known compression shrinkage process called Sanforization. Cotton can be drip dried or tumbled dry, but in both cases the dry cotton will be severely wrinkled, and ironing will be necessary. Cotton can be ironed safely at temperatures as high as 205 °C. Cotton, as well as all cellulosic fibers, is highly flammable and continues to burn after removal from a flame. After extinguishing of the flame the cotton will continue to glow and oxidize by a smoldering process called afterglow. The Limiting Oxygen Index of cotton is 18. A number of topical treatments have been developed to lower the flammability of cot ton and other cellulosics. The undesirable properties of cotton can be corrected to varying degrees through treatment of the Fibre with special finishes; however, the abrasion resistance of cotton is adversely affected. As a result, blends of cotton with the stronger man-made fibers have become important. Although man-made fibers have made inroads into applications previously reserved for cotton, cotton continues to be the major textile Fibre due to its great versatility, availability, and cost.

FLAX

Flax is a bats Fibre used to manufacture linen textiles. It is derived from the stem of the annual plant, Linumusitatissimum, which grows in many temperate and subtropical areas of the world. The flax plant grows as high as 4 feet tall, and flax fibers are found below the plant surface held in place by woody material and cellular matter. The flax fibers are freed from the plant by a fermentation process called retting.

Chemical retting using acids and bases has been employed with some success; however, such processes tend to be more expensive than natural fermentation techniques. The fibers are then removed from the plant through breaking of the woody core, removal of the woody material, and combing. The resulting fibers are ready for spinning and are 12-15 inches (30-38 cm) in length.

Structural Properties

Flax is nearly pure cellulose and therefore exhibits many properties similar to cotton. Strands of individual flax fibers may consist of many individual Fibre cells—fibrils held together by a natural cellulosic adhesive material. The molecular chains in flax are extremely long, and the average molccular weight of molecules in flax is 3 million. There are swellings or nodes periodically along the Fibre which show up as characteristic cross markings. The cell walls of flax are thick, and the Fibre cross section is polygonal with a large lumen in the center.

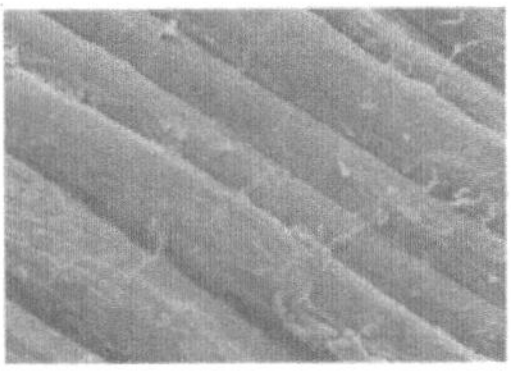

Fig. Flax

Physical Properties

Flax is a strong Fibre, and the average tenacity of dry flax is 2-7 g/d; wet flax is 2.5-9 g/d. Like cotton, the wet flax Fibre is about 20per cent stronger than the dry Fibre. Flax has a low elongation at break; however, the Fibre is fairly elastic at very low elongations. Flax has a higher regain than cotton, and a 12per cent regain is observed at standard conditions. Flax possesses good heat conductivity and is cool to the touch. The heat and electrical properties of flax are similar to those of cotton, and flax possesses solubility characteristics identical to cotton. Flax is a highly rigid Fibre, and it tends to crease on bending due to its poor resiliency. The specific gravity of flax is 1.54, the same as cotton. Flax is a dull Fibre, but more lustrous linen fabrics can be obtained by pounding the linen with wooden hammers. Linen fabrics have good dimensional stability unless stresses have been introduced during the weaving process.

Chemical Properties

The chemical properties of flax are similar to those of cotton. Flax has good resistance to acids, bases, and chemical bleaches. Flax is resistant to insects and microorganisms, and only under severe moist warm conditions will it be attacked by mildews. Flax is only slowly degraded by sunlight, and it decomposes at temperatures similar to that observed for cotton.

End-Use Properties

Flax finds only limited use in modern textiles. The methods used to produce flax involve extensive Labour and cost; thus flax usage is reserved for special luxury applications. The strength of flax makes it superior to cotton in certain applications. Flax is resistant to abrasion, but like cotton suffers from lack of crease retention and poor wrinkle recovery. Linen fabric's superior absorbency, coupled with its cool crisp hand, contributes to its desirability as a prestige Fibre. The thermal properties and chemical resistance of flax make it suitable for many consumer applications. Linen can be laundered repeatedly without deterioration and ironed safely at temperatures as high as 220°C. Flax is very flammable. Flax is used in prestige items including handkerchiefs, towels, tablecloths, sheets, and certain garments.

OTHER NATURAL CELLULOSIC FIBERS

A large number of natural cellulosic fibers from plant sources exist, including jute, hemp, ramie, kenaf, urena, sisal, henequen, abaca, pina, kapok, coir, and others. Each of the fibers is a plant or seed hair Fibre and has properties that make it suitable for selected applications. The major fibers—jute, hemp, and ramie—will be considered here. Hemp is a bast Fibre harvested from the hemp plant and processed in a manner similar to flax. It is a coarser Fibre than flax, darker in Colour and difficult to bleach. The Fibre is strong and durable, and the strands of hemp Fibre may reach six feet or longer. Individual hemp cells are 1/2-1 inch in length, and the Fibre cross section is polygonal. The Fibre is very stiff and contains considerable lignin. Although fine fabrics can be produced from select hemps, hemp is used mainly in coarse fabrics, including sack material, canvas, ropes, and twines.

Jute

Jute Fibre comes from a herbaceous annual plant which grows as high as 20 feet. The fibers are extracted from the plant stalk in a manner similar to flax. Jute tends to be brown in Colour due to about 20per cent lignin present in the Fibre, but it does have a silklike luster. The individual cells of jute are very short, being about 0.1 inch long, and the Fibre cross section has five or six sides. Jute is not as strong or as durable as flax or hemp. Jute fibers are stiff, but it does have an unusually high moisture regain. Jute's low cost, moderate strength, and availability make it an important Fibre for use in sacks, bags, and packing materials. Fabrics made of jute are called burlap.

Ramie

Ramie is a bast Fibre often referred to as china grass. Ramie Fibre is removed from the plan t by peeling or removing the bark and soaking the fibers in water, followed by scraping. The Fibre must be further degummed by treatment in base before spinning. Ramie is a white Fibre with excellent strength and luster. The Fibre is stiff and fairly coarse. The cells of the Fibre are very long, and the cross section irregular in shape. Ramie is useful in industrial applications and is being used in furnishings where rough, irregular fabrics are desired.

RAYON

Rayon was the first man-made Fibre to be produced commercially, and several types of rayon fibers are produced. Rayon is regenerated cellulosic material produced by solution of a cellulose source, followed by forcing of the solution through a spinneret and subsequent regeneration to form the Fibre. The Federal Trade Commission defines rayon as a manufactured Fibre composed of regenerated cellulose in which substituents have not replaced more than 15per cent of the hydroxyl hydrogens. Rayon is produced by three methods to give the viscose rayons, cuprammonium rayon, and saponified cellulose acetate. Of these, viscose rayon is the most important and the most inexpensive to produce.

Structural Properties and Formation of Rayon

Viscose Rayon

Viscose rayon fibers are produced by the viscose process, which involves formation of soda cellulose, reaction with carbon disulfide to form cellulose xanthate, and spinning into dilute acid to regenerate the cellulose as rayon Fibre. In practice the cellulose pulp is steeped in warm concentrated sodium hydroxide and pressed to remove excess sodium hydroxide. After aging, during which some degradation and breaking of cellulose chains occurs, the soda cellulose crumbs are mixed with carbon disulfide, and orange sodium cellulose xanthate is formed. The xanthate is then dissolved in dilute sodium hydroxide. At this point titanium dioxide is added to the solution if a delustered Fibre is desired. The solution is aged and ripened until a proper viscosity is reached; the solution is then forced through the spinneret into dilute sulfuric acid to decompose the xanthate and regenerate the cellulose by the wet spinning process. In addition to acid, the following additives are often found in the spinning bath:

- sodium sulfate,
- zinc sulfate, and
- glucose.

Initially, the outer portion of the xanthate decomposes in the acid bath and a cellulose skin forms on the Fibre. The sodium and zinc sulfates affect the rate of xanthate decomposition and Fibre formation, while the action of zinc sulfate is confined to crosslinking the outermost portion of the Fibre. The glucose tends to slow the rate of regeneration of the Fibre and tends to make the Fibre more pliable. The concentration and type of additives in the bath and the nature of the original cellulose will affect the physical properties of the rayon. By careful control of the coagulation bath followed by mechanical stretching, crimp can be produced in the Fibre. Slow regeneration and stretching of the rayon will tend to introduce greater areas of crystallinity within the Fibre. High-tenacity rayons are formed in this manner. Also, the Fibre can be modified through addition of other crosslinking agents to the spinning bath. High-wet-modulus rayons have been developed in recent years. These fibers are produced from high-grade cellulose starting materials, and the formation and decomposition of cellulose xanthate is carried out under the mildest of conditions to prevent degradation of the cellulose chains. After initial skin formation, the core of the Fibre decomposes, hardens, and shrinks, causing wrinkling on the Fibre skin. The final Fibre cross section of viscose rayon appears serrated and irregular, and the nature of the crystallites in the skin and core will differ.

Fig. Viscose rayon

This skin effect can be controlled sufficiently to give nearly round Fibre cross sections where carefully controlled coagulation occurs. The majority of viscose rayons have serrated and irregular cross sections due to the skin effect. The viscose rayon Fibre is long and straight unless the Fibre has been crimped, and striations due to the irregular cross section will run the length of the Fibre. If the Fibre has been delustered or spun dyed during Fibre formation, particles of pigment will appear in the Fibre. Normal viscose fibers will generally consist of 25per cent30per cent crystalline areas within the Fibre, and the average degree of polymerization of glucose units in the cellulose chains will be 200-700. The crystallites in viscose rayon are somewhat smaller than those found in cotton. Hi gher-tena city rayons and high-wet-modulus rayons tend to be more crystalline, and higher degrees of polymerization are found. The cross sections of these rayons tend to be more nearly round. The degree of crystallinity of high-wet-modulus rayons can reach 55per cent.

Cuprammoni urn Rayon

Cuprammoni urn rayon is produced by solution of cellulosic material in cuprammonium hydroxide solution at low temperature under nitrogen, followed by extrusion of the solution through a spinneret into water and then sulfuric acid to decompose the cuprammonium complex and regenerate the cellulose. Cuprammonium rayon is more silklike than any of the other celluloses, but the cost of production is correspondingly higher. Cuprammonium rayon has a smooth surface, and no markings or striations are found on the Fibre. The Fibre cross section is nearly round.

Fig. Cuprammonium rayon

Saponified Cellulose Acetate

Rayon can be produced from cellulose acetate yarns by hydrolysis in base, a process called saponification. Cellulose acetate molecules are unable to pack closely within the Fibre due to lack of hydrogen bonding and the presence of bulky acetate side chains within the molecule. Since cellulose acetate is more plastic in nature, it can therefore be drawn and extended to give a high degree of orientation to the Fibre. On hydrolysis the highly oriented cellulosic chains form hydrogen bonds with adjacent chains to give a rayon of high orientation and crystallinity. Saponified cellulose acetate has a nearly round but lobed cross section. The indentations caused by the lobes appear as light striations running the length of the filament.

Physical Properties

Since rayons are essentially pure cellulose, they would be expected to have physical properties similar to cellulosic fibers of natural origin. Differences in properties would be expected to depend on the degree of polymerization, crystallinity, and orientation within the Fibre. The dry and wet tenacities of the rayons vary over a wide range dependent on the degree of polymerization and crystallinity with tenacities of 2 to 6 g/d dry and 1 to 4 g/d wet. For the more crystalline and oriented rayons, the drop in tenacity observed on wetting of the Fibre is lower; however, none of the rayons exhibit higher wet strength than dry strength, as found with cotton. The great loss in strength of wet regular tenacity rayon makes it subject to damage during laundering. The percentage elongation at break varies from 10 to 30per cent dry and 15 to 40per cent wet. The recovery at 2per cent elongation ranges from 70 to 100per cent. In general, the rayons have significantly higher elongations at break than observed for cotton. As the degree of crystallinity and orientation of the rayon increases, the elongation at break is lower. Rayons generally have somewhat better elastic properties at low elongations than is found for cotton. The rayons possess greater luster than cotton, and often delusterants are added to rayon prior to spinning to give a more subdued Fibre. Since the rayons are less crystalline and oriented than cotton, they tend to swell more in water and undergo greater elongation under tension in both the wet and dry states. During weaving and wet finishing considerable stresses can be introduced into many rayon fabrics, and relaxation of these stresses will be necessary before a dimensionally stable fabric is obtained. Owing in part to the more extensive networks of amorphous regions found in rayons, the moisture regain of rayon is significantly higher than that of cotton. The regains of rayons under standard conditions range from 11per cent to 13per cent. The lower crystallinity and degree of polymerization of rayons also affect the way water acts on the fibers. Rayons as a consequence are swollen to varying degrees due to increased susceptibility to water penetration. During wetting rayon may increase up to 5per cent in length and swell up to double its volume. The heat conductivity and electrical properties are the same as those found for cotton. Rayon is cool to the touch, and static charges do not build up in the Fibre at humidities greater than 30per cent. Although rayon swells in water, it is not attacked by common organic solvents. It does dissolve in cuprammonium solutions. The rayons are moderately stiff fibers with poor resiliency and wrinkle recovery properties. As in the case of cotton, resin treatments will effectively increase the resiliency of rayon. Often such treatments will tend to be more effective on rayon than on cotton due to the greater accessibility of the interior of rayon to the resin. The specific gravities of viscose and cuprammonium rayons are the same as cotton and vary between 1.50 and 1.54. Only saponified cellulose acetate has a markedly different specific gravity.

Chemical Properties

The chemical properties of the rayons are essentially the same as those found for cotton. Cold dilute or concentrated base does not significantly attack rayon, but hot dilute acids will attack rayon at a rate faster than found for cotton. Although generally resistant to oxidizing bleaches, rayon is significantly attacked by hydrogen peroxide in high concentrations. Although silverfish are known to attack rayon, microorganisms which cause mildew do not readily attack rayon, except under more severe hot and moist conditions. Prolonged exposure of rayon to sunlight causes degradation of the cellulose chains and loss of strength of the fibers. The rayons are not thermoplastic, and they, like cotton, decompose below their melting point. Regular-tenacity rayon begins to decompose and lose strength at 150aC for prolonged periods and decomposes rapidly at 190°-210°C. High-tenacity rayons tend to decompose at slightly higher temperatures.

End-Use Properties

Rayon fibers found in consumer goods are known by numerous trade names. Regular- and high-tenacity viscose rayons are marketed as rayon or with names like Zantrel, Avril, Enkaire, and Fibro. Cuprammonium rayon and saponified cellulose acetate are not longer in production in the U.S. Regular- and medium-tenacity viscose rayons are among the least expensive of the man-made fibers, whereas high-strength, cuprammonium, and saponified cellulose acetate rayons are more expensive due to the greater care and additional steps necessary in manufacture of these fibers. Although the properties of rayons are very nearly those of cotton, a greater range of properties exist within the various types of rayons available. Being an inexpensive Fibre, rayon plays a role similar to that of cotton; however, rayon differs from cotton in that it can be modified to some degree during manufacture and is not as subject to world economic and climatic conditions. The dry strength of regula r- tenacityviscose and cuprammoni urn rayons are lower than that found for cotton, whereas high-tenacity viscose, polynosic, and saponified cellulose acetate rayons are significantly stronger than cotton. All rayons lose strength when wet and are more susceptible to damage while wet. The abrasion resistance of rayon is fair, and abrasion of rayon fabrics becomes noticeable after repeated usage and launderings. The rayons are resistant to pill formation. Rayon fabrics wrinkle easily and without chemical treatment show poor crease recovery and crease retention. Durable press and wash-and-wear finishes tend to cause less degradation of rayon than found with cotton. Rayon possesses excellent moisture absorbency characteristics, but the weaker rayons swell and deform somewhat in the presence of moisture. The hand of lower- strength viscose rayons is only fair, whereas the hand of cuprammonium and high-wet-modulus rayons is crisp and more desirable. Fabrics of rayon are generally comfortable to wear and are of acceptable appearance. The rayons are resistant to common household solvents and to light and heat except under extreme conditions or prolonged exposures.

Rayons can be dyed readily by a wide variety of dyes, and the colorfastness of dyed rayons is satisfactory. Rayons are moderately resistant to deterioration by repeated laundering, and a moderate degree of care is necessary for maintenance of rayon fabrics. The lower-strength rayons exhibit poor dimensional stability during laundering. Rayons are affected significantly by detergents or other laundry additives. Greater care must be taken when ironing rayon fabrics than would be necessary with cotton. Viscose and cuprammonium rayons may be safely ironed at 176°. The rayons possess poor flammability characteristics and ignite readily on contact with a flame, as is the case for natural cellulosics. The versatility of rayons, coupled with their lower price, makes them suitable for many textile applications. Rayon is used in clothing and home furnishings. Disposable Non-woven garments and products of rayon have been introduced to the consumer in recent years. The stronger rayons have been used in tire cord for several decades but have lost a significant portion of this important market in recent years. Rayon ha s been used more and more in blends with synthetic fibers, since rayon undergoes less degradation than cotton with durable press and wash-and-wear finishes.

CELLULOSE ESTER FIBERS

ACETATE AND TRIACETATE

The cellulose esters triacetate and acetate are the two major fibers of this type. The production of acetate fibers resulted from an attempt to find a new outlet for cellulose acetate used as aircraft "dope" in World War 1. By 1921, the first acetate fibers were being produced. Although small quantities of cellulose triacetate fibers were produced before World War I, it was not until after 1954 that cellulose triacetate fibers were produced commercially in large quantities. The Federal Trade Commission defines the acetate fibers as manufactured fibers in which the Fibre-forming substance is cellulose acetate. Triacetate fibers are those cellulose fibers in which more than 92per cent of the hydroxyl groups are acetylated. The term secondary acetate has been used in the past in reference to the acetates being less than 92per cent acetylated. In recent years, these fibers have simply gone by the designation acetate.

Although triacetate and acetate are very similar structurally, this slight difference in degree of esterification provides significant differences in properties for these fibers.

Structural Properties and Formation

The cellulose esters triacetate and acetate are formed through acetylation of cotton linters or wood pulp using acetic anhydride and an acid catalyst in acetic acid. At this point the cellulose is fully acetylated.

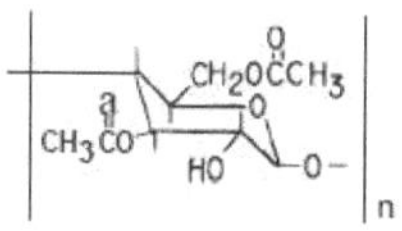

ACETATE

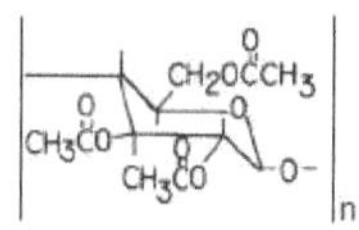

TRIACETATE

The triacetate is precipitated from solution by addition of water and dried. The cellulose triacetate is normally dissolved in methylene chloride to form about a 20per cent solution and forced through the spinneret into a stream of hot air to evaporate the solvent to form triacetate fibers by dry spinning. Acetate is formed by allowing the water-precipitated cellulose triacetate solution above to stand as long as 20 hours in aqueous solution. During this ripening process, some of the acetyl groups are hydrolyzed or removed from the cellulose. The acetate is then washed, dried, and dissolved in acetone to form the spinning "dope." The "dope" is forced through a spinneret into hot air to form the acetate Fibre by dry spinning.

Triacetate and acetate fibers can also be formed by wet spinning processes. In triacetate, essentially all hydroxyl groups of the cellulose have been acetylated, whereas 2.3-2.5 hydroxyl groups per anhydroglucose unit have been acetylated in cellulose acetate. In both triacetate and acetate, hydrogen bonding between cellulose chains for the most part is eliminated. Also, the bulky acetate group discourages close tight packing of adjacent cellulosic chains. As would be expected, these changes in molecular structure greatly affect the physical properties of triacetate and acetate. Although limited hydrogen bonding occurs between molecules within acetates, triacetate is incapable of forming hydrogen bonds. Van der Waals forces caused by the interaction of adjacent acetate and triacetate chains are the major associative forces between the acetylated cellulose molecular chains. The average number of acetylated anhydroglucose units in chains of acetates and triacetates usually varies between 250 and 300 units. The microscopic structures of triacetate and acetate are similar. The Fibre cross section of each Fibre is irregular, with as many as five or six lobes. The longitudinal views of these fibers show striations running the length of the Fibre. Small particles of delusterant will be visible in these fibers if they have been added to the Fibre prior to spinning. Acetate and triacetate fibers are very similar in appearance to the regular-tenacity viscose rayons.

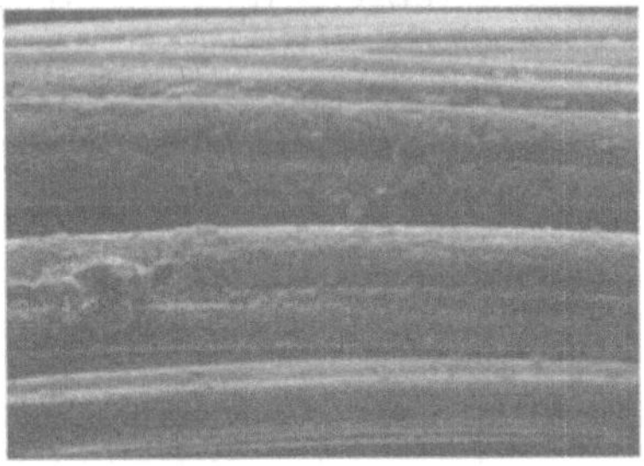

Fig. Acetate

Fig. Triacetate

Physical Properties

Since acetate and triacetate have only limited associative forces between molecular chains, they are considerably weaker than most of the cellulosic fibers. The tenacity of acetate and triacetate falls in the range 1-1.5 g/d. On wetting, their tenacities fall to 0.8-1.2 g/d. The elongations at break for acetate and triacetate are 25per cent-40per cent. On wetting, acetate breaks at 35per cent-50per cent elongation, whereas triacetate breaks at 30per cent-40per centelongation. These fibers possess good recoveries at low elongations. Above 5per cent elongation, plastic deformation of the fibers occurs and subsequent recovery is poor. Acetates and triacetates are moderately stiff fibers and possess good resiliency on bending and deformation, particularly after heat treatment. The specific gravities of acetate and triacetate range from 1.30 to 1.35. The moisture regains of these fibers are markedly lower than found for rayon and cotton. Acetate and triacetate have moisture regains of 6.0per cent and 4.5per cent, respectively. On heating, the moisture regain of triacetate drops further to 2.5per cent. Acetate and triacetate fibers conduct heat readily and are cool to the touch. Being thermoplastic, acetate and triacetate fibers soften at 200°C and 225°C and melt at 232°C and 300°C, respectively. Triacetate can be heat set more readily than acetate, and permanent pleats can be satisfactorily heat set into triacetate fabric. Heat treatment of triacetate will change the character of the Fibre and give a more dense molecular structure. Acetate and triacetate possess excellent electrical characteristics and can be used as insulating materials. As a consequence, static buildup is a problem on these fibers unless antistatic treatments are applied. Acetates and triacetates are attacked by a number of organic solvents capable of dissolving esters. Acetate swells or dissolves in acetone, ethyl acetate, phenol, chloroform, and methylene chloride, whereas triacetate dissolves in formic acid, acetic acid, methylene chloride, or chloroform.

Chemical Properties

Acetate and triacetate are resistant to dilute acids but are attacked by concentrated acids causing hydrolysis and removal of the acetate ester groups. While resistant to dilute alkalies, acetate is more susceptible to attack by alkalies than triacetate. Dilute alkali solutions attack acetate causing saponification and removal of the acetate groups on the fibers. Acetate is attacked by strong oxidizing agents, while triacetate has somewhat better resistance to attack by these reagents. Both fibers, however, can be bleached under proper conditions. Both acetate and triacetate are not attacked by insects or microorganisms. Acetate and triacetate are resistant to attack by sunlight, with greater resistance found for triacetate. Acetate and triacetate will withstand long periods at elevated temperatures below their melting points without significant loss of strength.

End-Use Properties

Acetates and triacetates are relatively inexpensive and have certain properties which are desirable in selected end-use applications. Acetate is marketed under trade names such as Airloft, Estrom, and Loftura, while triacetate fibers bear the trade names Tricel and Arnel. Both acetate and triacetate are weak fibers having low tenacities, and they cannot be used in applications where high strength is required unless they are blended with other fibers. The high elongations at break found for acetate and triacetate help compensate for these fibers' low strengths to some extent, and the fibers exhibit good recoveries at low extensions. The abrasion resistance of acetate and triacetate is poor, and these fibers cannot be used in applications requiring high resistance to rubbing and abrasion; however, the resistance of these fibers to pilling is excellent. While acetate and triacetate are moderately absorbent, their absorbencies cannot compare with the pure cellulosic fibers. The hand of acetate fabrics is somewhat softer and more pliable than triacetate, which possesses a crisp firm hand. Fabrics of both fibers possess excellent draping characteristics.

Fabrics of acetate and triacetate have a pleasing appearance and a high degree of luster, but the luster of these fabrics can be modified through addition of delusterants. Both acetate and triacetate are susceptible to attack by a number of household chemicals. Acetate and triacetate are attacked by strong acids and bases and by oxidizing bleaches. Acetate has only fair sunlight resistance, whereas the sunlight resistance of triacetate is superior. Both fibers have good heat resistance below their melting points. Acetate and triacetate cannot be dyed by dyes used for cellulosic fibers.

These fibers can be satisfactorily dyed with disperse dyes at moderate to high temperatures to give even, bright shades. The laundry colorfastness of triacetate is excellent; however, dyed acetate fabrics generally have moderate to poor colorfastness. The acetate fibers generally require a certain degree of care in laundering, although they exhibit greater wrinkle recovery on drying than cellulosic fibers. Both acetate and triacetate fabrics do not shrink or lose their shape readily during laundering and can be laundered satisfactorily under mild laundry conditions. Both acetate and triacetate may be satisfactorily dry cleaned in Stoddard solvents, but certain other solvents must be avoided due to Fibre solubility.

Acetate and triacetate dry quickly and may be tumble dried or drip dried. During tumble drying of these materials it is essential that the dryer be cool during the final tumbling. Both acetate and triacetate can be safely ironed, but direct contact of the fabric with the iron is not recommended. Acetate must be ironed at a lower temperature than triacetate, and the maximum safe ironing temperature for these fabrics is 175 °C and 200°C, respectively. The flammability characteristics of these fibers are similar to those of the cellulosic fibers, and they have LOIs of 18. Acetate is used in dresses, blouses, foundation garments, lingerie, garment linings, some household furnishings, and certain specialty fabrics. Triacetate is used in sportswear, tricot fabrics, and in garments where pleats and pleat retention is important, as well as in certain specialty fabrics.

PROTEIN FIBERS

The protein fibers are formed by natural animal sources through condensation of a-amino acids to form repeating polyamide units with various substituents on the a-carbon atom. The sequence and type of amino acids making up the individual protein chains contribute to the overall properties of the resultant Fibre. Two major classes of natural protein fibers exist and include keratin and secreted fibers. In general, the keratin fibers are proteins highly crosslinked by disulfide bonds from cystine residues in the protein chain, whereas secreted fibers tend to have no crosslinks and a more limited array of less complex amino acids. The keratin fibers tend to have helical portions periodically within protein sequence, whereas the secreted Fibre protein chains are arranged in a linear pleated sheet structure with hydrogen bonding between amide groups on adjacent protein chains. Keratin fibers are extremely complex in structure and include a cortical cell matrix surrounded by a cuticle sheath laid on the surface as overlapping scales.

The cell matrix of some coarser hair fibers may contain a center cavity or medulla. The keratin fibers are round in cross section with an irregular crimp along the longitudinal Fibre axis which results in a bulky texturized Fibre. Secreted fibers are much less complex in morphology and often have irregular cross sections.

Other fibrous protein materials include azlon fibers, spun from dissolved proteins, and a graft protein-acrylonitrile matrix Fibre. In general, protein fibers are fibers of moderate strength, resiliency, and elasticity. They have excellent moisture absorbency and transport characteristics. They do not build up static charge. While they have fair acid resistance, they are readily attacked by bases and oxidizing agents. They tend to yellow in sun light due to oxidative attack. They are highly comfortable fibers under most environmental conditions and possess excellent aesthetic qualities.

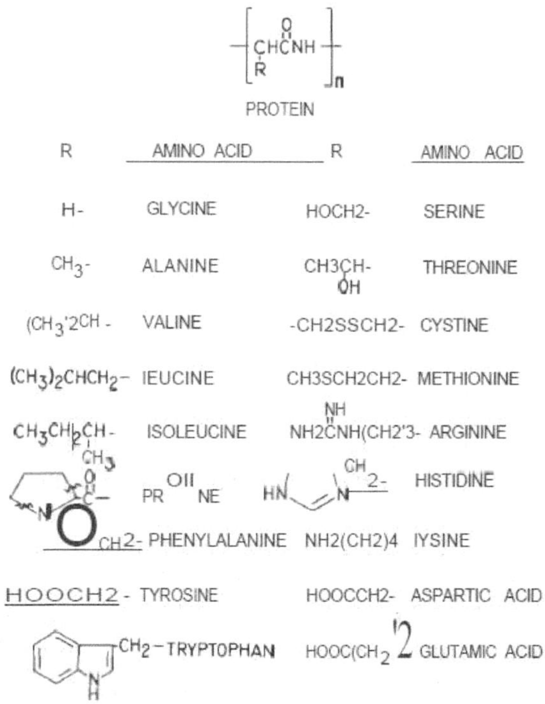

WOOL

Wool is a natural highly crimped protein hair Fibre derived from sheep. The fineness and the structure and properties of the wool will depend on the variety of sheep from which it was derived. Major varieties of wool come from Merino, Lincoln, Leicester, Sussex, Cheviot, and other breeds of sheep. Worsted wool fabrics are made from highly twisted yarns of long and finer wool fibers, whereas woolen fabrics are made from less twisted yarns of coarser wool fibers.

Structural Properties

Wool fibers are extremely complex, highly crosslinked keratin proteins made up of over 17 different amino acids.

The amino acid content and sequence in wool varies with variety of wool. The average amino acid contents for the major varieties of wool are given in table.

Table. Amino Acid Contents in Wool Keratins

Amino Acid	Content in Keratin
Glycine	5-7
Alanine	3-5
Valine	5-6
Leucine	7-9
Isoleucine	3-5
Proline	5-9
Phenylalanine	3-5
Tyrosine	4-7
Tryptophan	1-3
Serine	7-10
Threonine	6-7
Cystine	10-15
Methionine	0-1
Arginine	8-11
Histidine	2-4
Lysine	0-2
Aspartic acid	6-8
Glutamic acid	12-17

The wool protein chains are joined periodically through the disulfide crosslinked cystine, a diamino acid that is contained within two adjacent chains. About 40per cent of the protein chains spiral upon themselves and internally hydrogen bond to form an a-helix.

Near the periodic cystine crosslink s or at points where proline and other amino acids with bulky groups occur along the chain, the close packing of chains is not possible and a less regular Non-helical structure is observed. The crosslinked protein structure packs and associates to form fibrils, which in turn make up the spindle shaped cortical cells which constitute the cortex or interior of the Fibre.

The cortex is made up of highly and less crosslinked ortho and para cortex positions. The cortex is surrounded by an outer sheath of scalelike layers or cuticle, which accounts for the scaled appearance running along the surface of the Fibre.

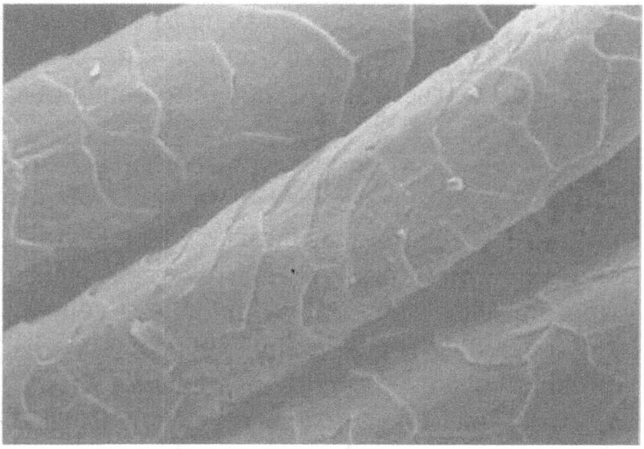

Fig. Wool

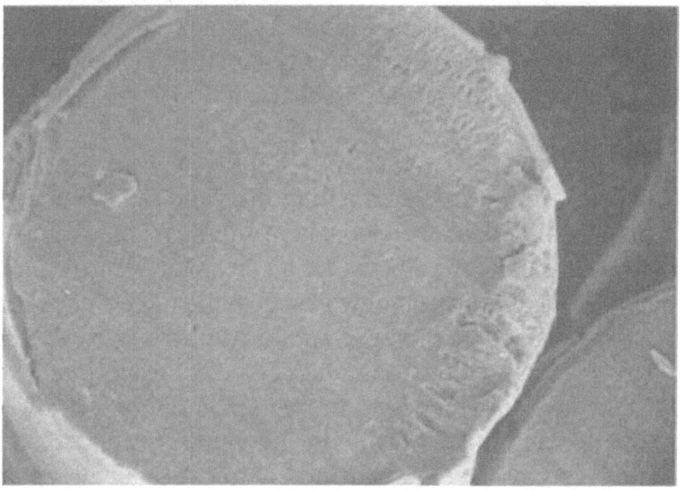

Fig. Wool cross-section

Physical Properties

Wool Fibre s possess low to moderate strength with tena cities of 1-2 g/ d dry and 0.8-1. 8 g/ d wet. Elongations at break vary from 25per cent to 40per centdry and 25per cent to 60per centwet. At 2per cent extension, wool shows 99per centrecovery, and even at 20per centextension a recovery as high as 65per centis observed. Wool fibers have excellent resiliency and recover readily from deformation except under high humidities. The stiffness of wool varies according to the source and the diameter of the individual fibers. The moisture regain of wool is very high and varies between 13per cent and 18per cent under standard conditions. At 100per cent RH, the regain approaches 40per cent. Wool fibers have specific gravities of 1.28-1.32. Wool is insoluble in all solvents except those capable of breaking the disulfide crosslinks, but it does tend to swell in polar solvents. Wool is little affected by heat up to 150°C and is a good heat insulator due to its low heat conductivity and bulkiness, which permits air entrapment in wool textile structures. At moderate humidities, wool does not build up significant static charge.

Chemical Properties

Wool is resistant to attack by acids but is extremely vulnerable to attack by weak bases even at low dilutions. Wool is irreversibly damaged and colored by dilute oxidizing bleaches such as hypochlorite. Reducing agents cause reductive scission of disulfide bonds within the wool, eventually causing the wool to dissolve. Under controlled conditions, reducing agents can be used to partially reduce the wool and flat set or set permanent pleats in the wool. Unless chemically treated, wool is susceptible to attack by several species of moths able to dissolve and digest wool fibers. Wool is quite resistant to attack by other biological agents such as mildew. Wool is attacked by short wavelength ultraviolet light, causing slow degradation and yellowing. On heating, wool degrades and yellows above 150 °C and chars at 300 °C.

End-Use Properties

Wool varieties include Merino, lincoln, leicester, Sussex, Cheviot, Ramboullett, and Shetland, as well as many others. Wool is a Fibre of high to moderate luster. Fabrics of wool posses s a soft to moderate hand and exhibit good drapability. Wool fibers are highly absorbent and have excellent moisture transmission properties. The low to moderate strength of wool fibers is compensated for by its good stretch and recovery properties. Wool is fairly abrasion resistant and does not tend to form pills due to its low strength. It resists wrinkling except under warm, moist conditions. Its crease retention is poor unless creases have been set using chemical reducing agents. Wool is attacked by alkalies and chlorine bleaches and is progressively yellowed by the short ultraviolet wavelengths in sunlight. Wool dyes readily, and the dyed wools exhibit good colorfastness. Owing to its felting action, wool cannot be laundered in hot water with agitation, but it can be dry-cleaned or washed in warm water with a mild detergent if no agitation is used. Due to its affinity for water, wool is slow drying. Wool may be ironed at 150 °C or below without steaming. Wool is a self-extinguishing Fibre and burns very slowly even in contact with a flame. It has an lOI of 25. Wool is extensively used in textile applications where comfort and aesthetics are important. It is used in men's and women's apparel, outer wear and cold weather clothing, suits, blankets, felts, and carpeting. It is often used in blends with cellulosic and man-made fibers.

SILK

Silk is a natural protein Fibre excreted by the moth larva Bombyx mori, better known as the common silkworm. Silk is a fine continuous monofilament Fibre of high luster and strength and is highly valued as a prestige Fibre. Because of its high cost, it finds very 1imited use in textiles. A minor amount of wild tussah silk is produced for specialty items. Attempts have been made to commercialize silk from spiders over the years, but all ventures have met with failure. Domestic and wild silks are essentially uncrosslinked and relatively simple in amino acid composition compared to the keratin fibers. The properties for silk 1isted here are for silk formed by Bombyx mori moth larvae. Liquid silk protein is extruded from two glands in the head of the silkworm. The fibers emerge from a common exit tube or spinneret and harden into a single strand by a protein gum called sericin. The completed silk cocoons are soaked in hot water to loosen the sericin, and the silk filaments are unwound. After unwinding, the silk filaments are washed in warm detergent solutions to remove the sericin. The fibroin silk fibers are more simple in structure than keratin and are composed predominantly of glycine, alanine, tyrosine, and serine. The average range of composition for silk is given in table.

Table. Amino Acid Contents in Fibroin

Amino	Acid Content in Fibroin (g/IOO g Fibroin)
Glycine	36-43
Alanine	29-35
Tyrosine	10-13
Serine	13-17
Valine	2-4
Leucine	0-1
Proline	0-1
Phenylalanine	1-2
Tryptophan	0-1
Threonine	1-2
Cystine	0
Methionine	0
Arginine	0-2
Histidine	0-1
Lysine	0-1
Aspartic acid	1-3
Glutamic acid	1-2

With no cystine present in the fibroin protein, little crosslinking is observed between protein chains. The degree of polymerization of silk fibroin is uncertain, with DPs of 300 to 3000+ having been measured in different solvents. In the absence of crosslinks and with limited bulky side chains present in the amino acids, fibroin molecules align themselves parallel to each other and hydrogen bond to form a highly crystal 1ine and oriented "pleated-sheet" or "beta" structure. Silk fibers are smooth surfaced and translucent with some irregularity in diameter along the Fibre. The fibers are basically triangular in cross section with rounded corners.

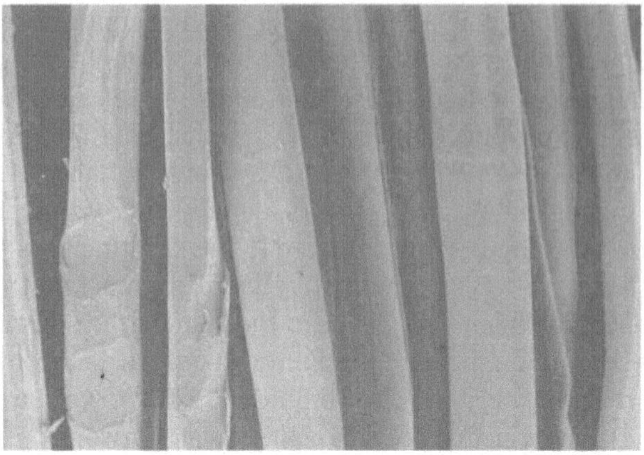

Fig. Silk

Physical Properties

Silk fibers are strong with moderate degrees of recovery from deformation. Silk has a dry tenacity of 3-6 g/d and a wet tenacity of 2.5-5 g/d. Silk exhibits a recovery of 90per cent from 2per cent elongation and of 30per cent-35per cent from 20per cent elongation. Silk fibers are moderately stiff and exhibit good to excellent resiliency and recovery from deformation, depending on temperature and humidity conditions. Silk has a specific gravity of 1.25-1.30 and a moisture regain of 11per cent under standard conditions. Silk is soluble in hydrogen bond breaking solvents such as aqueous lithium bromide, phosphoric acid, and cuprammonium solutions. It exhibits good heat insulating properties and is little affected by heat up to 150°C. Silk has moderate electrical resistivity and tends to build up static charges.

Chemical Properties

Silk is slowly attacked by acids but is damaged readily by basic solutions. Strong oxidizing agents such as hypochlorite rapidly discolor and dissolve silk, whereas reducing agents have little effect except under extreme conditions. Silk is resistant to attack by biological agents but yellows and loses strength rapidly in sunlight. Silk is often weighted with tin and other metal salts. These salts make silk even more sensitive to light-induced oxidative attack. Silk undergoes charring and oxidative decomposition when heated above 175°C in air over a prolonged period of time.

End-Use Properties

Silk possesses a combined set of aesthetic properties that make it useful for high-fashion luxury textile goods. Silk has a high luster and is translucent. Silk fabrics have pleasing appearance and drapability, and a characteristically pleasing crisp hand. Silk is highly moisture absorbent and has well to excellent resistance to wrinkling. It is a moderately strong Fibre with moderate recovery properties. It exhibits fair abrasion resistance and good resistance to pilling. Silk is sensitive to chlorine bleaches and to alkalies and is easily damaged by sunlight. The Fibre may be dyed with a wide variety of dyes to give dyed fibers with high colorfastness. Silk may be laundered under mild, Non-alkaline conditions and dry cleans readily. Because of its high affinity for water, it dries slowly but may be dried or ironed safely up to 150°C. Silk burns slowly and self-extinguishes when removed from a flame. Silk is used extensively in luxury fabrics and apparel and home furnishings, and in accessories such as scarfs.

OTHER NATURAL AND REGENERATED PROTEIN FIBERS

The other major hair fibers include:

- mohair,
- cashmere,
- llama,
- alpaca, and
- vicuna, as well as many others.

Regenerated Azlon fibers are derived from soluble proteins that can be spun into fibers, insolubilized, and regenerated. Soluble proteins also may be grafted to form a copolymer, dissolved, and then spun into fibers.

Mohair

Mohair is a very resilient hair Fibre obtained from the angora goat. The two primary classifications for mohair are the finer kid mohair and the coarser adult mohair. In many respects, mohair resembles wool in structure and possesses properties including the characteristic scale structure of the Fibre. The average length of mohair fibers is longer than wool, with 4-12 inch Fibre lengths being typical. It is a much stronger Fibre than wool, but its other tensile properties resemble wool. Mohair is remarkably resistant to wear and is used in applications where such durability is essential. Mohair possesses a beautiful natural luster and is used in blends to provide luster. Mohair fibers also provide a characteristic, resilient, and slightly scratchy hand in even small quantities blended with other fibers.

Cashmere

Cashmere is the fine, soft inner coat of down obtained from the cashmere goat found on the inner plateaus of Asia. In many ways the properties of cashmere resemble those of wool, but cashmere fibers are extremely fine and soft compared to wool. Cashmere is used in luxury applications where a soft, warm, fine Fibre with beautiful drape is desired.

Llama, Alpaca, and Vicuna

These fibers come from a group of related animals found in South America.

They are fine fibers that are white to tan and brown in Colour. They are longer than most wool fibers and generally stronger, with a finer scale structure. They are generally used only in the expensive luxury items of textiles and apparel.

Regenerated Protein Fibers

Azlon is the generic name given to manufactured fibers composed of a regenerated natural protein. Azlon is produced by dissolving proteins like casein from milk, soya bean protein, and zein from corn in dilute alkali and forcing these solutions through a spinneret into an acid-formaldehyde coagulating bath. Many of the properties of these fibers resemble the natural protein fibers, but they suffer from low dry and wet strength and sensitivity to alkalies. Although no longer produced in the U.S., azlon fibers are produced in Europe and used in blends with natural and man-made fibers.

Protein-Polyacrylonitrile Graft Copolymer

A Fibre consisting of a copolymer of casein protein grafted with 40per çent-75per cent acrylic monomers, of which at least half is acrylonitrile, has been developed in Japan under the tradename Chinon. The casein dissolved in aqueous zinc chloride and grafted with acrylonitrile is wet or dry spun into fibers. T

he Fibre has a tenacity of 3.5-5 g/d dry and 3-4.5 g/d wet and an elongation at break of 15per cent-25per cent wet or dry. It recovers 70per cent from 5per cent elongation. The Fibre has a moisture regain of 4.5per cent-5.5per cent and a specific gravity of 1.22. It dyes readily with acid dyes, but basic and reactive dyes can be used also. The Fibre is marketed as a substitute for silk.

POLYAMIDE FIBERS

The polyamide fibers include the nylons and the aramid fibers. Both Fibre types are formed from polymers of long-chain polyamides. In nylon fibers less than 85per cent of the polyamide units are attached directly to two aromatic rings, whereas in aramid fibers more than 85per centof the amide groups are directly attached to aromatic rings.

The nylons generally are tough, strong, durable fibers useful in a wide range of textile applications. The fully aromatic aramid fibers have high temperature resistance, exceptionally high strength, and dimensional stability. The number of carbon atoms in each monomer or comonomer unit is commonly designated for the nylons. Therefore the nylon with six carbon atoms in the repeating un it would be nylon 6 and the nylon with six carbons in each of the monomer units would be nylon 6,6.

NYLON 6 AND 6,6

Nylon 6 and 6,6 are very similar in properties and structure and therefore will be described together. The major s tructural difference is due to the placement of the amide groups in a continuous head-to-head arrangement in nylon 6, whereas in nylon 6,6 the amide groups reverse direction each time in a head-to-tail arrangement due to the differences in the monomers and polymerization techniques used:

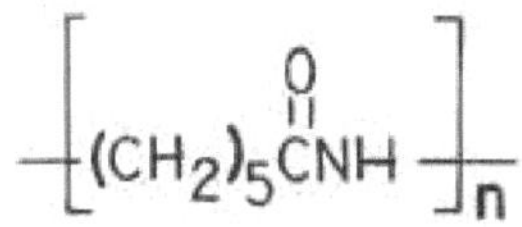

NYLON 6

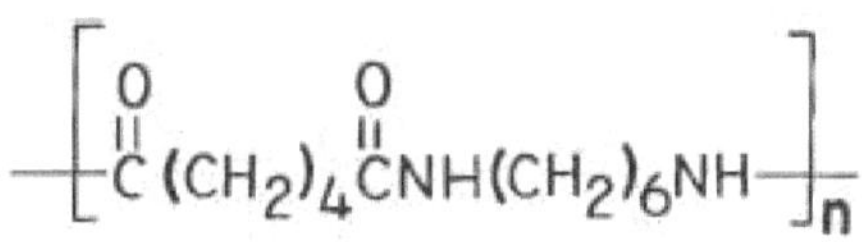

NYLON 6.6

Nylon 6,6 was developed in the United States, whereas nylon 6 was developed in Europe and more recently Japan. These nylon polymers form strong, tough, and durable fibers useful in a wide variety of textile applications. The major differences in the fibers are that nylon 6,6 dyes lighter, has a higher melting point, and a slightly harsher hand than nylon 6.

Structural Properties

Nylon 6 is produced by ring-opening chain growth polymerization of caprolactam in the presence of water vapor and an acid catalyst at the melt. After removal of water and acid, the nylon 6 is melt spun at 250°C260° C into fibers. Nylon 6,6 is prepared by step growth polymerization of hexamethylene diamine and adipic acid. After drying, the nylon 6,6 is melt spun at 280°C-290°C into fibers. Both nylon 6 and 6,6 are drawn to mechanically orient the fibers following spinning. The degree of polymerization of nylon 6 and 6,6 molecules varies from 100 to 250 units. The polyamide molecular chains lay parallel to one another in a "pleated sheet" structure similar to silk protein with strong hydrogen bonding between amide linkages on adjacent molecular chains. The degree of crystallinity of the nylon will depend on the degree of orientation given to the Fibre during drawing. Nylon fibers are usually rodlike with a smooth surface or trilobal in cross section. Multilobal cross sections and other complex cross sections are also found. Side-by-side bicomponent fibers with a round cross section are also formed. Following orientation and heating, these bicomponent fibers form small helical crimps along the Fibre to give a texturized Fibre useful in many applications including women's hosiery. Cantrece nylon is such a Fibre. Also, sheath-core nylons with differing melting characteristics are formed for use as self-binding fibers for Non-wovens.

Fig. Nylon 6,6, round.

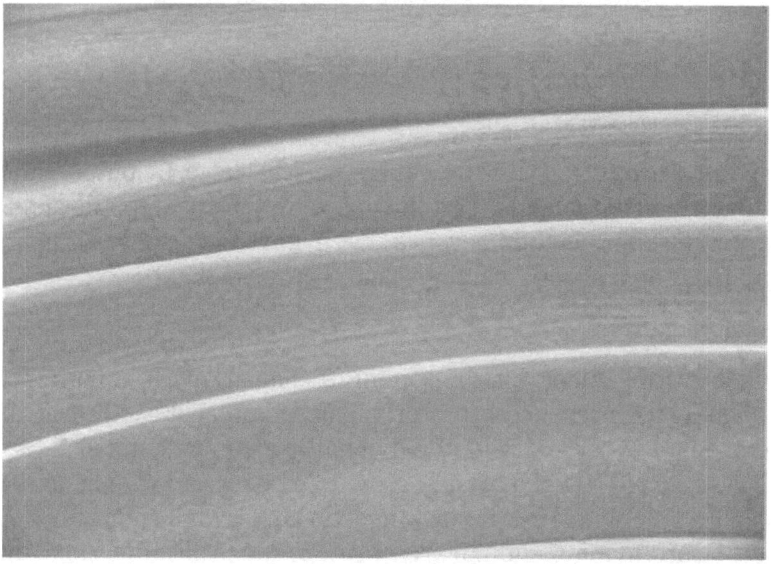

Fig. Nylon 6,6, trilobal

Physical Properties

Nylon 6 and 6,6 fibers are strong, with a dry tenacity of 4-9 g/d and a wet tenacity of 2.5-8 g/d. These nylons have elongations at break of 15per cent-50per cent dry, which increase somewhat on wetting. Recovery from stretch deformation is very good, with 99per cent recovery from elongations up to 10per cent. The nylons are stiff fibers with excellent resiliency and recovery from bending deformation. They are of low density, with a specific gravity of 1.14.

They are moderately hydrophilic with a moisture regain of 4per cent-5per cent under standard conditions and a regain of 9per cent at 100per cent relative humidity. Nylon 6 and 6,6 are soluble in hydrogen bond breaking solvents such as phenols, 90per cent formic acid, and benzyl alcohol. They have moderate heat conductivity properties and a re unaffected by heating below 150°C. The nylons have a high resistivity and readily build up static charge.

Chemical Properties

The nylons are fairly resistant to chemical attack. They are attacked by acids, bases, and reducing and oxidizing agents only under extreme conditions. They are unaffected by biological agents, but at elevated temperatures or in the presence of sunlight they will undergo oxidative degradation with yellowing and loss of strength.

End-Use Properties

Nylon 6 and 6,6 are marketed under several trade names. In fact, nylon was originally used as a trade name for duPont polyamide Fibre, but as a result of common usage the term nylon came to be the generic term for these polyamide fibers. Common trade names for these nylons include:

- Anso,
- Antron,
- Cadon,
- Cantrece,
- Cordura,
- Caprolan,
- duPont nylon, and
- Enkalure.

Nylon 6 and 6,6 are extremely strong fibers with excellent recovery and resiliency properties. Nylon fabrics have a fair hand, with nylon 6 having a somewhat softer hand than nylon 6,6. Nylon fabrics have high luster unless delustered. They have moderate to excellent draping properties depending on the denier of the Fibre. Unless bleached and dyed, nylons have a slightly yellow Colour. The nylons are wrinkle resistant and have good crease resistance if heat set. The Fibre is tough and has good abrasion resistance. In fact, nylon is reported to abrade other fibers in a Fibre blend. The Fibre has good resistance to household chemicals but exhibits poor resistance to attack by sunlight unless treated with antioxidants. The fibers have excellent dyeability with excellent colorfastness properties. Nylon 6 is somewhat deeper dyeing than nylon 6,6. The Fibre has good laundering and dry cleaning properties but tends to scavenge dyes bleeding from other fibers. The fibers have moderate moisture uptakes and dry readily at temperatures up to 150°C. Nylon 6 can be safely ironed up to 140°C, whereas nylon 6,6 can be safely ironed up to 180°C. The nylons are less flammable than cellulosics, with an LOI of 20. They melt and drip and tend to self-extinguish on burning. Nylons are extensively used in hosiery, lingerie, underwear, sweaters, and other knitted goods. They have been used extensively in light and sheer apparel articles such as windbreakers. Nylons are used in home furnishings and industrial applications including:

- carpets,
- upholstery,
- tie cord,
- parachutes,
- sails,
- ropes,
- thread, and
- outdoor wear.

ARAMID FIBERS

The aramid polyamide fibers are formed from a long chain of synthetic polyamides in which at least 85per cent of the amide linkages are attached to aromatic rings. These essentially fully aromatic polyamides are characteristically high melting and have excellent property retention at high temperatures and excellent durability. They are unaffected by moisture and most chemicals and are inherently flame retardant. The fibers have high strength and can be used in a number of unique high-strength applications.

Structural Properties

Aramids are formed through step growth polymerization of aromatic diacid chlorides with aromatic diamines in a polar aprotic solvent such as N,N-dimethylformamide (DMF) to a DP of 100-250. The meta- and para-substituted benzene dicarboxylic acid chlorides and diamines are characteristically used for aramid fibers presently in production, but other fully aromatic ring systems are possible future sources of aramid polymers for fibers:

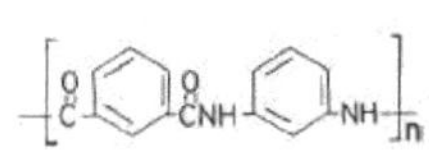

NOMEX

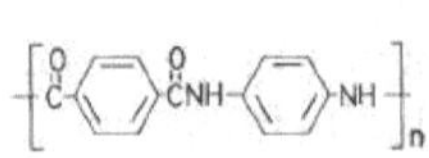

KEVLAR

ARAMID

The resultant aramid polymers are spun in suspension through a spinneret into hot air or a coagulating of both to form the fibers, followed by Fibre stretching and orientation. The aromatic units along aramid polyamide chains confer stiffness to the polymeric chains and limit their flexibility and mobility. Hydrogen bonding between amide groups on adjacent chains and extremely strong van der Waals interactions between aromatic rings planar to adjacent aromatic rings provide a tightly- packed, strongly held molecular structure and account for the strength and thermal stability of the aramids. The aramids are usually spun in round or dumbbell cross section.

Physical Properties

The aramids are the strongest of the man-made fibers, with the strength depending on the polymer structure, spinning method used, and the degree of orientation of the Fibre. Dry tenacities of 4-22 g/d are found for the aramids, and wet tenacities of 3-18 g/d have been recorded. The elongation at break of the fibers decreases with increasing tenacity over a range of 3per cent-30per centdry and wet. Recovery from low elongations of 5per cent or less is 98per cent-100per cent. The fibers are stiff, with excellent resiliency and recovery from bending deformation. The specific gravities of these fibers vary from 1.38 to 1.44. The aramids have a moisture regain comparable to the other polyamide fibers and are in the 3.5per cent-7per cent range. The fibers are swollen and dissolved in polar aprotic solvents or strong acids. The aramids have high heat and electrical resistivities, have excellent insulative capabilities, and are unaffected by heat up to 250°C.

Chemical Properties

The aramids are extremely resistant to chemical or biological attack. Only under extreme conditions and at elevated temperatures will concent rated acids or strong oxidizing agents attack aramids. Aramid fibers undergo initial oxidative attack in sunlight, causing discoloration and slight strength loss, but further exposure has little additional effect. The aramids also act as effective screens to high-energy nuclear radiation due to their ability to trap and stabilize radical and ionic species induced by radiation. The aramids undergo oxidative degradation on prolonged heating above 370°C.

End-Use Properties

Common trade names for aramid fibers include Nomex and Kevlar. Aramid fibers are extremely strong and heat resistant. Fabrics from the aramids have a high luster with a fair hand and adequate draping properties. The fibers are light yellow unless bleached and exhibit moderate moisture absorption characteristics. The fibers recover readily from stretching and bending deformation and are extremely abrasion resistant. They do tend to pill due to the high strength of the Fibre. The aramids are extremely resistant to attack by household chemicals and exhibit good resistance to sunlight. The fibers are difficult to dye except by special dyeing techniques with disperse dyes. Dyed aramids are reasonably colorfast. They have good launderability and dry-cleanability and are moderately easy to dry. The fibers are extremely heat stable and can be ironed up to 300°C. The fibers are of low flammability and self-extinguish on removal from a flame. They have an LOI of 30. Aramid fibers are used in fabric applications where high strength and low flammability are important, including protective clothing, electrical insulation, filtration fabrics, specialized military applications, and tire cord.

OTHER POLYAMIDES

Several other polyamides have been introduced for use as fibers in specialty applications where certain combinations of properties are desired. The major specialty nylons include Qiana, nylon 4, nylon 11, nylon 6,10, and biconstituent nylon-polyester. Qiana is a trade name for the luxury nylon Fibre formed through step growth polymerization of trans, transdimethane and a dibasic acid having 8-12 carbon atoms:

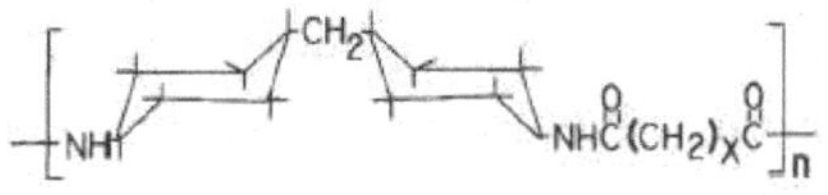

QIANA

Qiana resembles nylon 6 and 6,6 in many of its properties but also has a unique silklike texture, a lustrous appearance, excellent form retention, and an overall superior performance level which makes it useful in prestige textile products and in the sophisticated fashion apparel market. Qiana has a tenacity of 3-3.5 g/d and an elongation at break of 20per cent-30per cent. It is reported to recover from strain better than nylon 6,6 or polyester. It has a specific gravity of 1.03 and a moisture regain of 2per cent-2.5per cent, which are much lower values than nylon 6 or 6,6. Qiana melts at 275°C and is stable up to 185°C. The cross section of the Fibre is trilobal. It shows good chemical stability and is readily dyeable to fast colors. The excellent aesthetic qualities and drapability of Qiana have contributed greatly to its success.

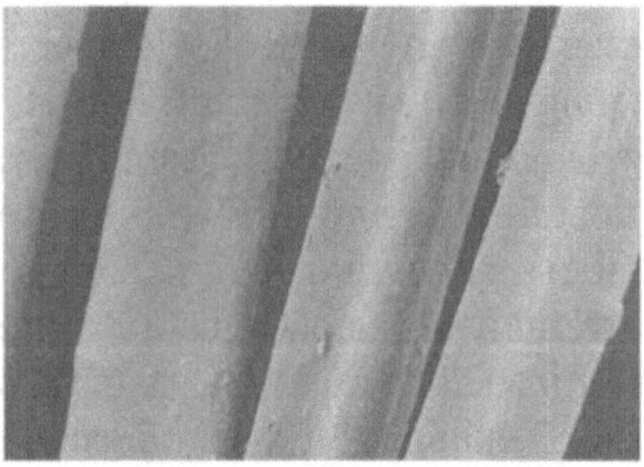

Fig. Qiana

Nylon 4

Nylon 4 is produced by polymerization of pyrrolidone using carbon dioxide catalyst. Nylon 4 is melt spun to give a Fibre of about 60per cent crystallinity. The Fibre is of moderate dry strength and slightly higher wet strength, with a higher specific gravity than nylon 6 and 6,6. The Fibre has excellent elongation and recovery properties and a water regain comparable to cotton. Nylon 4 is easily laundered and can be dyed easily to give colorfast shades but has limited wrinkle recovery properties and is sensitive to hypochlorite bleaches. Although a major attempt was made to commercialize nylon 4, market conditions limited the penetration of this new Fibre type and production was discontinued. Nylon 4 is no longer being produced.

Nylon 11

Nylon II is produced by self-condensation of ll-aminoundecanoic acid in Europe and marketed under the name Rilsanite. It is melt spun into Fibre and possesses essentially the same properties as nylon 6 and 6,6. It has high strength, low specific gravity, and low water regain. Nylon II has excellent electrical properties and is used in electrical products as well as in brush bristles, tire cord, lingerie, and hose.

Nylon 6,10

Nylon 6,10 is produced by condensation polymerization of he xame thylene diamine and sebacic acid and is melt spun into fibers. It resembles nylon 6 and 6,6 in many ways but has a lower moisture regain. It is primarily used in brush bristles.

Biconstituent Nylon-Polyester

Biconstituent Fibre of nylon 6 with polyester microfibrils dispersed throughout the Fibre matrix has been marketed under the trade name Source. The Fibre is reported to have unique optical and dyeing properties and a higher strength and lower regain than nylon 6,6 and is used primarily in carpets. A sheath-core bicomponent Fibre containing a nylon 6 sheath and a polyester core has been reported al so. It is said to have properties that are intermediate between both fibers.

POLYESTER FIBERS

Polyesters are those fibers containing at least 85per cent of a polymeric ester of a substituted aromatic carboxylic acid including but not restricted to terephthalic acid and f-hydroxybenzoic acid. The major polyester in commerce is polyethylene terephthalate, an ester formed by step growth polymerization of terephthalic acid and the diol ethylene glycol. Polyl, 4-cyclohexylenedimethylene terephthalate is the polyester of more limited usage and is formed through the step growth polymerization of terephthalic acid with the more complex diol 1,4-cyclohexylenedimethanol. A third polyester Fibre is actually the polyester ether PolY and is formed through step growth polymerization of £-hydroxybenzoic acid with ethylene glycol; it is no longer in commercial production, however. The polyester fibers all have similar properties, are highly resilient and resistant to wrinkling, possess high durabilily and dimensional stability, and are resistant to chemical and environmental attack.

POLYETHYLENE TEREPHTHALATE

Polyethylene terephthalate polyester is the leading man-made Fibre in production volume and owes its popularity to its versatility alone or as a blended Fibre in textile structures. When the term "polyester" is used, it refers to this generic type.

It is used extensively in woven and knitted apparel, home furnishings, and industrial applications. Modification of the molecular structure of the Fibre through texturizing and or chemical finishing extends its usefulness in various applications. Polyester is expected to surpass cotton as the major commodity Fibre in the future.

Structural Properties

Polyethylene terephthalate is formed through step growth polymerization of terephthalic acid or dimethyl terephthalate with ethylene glycol at 250°-300°C in the presence of a catalyst to a DP of 100-250:

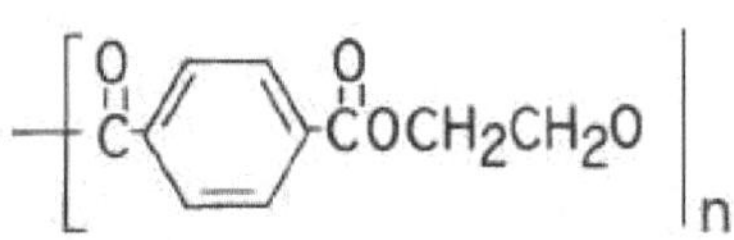

POLYETHYLENE TEREPHTHALATE
POLYESTER

The resultant polymer is isolated by cooling and solidification and dried. Polyester fibers are melt spun from the copolymer at 250°-300°C, followed by Fibre orientation and stretching. The polyester molecular chains are fairly stiff and rigid due to the presence of periodic phenylene groups along the chain. The polyester molecules within the Fibre tend to pack tightly and are held together by van der Waals forces. The polyesters are highly crystalline unless comonomers are introduced to disrupt the regularity of the molecular chains. Polyester fibers are usually smooth and rodlike with round or trilobal cross sections.

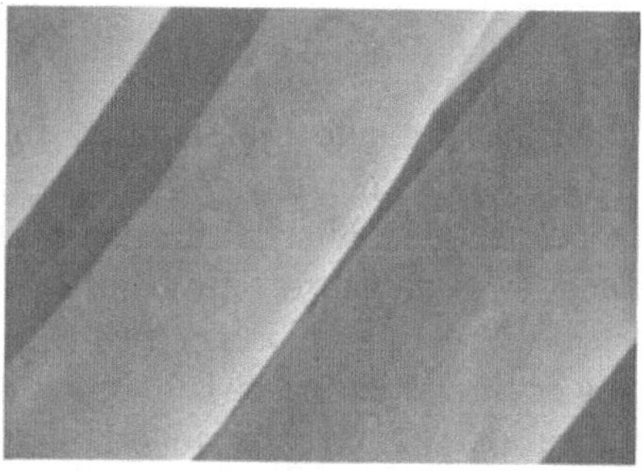

Fig. Polyester

Physical Properties

Polyester from polyethylene terephthalate is an extremely strong Fibre with a tenacity of 3-9 g/d. The elongation at break of the Fibre varies from 15per cent to 50per cent depending on the degree of orientation and nature of crystalline structure within the Fibre. The Fibre shows moderate recovery from low elongations. The Fibre is relatively stiff and possesses excellent resiliency and recovery from bending deformation. The Fibre has a specific gravity of 1.38. The Fibre is quite hydrophobic, with a moisture regain of 0.1per cent-0.4per cent under standard conditions and 1.0per cent at 21°C and 100per centRH. It is swollen or dissolved by phenols, chloroacetic acid, or certain chlorinated hydrocarbons at elevated temperatures. The Fibre exhibits moderate heat conductivity and has high resistivity, leading to extensive static charge buildup. On heating, the Fibre softens in the 210°-250°C range with Fibre shrinkage and melts at 250°-255°C.

Chemical Properties

Polyethylene terephthalate polyester is highly resistant to chemical attack by acid, bases, oxidizing, or reducing agents and is only attacked by hot concentrated acids and bases. The Fibre is not attacked by biological agents. On exposure to sunlight, the Fibre slowly undergoes oxidative attack without Colour change with an accompanying slow loss in strength. The Fibre melts at about 250°C with only limited decomposition.

End-Use Properties

Common trade names for polyethylene terephthalate polyester are:

- Monsanto,
- Dacron,
- Encron,
- Kadel IV,
- Polar Guard,
- Trevira,

- Fortrel, and
- Vycron.

This polyester is an inexpensive Fibre with a unique set of desirable properties which have made it useful in a wide range of end-use applications. It has adequate aesthetic properties and bright translucent appearance unless a delusterant has been added to the Fibre. The hand of the Fibre is somewhat stiff unless the Fibre has been texturized or modified, and fabrics from polyester exhibit moderate draping qualities. The Fibre is hydrophobic and Non-absorbent without chemical modification. This lack of absorbency limits the comfort of polyester fabrics.

Polyester possesses good strength and durability characteristics but exhibits moderate to poor recovery from stretching. It has excellent wrinkle resistance and recovery from wrinkling and bending deformation. Polyester shows only fair pilling and snag resistance in most textile constructions. Heat setting of polyester fabrics provides dimensional stabilization to the structure, and creases can be set in polyester by heat due to its thermoplastic nature. It has excellent resistance to most household chemicals and is resistant to sunlight-induced oxidative damage, particularly behind window glass. Due to its hydrophobicity and high crystallinity, polyester is difficult to dye, and special dyes and dyeing techniques must be used. When dyed, polyester generally exhibits excellent fastness properties.

Polyester has good laundering and dry-cleaning characteristics but is reported to retain oily soil unless treated with appropriate soil release agents. Owing to its low regain, polyester dries readily and can be safely ironed or dried at temperatures up to 160°C.

It is a moderately flammable Fibre that burns on contact with a flame but melts and drips and shrinks away from the flame. It has a LOI 21. Polyester is used extensively as a staple and filament Fibre in apparel and as a staple Fibre in blends with cellulosic fibers for apparel of all types. Polyester is also used extensively in home furnishings and as fiberfill for pillows, sleeping bags, etc. Polyester is used in threads, rope, and tire cord and in sails and nets as well as other industrial fabrics.

POLY-L,4-CYCLOHEXYLENEDIMETHYLENE TEREPHTHALATE

Poly-l,4-cyclohexylenedimethylene terephthalate resembles polyethylene terephthalate in most properties:

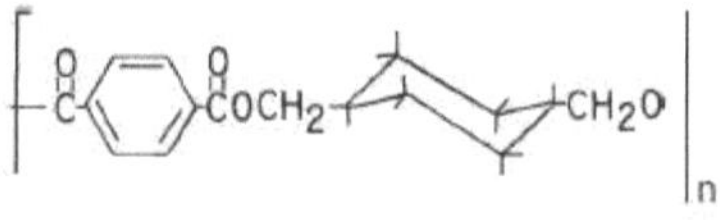

POLY-I.4-CYC LOHEXYLENEDIMETHYL ENE TEREPHTHALATE POLYESTER

The cyclohexylene group within this Fibre provides additional rigidity to the molecular chains, but the packing of adjacent polymer chains may be more difficult due to the complex structure.

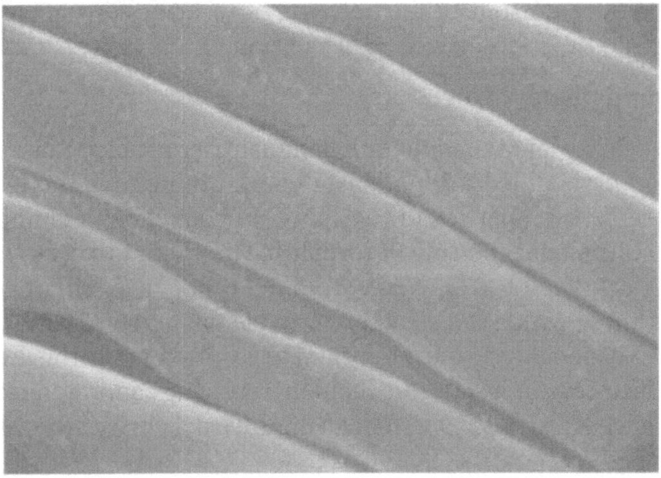

Fig. Poly-1,4-cyclohexylenedimethylene terephthalate polyester

As a result, the Fibre has a lower tenacity than polyethylene terephthalate. It has a tenacity of 2.5-3 g/d and exhibits lower elongations than polyethylene terephthalate. It has a lower specific gravity than polyethylene terephthalate. The Fibre melts at 290°-295°C and is attacked and shrunk by trichloroethylene and methylene chloride. The Fibre has good chemical resistance. The major trade name for this polyester is Kodel II. The Fibre is somewhat superior to polyethylene terephthalate in certain end-use properties including better recovery from stretch and better resistance to pillin9. The Fibre has superior resiliency and is particularly suited for use in blend with cellulosics and wool, as a carpet Fibre, and as fiberfill.

OTHER POLYESTERS

Poly-p-ethyleneoxybenzoate

This Fibre was introduced by a Japanese firm as A-Tell. The properties of the Fibre are very similar to the polyesters containing terephthalate units in most respects including tensile properties, specific gravity, melting point, chemical resistance, and sunlight resistance. As a result, the FTC expanded the polyester category to include the polymeric £ethyleneoxybenzoate esters. The Fibre is produced by reaction of ethylene oxide with £-hydroxybenzoic acid, followed by melt spinning. The Fibre has a tenacity of 4-5.5 g/d and an elongation at break of 15per cent-30per cent with nearly complete recovery from low elongation. Its specific gravity of 1.34 and moisture regain of 0.4per cent-5.0per cent are nearly the same as that of polyethylene terephthalate polyester. The Fibre melts at 224°C and softens at about 200°C. The Fibre is reported to be even more resistant to attack by acids and bases than terephthalate-based polyesters. The properties of this Fibre were not sufficiently different from other polyesters to achieve reasonable market penetration, and the Fibre has been discontinued.

Modified Terephthalate Polyesters

Poor dyeability and the moderate flammabi1ity of polyester have resulted in formulation of modified terephthalate esters to improve the dyeability and the flame retardant properties of the Fibre. Introduction of amino or sulfonic acid groups on the benzene ring of terephthalate leads to fibers that are more dyeable with cationic or acid dyes. Bromine, other halogen, or phosphonate groups substituted within the structure provide flame retardant characteristics to the Fibre, especially if antimony oxide is present in the Fibre matrix. Appropriate Non-terephthalate comonomers replacing some of the terephthalate groups introduced into the polyester can improve dyeability and/or flammability of the fibers.

ACRYLIC FIBERS

Acrylic fibers are synthetic fibers made from a polymer with an average molecular weight of ~100,000, about 1900 monomer units. To be called acrylic in the U.S, the polymer must contain at least 85per cent acrylonitrile monomer. Typical comonomers are vinyl acetate or methyl acrylate. The Dupont Corporation created the first acrylic fibers in 1941 and trademarked them under the name Orlon.

PRODUCTION

The polymer is formed by free-radical polymerization in aqueous suspension. The Fibre is produced by dissolving the polymer in a solvent such as N,N-dimethylformamide or aqueous sodium thiocyanate, metering it through a multi-hole spinnerette and coagulating the resultant filaments in an aqueous solution of the same solvent or evaporating the solvent in a stream of heated inert gas. Washing, stretching, drying and crimping complete the processing. Acrylic fibers are produced in a range of deniers, typically from 0.9 to 15, as cut staple or as a 500,000 to 1 million filament tow. End uses include sweaters, hats, hand-knitting yarns, rugs, awnings, boat covers, and upholstery; the Fibre is also used as a precursor for carbon Fibre. Production of acrylic fibers is centered in the Far East, Turkey, India, Mexico, and South America, though a number of European producers still continue to operate. All US producers have ended production.

Former U.S. brands of acrylic were:

- Acrilan (Monsanto),

- Creslan (American Cyanamid), and
- Orlon (DuPont).

Other brand names that are still in use include Dralon.

TEXTILE USES

Acrylic is lightweight, soft, and warm, with a wool-like feel. It can also be made to mimic other fibers, such as cotton, when spun on short staple equipment. Some acrylic is extruded in colored or pigmented form; other is extruded in "ecru", otherwise known as "natural," "raw white," or "undyed." Pigmented Fibre has highest light-fastness. Its fibers are very resilient compared to both other synthetics and natural fibers. Some acrylic is used in clothing as a less expensive alternative to cashmere, due to the similar feeling of the materials. Some acrylic fabrics may fuzz or pill easily. Other fibers and fabrics are designed to minimize pilling. Acrylic takes Colour well, is washable, and is generally hypoallergenic.

End-uses include socks, hats, gloves, scarves, sweaters, home furnishing fabrics, and awnings. Acrylic is resistant to moths, oils, chemicals, and is very resistant to deterioration from sunlight exposure. However, static and pilling can be a problem in certain fabrications. Acrylic has a bad reputation amongst some crafters who knit or crochet; acrylic yarn may be perceived as "cheap" because it is typically priced lower than its natural counterparts. The Fibre requires heat to "block" or set the shape of the finished garment, and it isn't as warm as alternatives like wool. Some knitters also complain that the Fibre "squeaks" when knitted. On the other hand, it can be useful in certain items, like garments for babies, which require constant washing, because it is machine-washable. Acrylic can irritate the skin of people with dermatological conditions such as eczema but this is unusual. Leading US spinners include National Spinning Co., Inc. Leading acrylic Fibre producers include:

- AkSa (Turkey),
- Montefibre (Spain),
- Dralon (Germany),
- Kaltex (Mexico), and
- Birla Acrylic (Thailand and Egypt).

POLYOLEFIN FIBERS

POLYETHYLENE AND POLYPROPYLENE

Polyolefin fibers are those fibers produced from polymers formed by chain growth polymerization of olefins and which contain greater than 85per cent polymerized ethylene, propylene, or other olefin units. In general, 1inear high-density stereoregular polyethylene and polypropylene are used in textile applications, with polypropylene predominating due to its superior temperature stability. These fibers have good strength and toughness, have good abrasion resistance, and are inexpensive. The fibers are difficult to dye and have relatively low melting points, but they are effectively used in a wide variety of textile applications.

Structural Properties

Linear polyethylene and polypropylene are polymerized from their corresponding monomers to a DP of 1000-3000 using complex metal catalysts at 40°-100°C and at moderate pressures.

Using this initiation technique, branching of the polyolefin due to free radical chain transfer is avoided and a linear unbranched structure is formed. With polypropylene, catalyst systems are selected that will lead to regular isotactic placement of the optically active methyl-substituted carbon on the backbone to give a structure capable of high crystallinity:

POLYETHYLENE POLYPROPYLENE

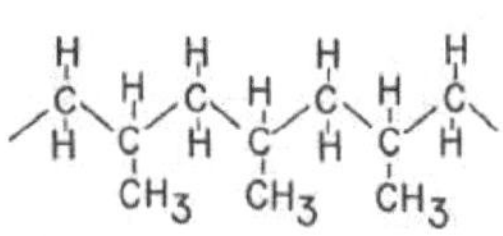

ISOTACTIC POLYPROPYLENE

The resultant polymers are dried, compounded with appropriate additives, melt spun into fibers, and drawn to orient.

The highly linear chains of these polyolefins can closely pack and associate with adjacent chains through van der Waals interactions and possess crystallinities in the 45per cent60per cent range, The surface of these fibers is usually smooth, and the Fibre cross section is round. Also, polyolefin films can be split using knife edges to form flat ribbonlike fibers with a rectangular cross section.

Fig. Polypropylene

Physical Properties

Polypropylene and polyethylene are strong fibers with good elongation and recovery properties. The tenacity of fibers varies from 3.5 to 8 g/d, with an elongation at break of 0per cent-30per cent. The fibers recover well from stress, with 95per cent recovery at 10per cent elongation. The fibers are moderately stiff and have moderate resiliency on bending. Moisture does not affect these properties, since polyolefins are hydrophobic and have a moisture regain of 0per cent. The polyethylene fibers have specific gravities of 0.95-0.96 and polypropylene specific gravities of 0.90-0.91. As a consequence these fibers float on water and are the lightest of the major fibers in commerce. The fibers are unaffected by solvents at room temperature and are swollen by aromatic and chlorinated hydrocarbons only at elevated temperatures. They exhibit excellent heat and electric insulation characteristics and are extensively used in these applications. The fibers are heat sensitive. Polyethylene softens at 130°C and melts at 150°C, while polypropylene softens at about 150°C and melts at about 170°C.

Chemical Properties

The polyolefins are extremely inert and resistant to chemical attack. They are unaffected by chemical and biological agents under normal conditions. They are sensitive to oxidative attack in the pre sence of sunlight due to formation of chromophoric keto groups along the hydrocarbon chain. These groups act as photo sensitizers for further decomposition. The Fibre only slowly undergoes oxidative decomposition at its melting point.

End-Use Properties

Common trade names for polyolefin fibers include:

- Herculon,
- Marvess, and
- Vectra.

Polyolefins are lustrous white translucent fibers with good draping qualities and a characteristic slightly waxy hand. They have excellent abrasion resistance and exhibit fair wrinkle resistance. They are essentially Non-absorbent, but reportedly exhibit wicking action with water. The polyolefins are strong, tough fibers which are chemically inert and resistant to oxidative attack except that induced by sunlight. Stabilizing chemicals must be incorporated into the fibers to lower their susceptibility to such light-induced attack. The polyolefins have high affinity for oil-borne stains, which are difficult to remove on laundering. The fibers are dry-cleanable with normal dry-cleaning solvents if the temperature is kept below SO°C. Since the Fibre s have no affinity for water, they dry readily, but care must be taken to keep the drying temperature below 70 °C for polyethylene and 120°C for polypropylene to prevent heat-induced thermoplastic shrinkage and deformation. The fibers are flammable, burn with a black sooty flame, and tend to melt and draw away from the flame. The polyolefins—particularly polypropylene—have found a number of applications particularly in home furnishing s and industrial fabrics. Uses include indoor-outdoor carpeting, carpet backing, upholstery fabrics, seat covers, webbing for chairs, non-wovens, laundry bags, hosiery and knitwear, fishnet, rope, filters, and industrial fabrics.

VINYL FIBERS

Vinyl fibers are those man-made fibers spun from polymers or copolymers of substituted vinyl monomers and include:

- vinyon,
- vinal,
- vinyon-vinal matri x,
- saran, and
- polytetrafluoroethylene fibers.

Acrylic, modacrylic and polyolefin are also formed from vinyl monomers, but because of their wide usage and particular properties they are usually considered as separate classes of fibers. The vinyl fibers are generally specialty fibers due to their unique properties and uses. All of these fibers have a polyethylene hydrocarbon backbone with substituted functional groups that determine the basic physical and chemical properties of the Fibre.

VINYON

Vinyon is defined as a Fibre in which at least 85per cent of the polymerized monomer units are vinyl chloride. Vinyon fibers have high chemical and water resistance, do not burn, but do melt at relatively low temperatures and dissolve readily in many organic solvents, thereby limiting their application.

Structural Properties

Vinyon is formed by emulsion polymerization of vinyl chloride or by copolymerization of vinyl chloride with less than 15per centvinyl acetate in the presence of free radical catalysts to a DP of 200-450:

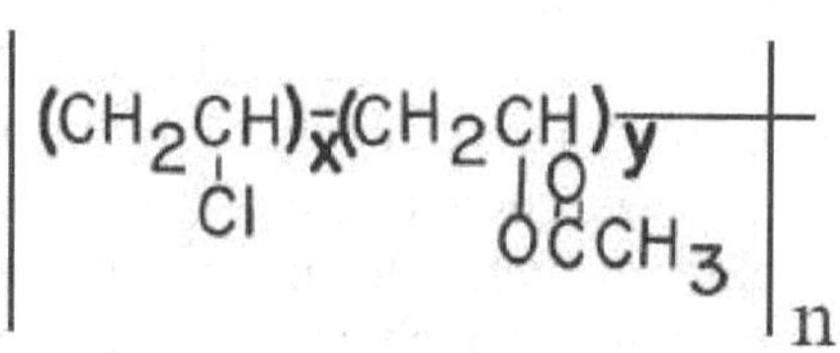

$x > 85\%$. $y < 15\%$

VINYON

The polymer is precipitated and isolated by spray drying and then dissolved in acetone; the polymer solution is dry spun to form the Fibre. Comonomers such as vinyon cetate are added to reduce the crystal linity of the drawn Fibre and to increase the amorphous areas within the Fibre. Weak hydrogen bonding between chlorine and hydrogen on adjacent vinyon chains would be expected with tight packing of the molecular chains in the absence of comonomer. The fibers are spun in nearly round or dog-bone cross section.

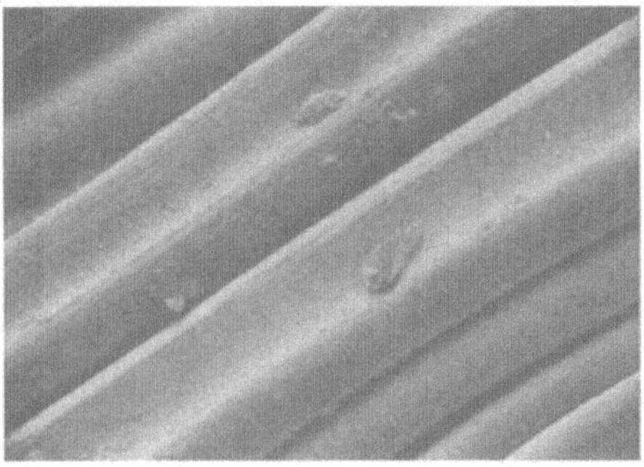

Fig. Vinyon.

Physical Properties

Vinyon fibers have a strength of 1-3 g/d in both the wet and dry states; elongations at break vary between 10per cent and 125per cent. At low elongations vinyon fibers recover completely from deformation. Vinyon fibers are soft and exhibit good recovery from bending deformation. The Fibre has moderate density, with a specific gravity of 1.33-1.40. Vinyon is extremely hydrophobic, having a moisture regain of 0.0per cent-0.1per cent under standard conditions. Vinyon is readily dissolved in ketone and chlorinated hydrocarbon solvents and is swollen by aromatic solvents. The Fibre is a poor heat and electrical conductor and possesses potential in insulation applications. Unfortunately, the oriented Fibre softens and shrinks at temperatures above 60 ° C, thereby limiting the applications in which it may be used.

Chemical Properties

Vinyon Fibre is chemically inert and possesses chemical properties similar to polyolefin Fibre. Vinyon is only slowly attacked by ultraviolet rays in sunlight. Vinyon Fibre melts with decomposition at 135°-180°C, with vinyon containing comonomer having a lower melting/decomposition temperature.

End-Use Properties

Vinyon as pure polyvinyl is marketed as PVC-Rhovyl, while vinyon HH is a copolymer. The Fibre is of low strength but has properties that make it useful in apparel where heat is not a factor. It is difficult to dye. It may be laundered readily, but it is attacked by convnon dry-cleaning solvents. Since the Fibre may not be heated above 60°C, it may be tumble dried only at the lowest heat settings.

The Fibre is Non-flammable and does not support flaming combustion. Vinyon finds its major use in industrial fabrics including filters, tarps, and awnings, in protective clothing, and in upholstery for outdoor furnishings. It is also used as a bonding Fibre in heat bonded Non-wovens and as tire cord in specialty tires.

VINAL

Vinal fibers are made from polymers containing at least 50per cent vinyl alcohol units and in which at least 85per cent of the units are combined vinyl alcohol and acetal crosslink units. The Fibre is inexpensive, resembles cotton in properties, and is produced in Japan.

Structural Properties

Vinal fibers are formed from inexpensive starting materials through a complex process. Vinyl acetate monomer is solution polymerized to polyvinyl acetate using free radical initiation techniques. The polyvinyl acetate is hydrolyzed to polyvinyl alcohol in methanol under basic conditions. Aqueous polyvinyl alcohol solution is wet or dry spun followed by exposure to an aldehyde such as formaldehyde or benzaldehyde and heat to form acetal crosslinks between chains to insolubilize the Fibre. The polyvinyl alcohol chains are tightly packed except near the juncture of the acetal crosslinks. Extensive hydrogen bonding between hydroxyl groups on adjacent chains occurs, which contributes to its highly packed structure, and the Fibre is about 50per cent crystalline:

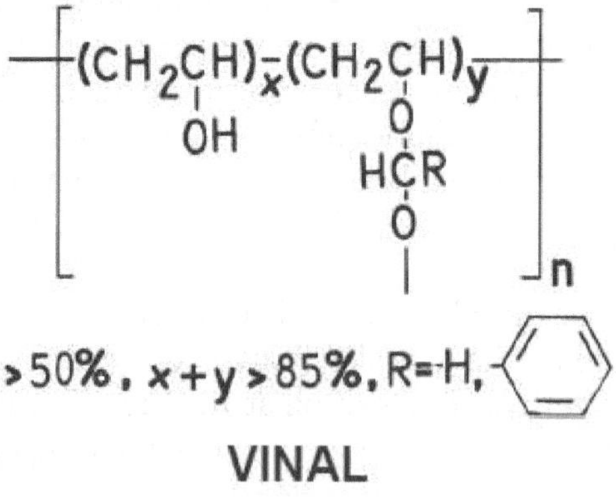

VINAL

The Fibre surface is somewhat rough with lengthwise striations and possible periodic twists. The Fibre may be nearly round or U-shaped in cross section with a noticeable skin and core.

Physical Properties

Vinal Fibre is moderately strong with a dry tenacity of 3-8.5 g/d and possesses a moderate extension at break of 10per cent-30per cent. On wetting the strength of the Fibre decreases and the elongation at break increases slightly. The Fibre exhibits poor to moderate recovery from elongations as low as 2per cent. The Fibre is moderately stiff, but does not recover readily from dry or wet deformation. The Fibre has a specific gravity of 1.26-1.30. Vinal Fibre has a moisture regain of 3.0per cent-5.0per cent under standard conditions, and is swollen and attacked by aqueous phenol and formic acid solutions. The Fibre is a moderate heat and electrical conductor. On heating to 220°-230° the Fibre undergoes 10per cent or more heatinduced shrinkage.

Chemical Properties

Vinal is somewhat sensitive to acids and alkalies. It is attacked by hot dilute acids or concentrated cold acid solutions with Fibre shrinkage and is yellowed by strong aqueous solutions of alkali. Oxidizing, reducing, and biological agents have little effect on vinal. Sunlight causes vinal to slowly lose its strength with perceptible changes in Colour. At 230 °-250° vinal shrinks and softens with decomposition.

End-Use Properties

Vinal resembles cotton and other cellulosics in end-use properties. Kuralon and Manryo are names under which vinal fibers are marketed. Vinal Fibre has good strength and excellent abrasion and pilling resistance. Like cellulosics, vinal breaks at low elongations, exhibits poor recovery from small deformations, and wrinkles readily unless treated with durable press resins of the type used for cellulosics. Fabrics of vinal have a warm comfortable hand, are absorbent, and exhibit good drapabi1ity. The Fibre has a silklike appearance and luster. It has excellent sunlight resistance and fair heat resistance. It dyes readily with dyes for cellulosics. Its launderability and dry-cleanability are very good, and it dries more readily than cellulosics. Vinal may be ironed dry up to 120°C, but undergoes severe deformation if ironed wet below 120°C. Vinal burns readily like a cellulosic Fibre and has a LOI of 20. Vinal is used in apparel as well as industrial applications where an inexpensive Fibre of moderate strength and properties is desired. It is useful in Fibre blends to increase comfort and aesthetic properties.

VINYON-VINAL MATRIX FIBRE

In response to the need for a Fibre of low flammability and low toxic gas formation on burning, Kohjin company developed and marketed a vinyon-vinyl matrix Fibre under the trade names Cordela and Cordelan. The Fibre is believed to be formed through grafting of vinyl chloride to polyvinyl alcohol followed by mixing of the resultant copolymer with additional polyvinyl alcohol. The polymer mixture is wet spun, oriented, and crosslinked with aldehydes. The Fibre has a kidney-shaped cross section, and no outer skin is evident. The Fibre is moderately strong with a dry tenacity of about 3 g/d and wet tenacity of 2.2 g/ d. The elongation at break is 15per cent-20per centwet or dry, with poor recovery from even low elongations. It has a specific gravity of 1.32 and moisture regain of 3.0per cent. Besides its low flammability, it has good chemical resistance except under extreme conditions. The Fibre is heat sensitive above 100 °C, but it may be dyed at lower temperatures with a large range of dyes for cellulosics as well as man-made fibers. Vinyonvinyl matrix Fibre is used primarily in applications where low flammability in apparel is desired and has been used extensively in children's sleepwear.

SARAN

Saran is the generic name for fibers made from synthetic copolymers that are greater than 80per cent vinylidene chloride. The Fibre is formed through emulsion copolymerization of vinylidene chloride with lesser amounts of vinyl chloride using a free radical catalyst, and the precipitated copolymer is melt spun into the Fibre:

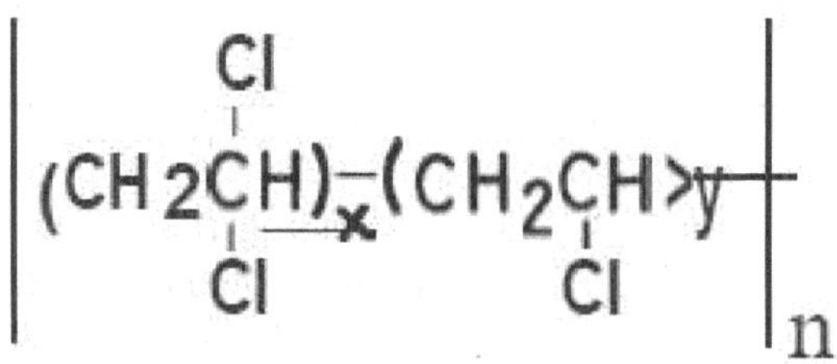

$x > 80\% \cdot y < 20\%$

SARAN

The Fibre is of moderate crystallinity and resembles vinyon in properties. The Fibre has wet and dry strength of 1-3 g/d, elongation at break of 15per cent-30per cent, and good recovery from less than 10per cent elongation. The Fibre has a specific gravity of 1.7 and has excellent resiliency. It has essentially no affinity for water, but it is soluble or attacked by cyclic ethers, ketones, and aromatic solvents. It is a good heat and electrical in sulator, but it softens at 115°-150°C and melts at 170°C. The Fibre is chemically inert and only slowly affected by the ultraviolet rays in sunlight. The Fibre is of low flammability. Saran is marketed as saran or under the trade name Rovana. The Fibre has high resiliency and low water absorbency and is chemically inert and used in applications where these qualities are desirable and where heat sensitivity is not a problem. The Fibre is generally not used in apparel but rather in automobile upholstery, outdoor fabrics, home furnishings, and industrial applications.

POLYTETRAFLUOROETHYLENE

Tetrafluoroethylene is better known by its trade name—Teflon—and is widely used in many applications including specialty fibers. Poly tetrafluoroethylene Fibre is extremely hydrophobic and chemically and thermally stable and is used in applications where such stability and inertness is needed. Polytetrafluoroethylene fibers are produced through polymerization of tetrafluoroethylene under conditions of high temperature and pressure using peroxide catalysts.

The polymer is spun as a dispersion to form a weak Fibre which is then heated at 385°C to fuse the individual Fibre particles. The Fibre must be bleached to give a white Fibre. The highly electronegative fluorine atoms in the Fibre result in an extremely tight packing of adjacent molecular chains tightly held by van der Waals forces and with a high degree of crystallinity. The fibers are smooth with a round cross section. The Fibre has a tenacity of 1-2 g/d and elongations at break vary from 15per cent to 30per cent. It is a soft flexible Fibre of high density, and it is totally hydrophobic. It is unaffected by solvents except for perfluorinated hydrocarbons above 300°C. It is an excellent electrical and heat insulator and is not affected by heat up to 300°C. It is chemically inert and may be used in numerous industrial end-use applications including protective clothing. Its hydrophobicity has permitted is to be used to form breathable but water repellant composite materials for textile usage, particularly in outdoor and rainwear.

Elastomeric Fibers

Elastomeric fibers are those fibers that possess extremely high elongations at break and that recover fully and rapidly from high elongations up to their breaking point. Elastomeric fibers are made up of molecular chain networks that contain highly amorphous areas joined together by crossl inks. On elongation, these amorphous areas become more oriented and more crystalline in nature. Elongation continues until the crosslinks in the structure limit further molecular movement. If additional force is placed on the elastomeric fibers at this point, molecular scission occurs, causing a reduction in properties and ultimate breaking of the Fibre.

Elastomeric fibers include:

- the crosslinked natural and synthetic rubbers,
- spandex fibers,

- anidex fibers and the side-by-side biconstituent Fibre of nylon and spandex.

The fibers are all used in specialized applications where high elasticity is necessary within the textile structure.

RUBBER

Rubber fibers from natural sources have been known for over 100 years. Natural rubber in commerce is derived from coagulation of Hevea brasiliensis latex and is primarily polyisoprene, a diene polymer. Most synthetic rubbers were developed during and following World War II.

They are crosslinked diene polymers, copolymers containing dienes, or amorphous polyolefins. Both the natural and synthetic rubbers must be crosslinked with sulfur or other agents before true elastomeric properties are introduced. In addition, accelerators, antioxidants, fillers, and other materials are added to the polymeric rubber prior to Fibre formation. Rubber fibers exhibit excellent elastic properties but are sensitive to chemical attack, thereby limiting their usefulness.

Structural Properties

Rubber fibers are derived from the sources outlined above and structurally are:

- crosslinked polyisoprene,
- polybutadiene,
- diene-monomer copolymers, or
- amorphous polyolefins.

Less common or special use rubbers include the acrylonitrile-diene copolymer Lastrile and the chloroprene polymer Neo-prene. Structural formuli of typical rubbers follow:

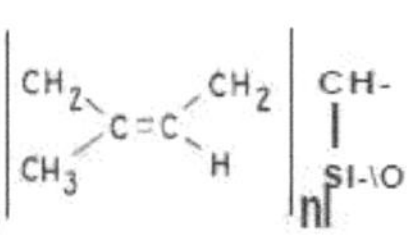

CROSSLINKED cis-POLYISOPRENE (VULCANIZED NATURAL RUBBER)

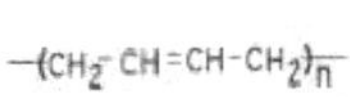

POLYBUTADIENE (BUTYL RUBBER)

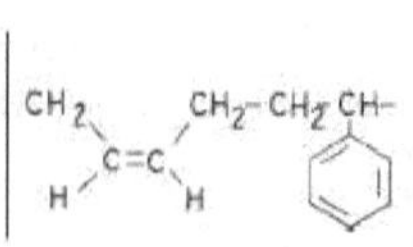

cis-POLYBUTADIENE-STYRENE COPOLYMER RUBBER

The rubber polymers emulsified as a latex are blended with vulcanizing agents, chemical accelerators, antioxidants, fillers, and other materials. The polymer blend is extruded or coagulated and cut into fibers and heated to crosslink the structure. Of necessity, the rubber fibers are large in cross section and have a higher linear density compared to other man-made fibers. The hydrocarbon chains in the crosslinked rubbers are in a highly folded but random and amorphous configuration, but they are attached periodically to adjacent polymer chains through multiple sulfur or other crosslinks. On stretching, the molecular chains in the amorphous region untangle and straighten and orient into parallel, more crystal line structures, up to the elastic limit determined by the crosslinks present. On relaxation, the molecular chains return to their lower-energy amorphous state. Rubber fibers tend to possess a square or round cross section depending on whether the fibers were extruded or cut.

Physical Properties

Rubber fibers possess tenacities of only 0.5-1 g/d wet or dry. Elongation at break for these elastomeric fibers varies between 700per cent and 900per cent, with nearly complete elastic recoveries even at higher elongations. Because of their size, rubber fibers are moderately stiff but resilient. They are light, having a specific gravity of only 0.95-1.1. Certain environmental properties of rubber limit its usefulness. Being hydrocarbon material, rubber fibers do not absorb moisture and have a regain of 0.0per cent under standard conditions. The Fibre is a poor heat conductor and a good heat and electrical insulator. The Fibre is readily swollen by ketones, alcohol, hydrocarbons, and oils and softens if heated above 100°C.

Chemical Properties

Rubber fibers are quite susceptible to attack by chemical s both in solutions and in the environment due to the presence of extensive unsaturation in the molecule. Rubber is readily attacked by concentrated sulfuric and nitric acids and by oxidizing agents such as hydrogen peroxide and sodium hypochlorite at elevated temperatures. Al though biological agents don't readily attack rubber, heating above 100°C or exposure to sunlight accelerates its oxygen-induced decomposition. Rubber fibers are particularly susceptible to attack by atmospheric ozone, a major component of smog.

End-Use Properties

Rubber fibers are manufactured by several companies as rubber Fibre or under trade names such as:

- Buthane,
- Contro,
- Hi-Flex,
- Lactron,
- Lastex, and
- Laton.

Rubber fibers show good elastomeric properties and reasonable aesthetic properties particularly as the core of a textile yarn structure. Rubber fibers usually have a dull luster due to the additives and fillers within the Fibre. The Fibre has poor resistance to household chemicals, sunlight, heat, and atmospheric contaminants. It cannot be dyed readily, and the polymer must be colored prior to Fibre extrusion and curing. The Fibre possesses only fair launderability and poor dry-cleanability. Owing to its low moisture absorbance, it dries readily, but care must be taken to dry or iron the Fibre below 100°C. Rubber fibers burn readily but tend to be self-extinguishing. Rubber fibers are used extensively in applications where elastomeric fibers are desired, such as in swimwear, foundation garments, outerwear, and underwear, hosiery, and surgical bandages. The use of rubber fibers has decreased in recent years as more stable elastomeric fibers such as spandex have been introduced.

SPANDEX

Spandex fibers are elastomeric fibers that are >85per cent segmented polyurethane formed through reaction of a diisocyanate with polyethers or polyesters and subsequent crosslinking of polyurethane units. The spandex fibers resemble rubber in both stretch and recovery properties, but are far superior to rubber in their resistance to sunlight, heat, abrasion, oxidation, oils, and chemicals. They find the widest use of any of the elastomeric fibers.

Structural Properties

Spandex is a complex segmented block polymer requiring a complex series of reactions for formation. Initially, low molecular weight polyethers and polyesters containing reactive terminal hydroxyl and/or carboxyl groups are reacted with diisocyanates by step growth polymerization to form a capped prepolymer. This polymer is melt spun or solvent spun from N,N-dimethylformamide into a Fibre; then the Fibre is passed through a cosolvent containing a reactive solvent such as water that reacts with the terminal isocyanite groups to form urethane crosslinks. Typical structures for spandex with urethane crosslinks follow:

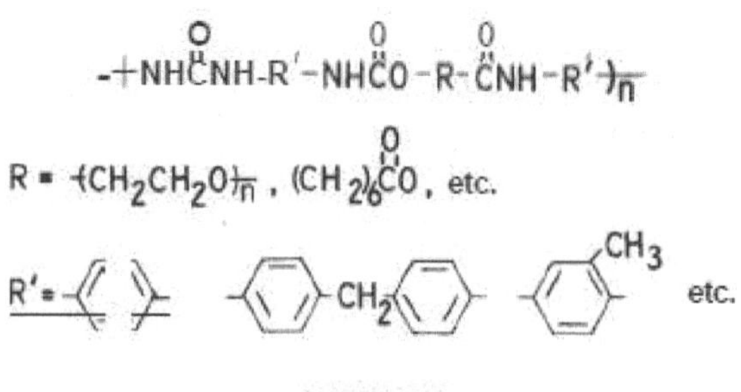

SPANDEX

The polyether or polyester segments in spandex are amorphous and in a state of random disorder, while urethane groups segmenting the polyether or polyester segments can form hydrogen bonds and undergo van der Waals interactions with urethane groups on adjacent chains.

Chain ends will be crosslinked or joined to other chains through urea groups. On stretching, the amorphous segments of the molecular chains become more ordered up to the limit set by the urea linkages. On relaxation, the Fibre returns to its original, less ordered state. Spandex fibers are long, opaque fibers available in many cross sections including round, lobed, or irregular.

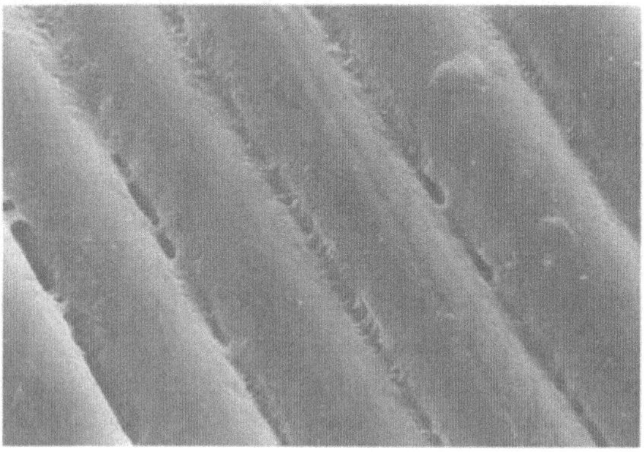

Figure. Spandex.

Physical Properties

Spandex fibers are weak but highly extensible fibers with tenacities of 0.5-2 g/ d dry with slightly weaker wet tenacities, and elongations at break of 400per cent-700per cent. Spandex exhibits nearly complete recovery from high elongations and is quite resilient.

The fibers have specific gravities of 1.2-1.4 and are somewhat more dense than rubber fibers. Owing to the more hydrophilic nature of spandex, its moisture regain varies between 0.3per cent and 1.3per cent under standard conditions.

Although spandex is swollen slowly by aromatic solvents, it is generally unaffected by other solvent s. Spandex is a heat insulating material and shows poor heat conductivity. It softens between 150°C and 200 °C and melts between 230°C and 290° C. It has moderate electrical resistivity and builds up some static charge under dry conditions.

Chemical Properties

Spandex fibers are much more stable to chemical attack than rubber. Spandex is attacked by acids and bases only at higher concentrations and under more extreme reaction conditions. Reducing agents do not attack spandex, but chlorine bleaches such as sodium hypochlorite slowly attack and yellow the

Fibre. Spandex generally resists attack by organic solvents and oils and is unaffected by biological agents. Sunlight causes slow yellowing and deterioration of spandex, and heat causes slow deterioration at elevated temperatures.

End-Use Properties

Spandex is widely used and sold as an elastomeric Fibre under that designation or under such trade names as:

- Cleerspan,
- Glospan,
- Lycra, or
- Numa.

Spandex possesses excellent elastomeric properties and acceptable aesthetics for use in constructions requiring such a Fibre. Spandex is dull, but luster may be improved with brighteners. Although it has low water absorbency, spandex has good resistance to abrasion and wrinkling and may be heat set if desired. It has good resistance to household chemicals other than bleach and is resistant to sunlight or heat-induced oxidation. It can be dyed relatively easily to reasonable degrees of colorfastness.

It stands up well to repeated laundering and dry-cleaning, and it dries readily without damage at moderate temperatures. Spandex burns readily but melts and shrinks away from the flame. Spandex can be used effectively in all elastomeric Fibre applications.

OTHER ELASTOMERIC FIBERS

Anidex

In 1970, anidex fibers were introduced as an elastomeric Fibre by Rohm and Haas with the trade name Anim. Anidex fibers are defined as fibers containing polymers that are at least 50per centof one or more polymerized acrylate esters. Anidex fibers are formed through emulsion copolymerization of acrylate esters with reactive crosslinkable comonomers. The resulting copolymer ernulsion is mixed with a filler and wet spun to form a Fibre which is heated to crosslink the polymer chains and provide the necessary elastomeric properties.

The morphology and elastomeric action of the Fibre resemble spandex and rubber, but anidex generally has a lower elongation at break than these fibers. It has a round cross section. The Fibre has a specific gravity of 1.22 and a moisture regain of 0.5per cent under standard conditions. An index is reported to be more resistant to heat, light, and chemicals than either spandex or rubber. Otherwise, the Fibre possesses end-use properties very much like those of spandex. Anidex fibers apparently did not have sufficient differences in properties to become an economic success and are no longer being produced.

Nylon-Spandex Biconstituent Fibre

An inherent problem with elastomeric fibers is their low strength and limited abrasion resistance. A new biconstituent elastomeric Fibre-Monvelle—has been introduced to answer these problems. The Fibre is a side-by-side bicomponent Fibre containing nylon and spandex. The Fibre is melt spun by special spinnerets to form the Fibre.

On drawing, the nylon portion of the Fibre stretches and becomes more molecularly oriented, whereas the spandex simply elongates as would be expected for an elastomeric Fibre. On relaxation, the spandex portion contracts, whereas the nylon portion does not, and the Fibre coils into a tight coil which will have excellent elastomeric properties and high strength even as the spandex approaches its elastic limit. Sheer fibers may be formed from this bicomponent Fibre, and the Fibre is readily dyeable due to its nylon content. The Fibre is particularly useful in support hose and in pantyhose and other constructions where elasticity and strength are important.

MINERAL AND METALLIC FIBERS

A number of fibers exist that are derived from natural mineral sources or are manufactured from inorganic and mineral salts. These fibers are predominantly derivatives of silica or other metal oxides. In addition, metal fibers are produced. The common feature of these fibers is their inorganic or metallic composition and tendency to be heat resistant and Non-flammable, with the exception of polymer-coated metallic fibers.

GLASS

Fibers spun from glass are completely inorganic in nature and possess unique properties that cannot be found in organic textile fibers. Glass fibers have some deficiencies in properties that severely limit their use in apparel. Glass fibers are used in a number of industrial and aerospace applications and in selected home furnishing uses where heat and environmental stability are of primc importancc.

Structural Properties

Glass fibers are formed from complex mixtures of silicates and borosilicates in the form of mixed sodium, potassium, calcium, magnesium, aluminum, and other salts. The silicate-borosilicate prepolymer is prepared through mixing and fusing the following inorganic salts in the range of concentrations indicated:

- Silica: 50per cent-65per cent
- Calcium oxide: 15per cent-25per cent
- Alumina: 2per cent-18per cent
- Boron oxide: 2per cent-15per cent
- Other oxides: 1per cent-10per cent

The exact composition of the manufactured glass polymer will affect the ultimate properties of the glass Fibre formed. The glass thus formed as small marbles is melted in an electric furnace at high temperatures and melt spun to form fine smooth glass Fibre filaments.

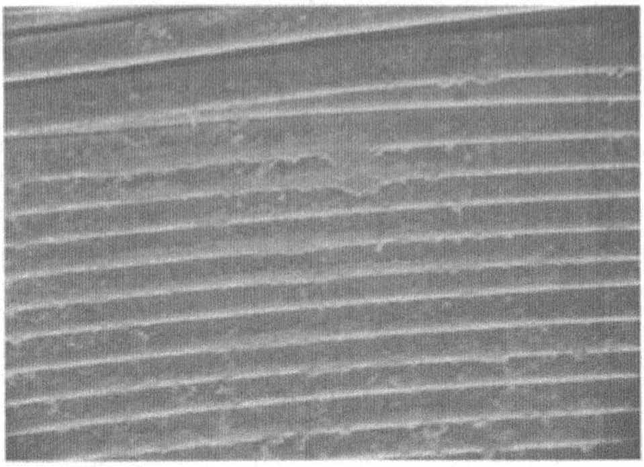

Fig. Glass.

The resultant silicate-borosilicate chains within the Fibre are not ordered, and the glass structure is totally amorphous. Glass is actually a supercooled liquid that exhibits extremely slow but observable flow characteristics with time. The polymer chains within a glass Fibre can be represented as follows:

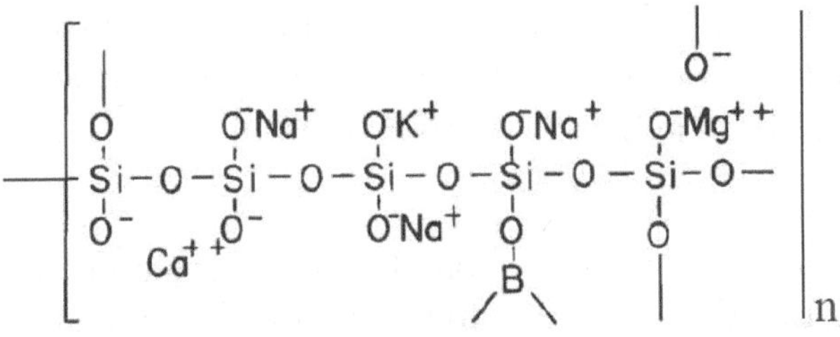

GLASS

Physical Properties

Glass fibers are extremely strong, with tenacities of 6-10 g/d dry and 5-8 g/d wet. They possess elongations at break of only 3per cent or 4per cent but are perfectly elastic within this narrow deformation range up to their breaking point. Glass fibers are quite stiff and brittle and break readily on bending. As a result, they exhibit poor resistance to abrasion, although appropriate organic sizings can alleviate this problem to some extent. Glass fibers are extremely dense, having a specific gravity of about 2.5.

The surface of glass fibers can be wetted by water, but otherwise the fibers have essentially no affinity for water. As a result the moisture regain of these fibers is 0.5per cent or less. Glass fibers are not soluble in common organic solvents but can be slowly dissolved by concentrated aqueous base and more rapidly by hydrofluoric acid. Glass fibers exhibit excellent heat and electrical insulation properties and are not affected by heat up to their melting point of 750°C or above.

Chemical Properties

Glass fibers are chemically inert under all but the most extreme conditions. Glass fibers are attacked and etched by dilute hydrofluoric acid and by concentrated alkalies over long periods of exposure. Oxidizing and reducing agents and biological agents have no effect on glass fibers under normal conditions, and glass fibers are unaffected by sunlight or heat.

End-Use Properties

Glass fibers are manufactured for industrial and consumer use under a number of names including fiberglas, Beta glass, J-M fiberglass, PPG fiberglass, and Vitron. Glass fibers are strong, but they exhibit poor abrasion resistance, which limits their use in textile structures for consumer goods. Heat setting of glass fabrics imparts good wrinkle resistance to fabrics made from these fibers. Glass fabrics have low moisture absorbency and good hand and luster when used in home furnishing applications. The high density of glass provides reasonable draping properties for most applications, although glass fabric constructions are extremely heavy in weight. Its sunlight and heat resistance make glass fibers useful in window coverings. Glass fibers are extremely difficult to dye, and pigmentbinder or pigment-melt coloration must be used. Glass substrate must be laundered or dry-cleaned using minimum mechanical action and drip drying. Glass substrates do not need ironing, although careful ironing up to 400°C is possible. Glass and related inorganic fibers are the only truly Non-flammable man-made fibers. They are used extensively in curtains and draperies, electrical and thermal insulation, tire cord, reinforced plastics, industrial filters, and protective clothing and accessories. Apparel applications are limited to very fine Beta glass fibers.

INORGANIC FIBERS

A series of man-made inorganic fibers other than glass exist that are Non-flammable, heat stable amorphous materials useful in industrial fabric constructions, including refractory materials. These inorganic fibers include pure silica, potassium titanates, aluminum borosilicates, and aluminum oxide-zirconium oxide polymers, Most of these fibers have high strength, are less susceptible to chemical attack than glass, and melt above 1000°C. They may be used in higher-temperature applications than possible with most glass fibers.

ASBESTOS

Asbestos is the name given to several natural minerals which occur in a fibrous crystalline form. The asbestos is initially crushed to open up the Fibre mass, followed by carding and spinning to yield fibers of circular cross section 1-30 cm in length. Asbestos is very resistant to heat and burning, to acids and alkalies, and to other chemicals. Although it has low strength, asbestos Fibre does not deteriorate in normal usage, and it is not attacked by insects or microorganisms. Asbestos is used in fireproof clothing, conveyor belts, brake linings, gaskets, industrial packings, electrical windings, insulations, and soundproofing materials. Inhaled asbestos fibers have been shown to be a serious health hazard, and it has been removed from the textiles market.

METALLIC FIBERS

Metallic fibers are defined as fibers composed of metal, plasticcoated metal, or metal-coated plastic. Single-component metallic fibers for textile usage are fine drawn filaments of metal which can be spun and woven on normal textile machinery. These metallic fibers possess the properties of the metal from which they are formed. Multicomponent metallic fibers are more commonly used in textiles and are usually made from flat aluminum filaments surrounded with or bonded between clear layers of polyester, cellophane, or cellulose ester or from polyester film which has been metallized through vacuum deposition of aluminum and then encapsulated in polyester. In general, the properties of these fibers resemble the properties of the plastic film used to form the multi component Fibre. The fibers are generally weak and easily stretched but can be used for decorative purposes and for applications where electrical conductivity and heat resistance are important. Trade names for metallic fibers include Brunsmet and Lurex.

MISCELLANEOUS FIBERS

In this stage, fibers which do not logically "fit" under other classifications are listed—novaloid, carbon, poly-.!!!-phenylenedibenzimidazole and polyimide fibers. These fibers were developed for specific industrial applications and do not find wide use in consumer goods. For the present, it appears unlikely that new generic Fibre types will be introduced in the marketplace unless they have distinguishing characteristics that make them uniquely suited to particular applications. All of these fibers possess such properties.

NOVALOID

Novaloid is the designation assigned by the FTC for a class of flame retardant fibers made from crosslinked phenol-formaldehyde polymer. The fibers of this class in U.S. production are called Kynol and are manufactured by American Kynol, Inc. The Fibre is golden yellow in Colour and possesses good physical and chemical properties. The Fibre is thought to be formed through spinning phenol-formaldehyde prepolymer, followed by heating the Fibre in formaldehyde vapor to crosslink the structure. The Fibre has low crystallinity with a tenacity of 1.5-2.5 g/d and an elongation at break of about 35per cent. The Fibre has a specific gravity of 1.25 and a moisture regain of 4per cent-8per cent. It is inert to acids and organic solvents but is more susceptible to attack by bases. The Fibre is totally heat resistant up to 150°. This characteristic yellow Fibre is dyeable with disperse and cationic dyes. Novaloids are characteristically used in flame retardant protective clothing and in apparel and home furnishings applications where low Fibre flammability is desired.

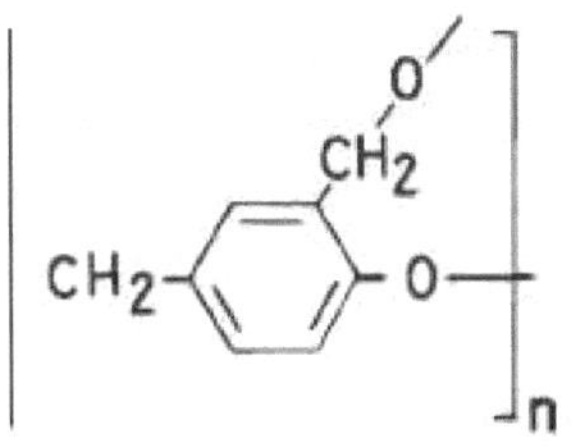

NOVALOID

CARBON

Carbon or graphite fibers have been developed recently for use in industrial and aerospace applications. The carbon fibers are prepared from rayon, acrylic, or pitch fibers by controlled oxidation under tension in limited oxygen atmosphere at 300°-400°C. At this stage of oxidation, carbon fibers have sufficient flexibility to be used in apparel applications such as flame retardant clothing. Further oxidation at temperatures near or exceeding 1000°C under tension results in a high strength Fibre consisting of a continuous condensed aromatic network similar to graphite.

Fully oxidized carbon fibers have tenacities from 10 to 23 g/d dry or wet and elongations at break of just 0.4per cent to 1.5per cent. They have specific gravities from 1.77 to 1.96 and are mildly electrically conducting. Although the fibers are of high strength, they exhibit poor abrasion characteristics and must be sized with epoxy or other resins before formulation into textile substrates. Carbon fibers are inert to all known organic solvents and to attack by acids and bases under normal conditions. These black fibers have excellent resistance to sunlight and biological agents, are inherently flame retardant and are resistant to oxidation at high temperatures. Carbon fibers are extensively used in reinforcing fibers for resins and plastics in high performance Fibre-polymer composites and are marketed as:

- Celion,
- Hi-Tex, and
- Thornel.

POLY-M-PHENYLENEDIBENZIMIDAZOLE (PBI)

PBI was developed by the U.S. Air Force and Celanese as a flame retardant Fibre for use in aerospace applications. The Fibre is spun from N'N. dimethylacetamide followed by derivatization with sulfuric acid to form a golden Fibre. It has moderate strength and good elongation at break, a moderate density, and a high moisture regain. Fabrics from the Fibre possess good hand and drape and are stable to attack by ultraviolet light. PBI does suffer from poor dyeability, however. In addition to its low flammability, the Fibre has a high degree of chemical and oxidation resistance, does not show appreciable heat shrinkage up to 600°C and generates only small quantities of smoke and toxic gases on ignition. In recent years, it has shown potential as a replacement for asbestos, as a flue gas filter material, and as an apparel fabric in specialized applications.

POL YIMIDE

An aromatic polyimide has been introduced by Upjohn Company for use in flame retardant, high-temperature applications. The Fibre is spun from the polymer by wet or dry processing techniques using a polar organic solvent such as N,N-dimethylformamide to give a Fibre with a round or dog-bone cross section. The highly colored Fibre may be crimped at 325°C using Fibre relaxation. The Fibre has a tenacity of 2-3 g/d and a 28per cent-35per cent elongation at break. The Fibre has a moisture regain of 2.0per cent-3.0per cent and melts at about 600°C.

9

Yarn and Textile Substrate Formation

YARN FORMATION

Yarn formation methods were originally developed for spinning of natural fibers including cotton, linen, wool and silk. Since the overall physical characteristics of the fibers and processing factors needed differed from Fibre to Fibre, separate processing systems were developed. As synthetic fibers were introduced, synthetic spinning systems for texturized and untexturized cut staple were developed as modifications of existing staple systems, whereas spinning systems for texturized and untexturized filament were developed separately. Staple yarn formation involves multiple steps and can include:

- Fibre cleaning and opening (as needed for natural fibers);
- Fibre blending (to assure uniform mixing in natural fibers or in Fibre blends);
- carding (to align fibers and to remove short fibers);
- combing (if highly aligned fibers are desired);
- drawing and spinning (to reduced the denier of the yarn, to provide twist and to give cohesion to the yarn); and
- doubling or plying and twisting of the yarns (as needed to provide greater uniformity).

In recent years a number of staple spinning processes other than ring spinning have been developed that reduce or shorten the number of steps necessary for formation of yarns suitable for textile substrate formation and are discussed separately following conventional ring spinning techniques. Yarn preparation from Fibre filaments is much less complex and often no or only limited twist is imparted prior to use in the textile substrate. The steps involved in yarn formation are outlined in figure.

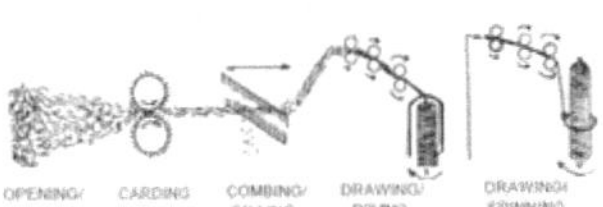

Fig. Spinning

COTTON SYSTEM

When cotton bales arrive at the textile mill, they are highly compressed and have been ginned to remove seeds and some of the impurities present. On removing the cotton from the bales, the cotton is opened up, blended and mixed with cotton from other bales. During the opening and blending process, the fibers are separated and loosened from each other, trash is removed from the fibers, and the fibers are more randomly mixed to assure greater uniformity. Finally the fibers are formed into a thin partially oriented continuous web of intertwined fibers called a picker lap. The picker lap in turn undergoes carding to remove short fibers and remaining trash and to provide additional orientation to the fibers. Carding involves pulling, separating, and orienting the Fibre s by passing the lap between successive cylinders moving at different speeds and containing fine bent wire bristles that catch the fibers. The carded lap is removed by a doffer cylinder in the form of sliver (a rope-like Fibre mass) and coiled into a rotating can. The sliver may undergo additional blending to improve its uniformity and density so that it may be more effectively drawn and spun into yarn. At this point, the sliver is suitable for drawing and spinning into medium and coarse yarns. Before drawing and spinning into fine yarns, the cotton sliver must first be combed to further straighten and orient the fibers and to remove additional short tangled fibers. The drawing and spinning process involves passing the sliver between and through a series of rollers moving at progressively higher speeds to draw the sliver to a finer, more oriented, and uniform structure followed by twisting as the sliver is played onto a turning spindle. The drawing portion of the operation is referred to as the drafting process. The degree of twist will depend on the speed of the turning spindle with each complete turn of the spindle providing a single complete twist in the yarn. In the initial stages where little twist is present in the drawn sliver, the sliver is fed through a tube and onto the spindle. This process is called roving. Subsequent drawing and high speed twisting is carried out by ring spinning in which the drafted, lightly twisted sliver (roving) is fed from the drafting unit onto a high speed spindle via a traveller holding the spun yarn to a ring surrounding the reciprocating spindle. The traveller can move easily around the ring and provides a slight drag on the yarn as it is fed onto the spindle. Ring spinning proceeds at 5,000 to 10,000 revolutions per minute.

WOOLEN AND WORSTED SYSTEMS

Wool is spun into yarns by either the woolen or worsted system. The woolen spinning system is less complex than the worsted system and utilizes shorter wools of a wider range of lengths and diameters to give a low twist bulky yarn. The worsted system produces highly twisted fine yarns utilizing fine fibers of a narrow distribution of length and size. Fabrics made from woolen yarns tend to be bulky and contain more entrapped air; whereas fabrics from worsted yarns tend to be tightly woven and fine with a hard flat surface. On arrival at the mill as bales, raw wool contains large amounts of grease, swint (salts from the body of the sheep), dirt, vegetable material and other impurities. The wool must be washed in successive baths of detergent solution to remove these impurities. The process is called scouring, and the weight of the raw wool can be reduced by as much as 50per cent by the scouring process. Wool grease (lanolin) is effectively recovered from the scouring liquor as a commercial product. Vegetable matter remaining in the wool can be removed by passing the wool through concentrated sulfuric acid to chemically destroy the cellulosic matter, a process called carborization. After washing and drying, the cleaned wool is blended and carded. At this point the sliver can be drawn and slightly twisted to form a roving which can then be spun into a woolen yarn. The sliver must undergo additional straightening, orientation, and removal of short fibers to be used in the worsted system. This process involves several successive steps including gilling (a form of pin orientation) and combing to give wool top. The wool top is drawn and slightly twisted in several stages to form a roving which is finally spun into a highly twisted worsted yarn.

OTHER STAPLE SYSTEMS

Other natural staple and synthetic fibers can be spun on cotton and wool systems if these systems are modified to take into account unique factors such as Fibre length, crimp and linear density. Cut staple manmade fibers arrive in boxes at the mill and are ready for carding and processing into yarn. When two or more different staple fibers are mixed, it is critical to provide extensive blending before carding and repeated doubling of the sliver to assure intimate blending prior to roving and spinning. Other spinning methods that take fewer steps have been developed for staple spinning and are discussed under other yarn forming methods.

FILAMENT SYSTEMS

Filament spinning systems are much less complex because the fibers are continuous and do not need to be highly twisted to give a cohesive strong yarn. Filament yarn spinning usually involves man-made fibers and only the portion of the ring spinning system that involves twisting and winding onto spindles is used.

THER YARN-FORMING SYSTEMS

Since conventional ring spinning systems take several steps to reduce fibers to a spun yarn, several alternative techniques have been developed that reduce the number of these steps, Most of these processes have more limited usefulness with regard to the range of linear densities and type and distribution of Fibre lengths that can be processed by these systems,

Open-End or Break Spinning

Open-end or break spinning systems have come into wide use for spinning of short and medium staple fibers directly from sliver to yarn in one step without carding or roving. Sliver is fed into a breaking or opening unit to separate the fibers from one another, and the fibers are then forced by air pressure into a hollow rotor rotating at high speeds (up to 50,000 rpm).

The fibers are deposited by centrifugal force on the sides of the hollow rotor, and the resulting yarn is removed continuously by a stationary tube mounted with in the rotor. The rotating rotor provides twist to the yarn and produces a yarn with somewhat greater higher pitch and bulk and somewhat lower strength than ring spun yarns.

Open-end spinning operates at a rate up to five times that of ring spinning and can be effectively used for cotton, polyester-cotton blends, as well as other short and medium staple systems. Synthetic staple fibers such as polyester alone can not be effectively open end spun due to dusting of oligomer from the fibers that interferes with the spinning action of the rotor.

Friction Spinning

Friction spinning is a variation of the open-end spinning system and often referred to as the DREF system. In this system, one or more slivers are fed onto a rapidly rotating card drum which opens the sliver to form single fibers. The separated single fibers are blown from the card drum by a stream of air onto the junction of two parallel perforated drums turning in the same direction. The rotating perforated drums under suction cause the fibers to be compressed and twist around one another to form a uniform yarn which is continuously removed onto a spool. The resultant yarns are bulky and have properties similar to woolen yarns.

Air-Vortex Spinning

Air-vortex spinning is similar to open-end spinning but utilizes a stationary tube rather than a rotor. A high speed air vortex is created in the tube that deposits fibers with in the tube and simultaneously provides twist due to the vortex. The yarn is continuously withdrawn as in the case of open end spinning.

Fasciated Spinning

In fasciated spinning long staple sliver without twist is introduced into a limited space and subjected to a torque jet operating at right angles to the flow of the sliver thereby imparting a false twist to the sliver. As the sliver exits the torque jet it rapidly untwists, and the outer fibers tend to break away from the sliver and wrap around the inner sliver to give a strong yarn consisting of mostly parallel fibers with some fibers tightly twisted around the outside.

Self-Twist Spinning

In self-twist spinning, two parallel slivers are fed between two reciprocating rollers which form identical left hand or right hand twists in each of the slivers alternating down the length of the sliver. The resulting false-twisted yarns are then brought together so that the right hand twist segment of one sliver is phased with the left hand twist of the other sliver. On relaxation, the slivers untwist over one another to form a stable yarn. The process was originally developed for spinning wool, but has been used extensively for acrylic yarns. In the related Selfil spinning method the self-twist yarn is wrapped by alternating phases of continuous filaments to form a highly stable wrapped yarn containing less than 10per cent filament.

Covers pun Spinning

Covers pun is both the name of a spinning system and a yarn. In this system staple rovings are drafted in a conventional manner, and then the roving is passed into a hollow vertical spindle. On the outside of the spindle a filament yarn, from a cylindrical spool rotating at 20,000 to 30,000 rpm, is fed into the top of the hollow vertical spindle with the roving. The rotating filament spindle causes the filament to wrap around the core of the staple to produce a wrapped yarn consisting of 80per cent to 95per cent staple. Polyester filaments are usually used to wrap pure yarns or blends of cotton, wool, nylon, and acrylic staple to form sewing threads or yarns for textile substrate production.

Integrated Composite Yarn Spinning

A composite yarn is formed by melt extruding fibers from a spinneret or by coating filaments with a molten polymer followed by coating the emerging fibers with short staple fibers. The resulting matrix is immediately twisted causing the staple fibers to imbed in the extruded fibers before cooling. The resulting composite yarn is formed at a very rapid rate on the Bobtex spinning apparatus developed for this process.

Twistl ess Systems

In twistless systems liquid or powdered polymer adhesives are applied to sliver or filament tow and the adhesive activated by heating or steam to cause the individual fibers to adhere to one another. In some systems, after textile substrate formation from the twistless yarns the adhesive is removed to improve the aesthetics of the resulting textile. Adhesives used include polyvinyl acetate, polyvinyl alcohol and starch.

TEXTILE SUBSTRATE FORMATION

PREPARATION

Yarns often must undergo additional processing before they are ready for use in forming of a textile substrate. The yarns may need to be rewound onto appropriate packages, reinforced by application of size, lubricated by application of spinning oil, and/or drawn-in and tied into the machine used in fabric forming. The combined process used for size application and lubrication of warp yarns in weaving and warp-knitting is called slashing.

Winding

Winding processes involve movement of yarn from one package to another and often conversion of the overall size, shape and tightness of the packages. These processes also serve other important functions. Winding allows clearing of the yarn to eliminate thin spots, thick spots, knots, and other imperfections, and makes it possible to regulate tension within the package, combine or segment yarn packages, and prepare packages for dyeing prior to substrate formation. In shuttle weaving, it is necessary to prepare small packages referred to as quills or pirns that fit within the shuttle. The yarn is wound onto the pirn sequentially in such a way to assure steady and even release of yarn from the pirn during the weaving process.

Warping and Slashing

A specialized type of package formation is involved in preparing warp beams for weaving or warp knitting. A high degree of tension is placed on the warp during these processes, therefore the yarns must be lubricated to minimize friction between yarn and machine parts and adhesive must be applied to the yarn to strengthen and reduce the hairiness of the yarn. Warping involves winding yarns from several thousand packages pl aced on creels onto a flanged beam passing through a reed (a comb-like device). The reed maintains the yarns parallel to one another as they are wrapped onto the beam under as even a tension as possible. Warping of small sections of warp (tape warping) is also often carried out, and the tape warps are later placed parallel to one another to provide a full width warp for use in the loom or warp knitting machine. Staple yarns and some filament yarns must undergo slashing. Slashing involves simultaneous application of sizing and lubricant to the warp from one bath called a size box, followed by drying to remove water or solvent, breaking the slashed warp yarns away from one another using least rods, and rewinding of the warp. Sizes and lubricants used on warps will vary with Fibre type. Sizes used include starches and gums, cellulose derivatives such as carboxymethyl cellulose, proteins, polyvinyl alcohol, polyvinyl acetate and acrylic copolymers, while the lubricants used are similar to spinning oils and include mineral and vegetable oils and waxes as well as derivatives of these materials.

Drawing-In and Tying-In

After warp beams are prepared, the warp yarns must be drawn through certain elements in the loom or warp knitting machine before fabric can be produced. This process in the past was carried out by hand using a special hooked wire to draw each yarn through the elements of the loom or wrap knitter followed by hand-knotting of the yarn to the corresponding yarn on the take-up warp beam. Machines are now primarily used to perform the function of drawing-in and tying-in at a high rate. Similar drawing-in of yarns for fill knitting and tufting is also necessary, but the process is not as complex as drawing-in a warp beam. In Non-woven formation, a sliver or a random or plied Fibre web is used.

TEXTILE SUBSTRATE FORMATION

Textile substrates are formed from yarns or Fibre webs by several techniques including weaving, knitting, tufting, and Non-woven formation. In addition, composites of textile substrates are formed by methods such as adhesive bonding, formation of back coatings on fabric substrates, and flocking. Weaving involves interlacing two sets of yarns usually at right angles to one another using a 100m. The warp yarns are fed into the loom and filling (weft) yarns inserted into the warp using a shuttle or an alternative insertion technique. Knitting involves interconnecting yarns by looping them around one another. In warp knitting the yarns in a warp beam are looped over adjacent yarns in a zig-zag repeating pattern to form a fabric, while in fill (weft) knitting a fill yarn is formed into a series of loops that are passed through the loops previously formed in the fill direction. In tufting, yarns threaded through needles are punched through a backing fabric and the loops thus formed are held in place as the needles are withdrawn from the backing followed by formation of the next tuft in the same manner. In non-woven formation, a Fibre web or yarns are entangled or bonded to adjacent fibers through use of mechanical or chemical bonding techniques to make a continuous interconnected web. Composites of textile substrates are formed by bonding two fabrics together by use of an adhesive to form a bonded substrate or backed substrate or by application of cut fibers to an adhesive-coated substrate to form a flocked substrate.

WEAVING

Weaving has been traditionally conducted on looms using a shuttle carrying a package (pirn) containing fill (weft) yarn which inserts the fill yarn into the warp which ha s been drawn-in and tied-in to the 100m. In recent years, many shuttle looms have been replaced with shuttleless systems particularly for simple fabric constructions. The shuttle 100m continues to be the most versatile weaving machine capable of weaving the widest range of yarns into fabric. The basic components of a loom are presented in figure. The 100m functions in the following manner. The warp beam is connected to a let-off mechanism that meters the warp yarns off the beam as fill insertion proceeds. Each yarn in the warp passes through metal warp stop mechanisms that can detect broken warp yarns, through the eyes of heddles that are contained in the various harnesses used to lift the warp yarns, through the reed used to beat-up the filling yarn, and finally onto the take-up beam.

Each warp yarn passes only through the eyes of heddles within harnesses that will be used to raise that particular warp yarn. The more harnesses that are used in raising and lowering the yarns, the more complex the weave that is possible. The pattern of the warp is determined by which of the harnesses each of the warp yarns is passed. The actual raising and lowering of the harnesses within the 100m is referred to as shedding, and the space between the separated warp yarns is called the shed. The harnesses are raised and lowered by use of cams or a dobby attachment or can also be raised and lowered individually by use of a Jacquard mechanism.

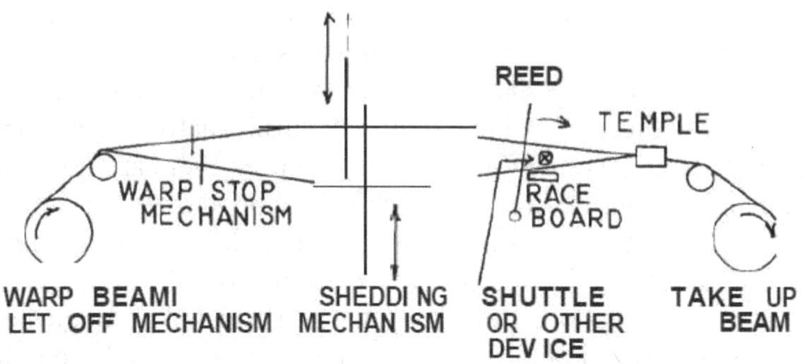

Fig. Basic components of a 100m

The fill yarn is inserted at right angles to and through the shed by use of a shuttle or an alternative insertion mechanism by a process called picking. Below the insertion mechanism is the race board that helps provide support for the shed and the fill insertion area. The reed is placed between the open shed and the fill insertion area of the shed. After the fill yarn has been inserted, the reed is used to push the fill yarn tightly into place in the fabric by a process called beat-up.

The temple provides uniform tension on the formed fabric to prevent loosening of the inserted fill yarns, and a take-up mechanism keeps proper tension on the fabric as it is formed and taken up. Therefore, the basic repeating sequence of actions in the loom is shedding to open the warp, picking to insert the fill yarn between the separated warp, and beat-up to push the fill tightly into place. The sequence must be carefully timed and synchronized to assure proper insertion. The ratio of fill insertion in shuttle loom varies from 75 to 300 picks per minute, while shuttle-less insertion systems operate at 150 to 600 picks per minute. Fabrics woven from staple and filament yarns are found in figures, respectively.

Fig. Woven staple fabric

Fig. Woven filament fabric

Shedding Mechanisms

The cam system is most limiting in the complexity of weaves possible since only 8 to 10 harnesses can be effectively raised and lowered by this method. The cams are genera 11 y positioned be low the harnesses and raise and lower the harnesses by use of mechanical tappets. For more complex weaves the dobby mechanism or the Jacquard system must be used. The dobby mechanism is actually mounted on the side of the loom and is capable of raising as many as 20 to 28 harnesses. The dobby mechanism uses a slotted drum and continuous pattern chain containing patterns of pegs on each bar of the pattern chain to lift selected harnesses during shedding. As a bar of the pattern chain is presented to the slotted drum, the pegs present on that bar enter and block the corresponding slots in the drum. In turn, this prevents hooks attached to the harnesses from entering the slots, engaging, and raising that harness. Where the slots are not occupied by pegs, the hooks attached to individual harnesses enter the slots, engage, and that harness is raised. When the next bar on the pattern chain is presented to the slotted drum, a new series of harnesses are raised depending on the pattern of pegs on that bar. In the Jacquard system, each warp yarn is attached by an individual heddle to a draw string mounted above the loom and may be raised and lowered independently. The draw strings from the warp yarns pass parallel to each other through separate holes in a perforated board. A series of horizontal needles mounted above the perforated board are pressed against a continuous series of cards having a pattern of perforations. Each cord is in turn attached to a rod that passes through the eye of one of these needles. The upper ends of these rods are hooked and rest over a series of bars called griffes. Where the card has a perforation the needle passes through the perforation, the hook for that warp remains on the bar, and as the bar is raised the warp yarn is lifted. Where no perforation is present, the hook is disengaged and the warp yarn is not lifted. By this method extremely complex and intricate patterns can be developed. Due to the complexity of the mechanism, the weaving rate is much slower with Jacquard than with dobby or cam mechanisms.

Fill Insertion

Until recent years fill (weft) insertion was carried out by traverse of a shuttle containing a package of fill (pirn) within the shed back and forth a cross the width of the warp. As a result, fabrics produced by shuttle weaving have a selvedge (edge) in which the fill turns in a U at the edge of the fabric to return as the next row of fill in the fabric. Different filling yarns can be inserted by use of multiple shuttles in a magazine arrangement. Because movement of the shuttle back and forth across the fabric is necessary in a shuttle 100m, a mechanism for projection (picking) and checking of the shuttle at both sides of the loom is required. The picking stick strikes the shuttle to provide the force necessary to accelerate the shuttle to sufficient velocity to rapidly travel across the width of the loom (picking). As the shuttle nears the other side of the loom, damping mechanisms slow or decelerate and ultimately stop the shuttle (checking), so that it is ready for rapid return across the loom width. The picking and checking action of a conventional loom requires large amounts of energy, and a high degree of vibration is inherent in shuttle systems. In order to minimize energy consumption and machine vibration and to increas8 the rate of fill insertion, a number of shuttle-less systems of fill insertion have been developed. Although these systems are not as versatile as shuttle fill insertion, the improved efficiency and reduced noise levels of these insertion methods make them quite suitable for weaving of less complex weaves. Ultimately, shuttle-less systems are expected to be used in production of 80per centof all woven fabric. The shuttle-less systems can be divided into two major categories, mechanical systems and fluid systems. Mechanical systems include the use of grippers or rapiers (single and double), while fluid systems use an air jet or water jet. The major fill insert methods are presented in figure.

SHUTTLE

Fig. Fill insertion methods in weaving

The shuttleless systems all insert individual pre-measured lengths of fill yarn. Therefore a fill package does not have to be carried across the shed thereby greatly reducing the energy required for fill insertion. The pre-measured fill is generally only introduced from one side of the 100m, and a traditional stable selvedge with the fill yarn turning back on itself is not produced. The edge of the fabric produced by the shuttleless systems is normally fringed. However, the fringed selvedge can be reinforced by use of a higher density of warp yarns at the selvedge or by attachments producing tucking-in of the selvedge or a leno selvedge. In shuttle less looms, the fill yarn is usually premeasured and cut and held in place by a vacuum tube or on a storage drum prior to insertion. The gripper system is most closely related to the shuttle system. The gripper is fired as a projectile across the width of the 100m carrying a single length of fill. The gripper is much smaller and lighter than a shuttle, since it does not need to carry a fill package. Multiple grippers are used, and a gripper after insertion of fill yarn is returned by a conveyor system back across the 100m. A multiphase gripper system is also used in which a series of grippers each carrying fill yarns are conveyed across that loom in sequence by use of a magnetic or mechanical drive mechanism.

This method must use phased shedding and beat-up motions to permit simultaneous movement of several grippers across the face of the 100m. Although the velocity of the multiple grippers across the loom is much slower than use of a single gripper, the composite rate of fill insertion is much faster than more conventional looms. In the rapier systems, the fill yarn is carried across the warp by a single or two mechanical arms. The rapiers must be removed from the shed prior to beat-up. In the single rapier system the end of the arm contains a clip to hold the fill yarn that releases the yarn after the fill yarn is completely inserted. In the double rapier system, one arm equipped with a clip (giver) conveys the yarn to the middle of the shed, and the taker on the other arm simultaneously is inserted from the other side and takes the yarn across the rest of the shed. Since rigid rapiers effectively double the width of the loom, flexible rapiers that uncoil on fill insertion have been developed that reduce the loom width. Fluid fill insertion systems do not use a device to carry the fill yarn across the loom.

They operate by impinging sonic velocity water or air jets onto the end of the yarn which accelerates and carries the yarn across the loom. Since liquid water is more cohesive than air and the energy conferred to the water is not as readily dissipated as the energy conferred to air, the water jet is capable of conveying the fill yarn greater distances than an air jet. The major disadvantage of water jet fill insertion methods is related to the hydrophilic character of the water and its ability to dissolve many sizes and to wet out hydrophilic fibers. Therefore, water jets can only be effectively used on hydrophobic fibers such as polyester that are unsized filaments or that contain sizings unaffected by water. To enhance the projection distance of air jet systems, guides are mounted across the loom that are inserted through the warp during fill insertion to provide a turbulence-free path across the loom. Also booster jets are often mounted periodically across the loom and fired sequentially as the fill yarn is inserted to assist in carrying the yarn across the width of the loom.

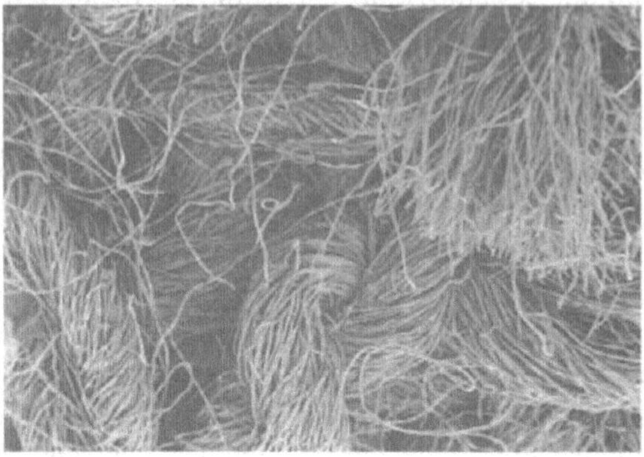

Fig. Loop pile woven staple fabric

Special Weaving Methods

When pile type fabrics are produced on a loom an additional warp is necessary. The tension on one of the warps is released before beat-up permitting warp loops to be formed. Also, two warps can be used to weave a double fabric with yarns connecting the two fabric sections. When these yarns are cut, two cut pile fabric pieces are then formed. A looped pile and sheared pile-woven fabric are found in figures. Carpets can be woven on looms using more than one warp and a complex fill insertion system. The loops formed in the warp direction are held in place by wires inserted through them from the sides of the loom until the loop is firmly locked in place by the weaving of the backing and then removed. Carpet looms are very complex and carpet formation is very slow. Therefore, most carpeting is produced by tufting.

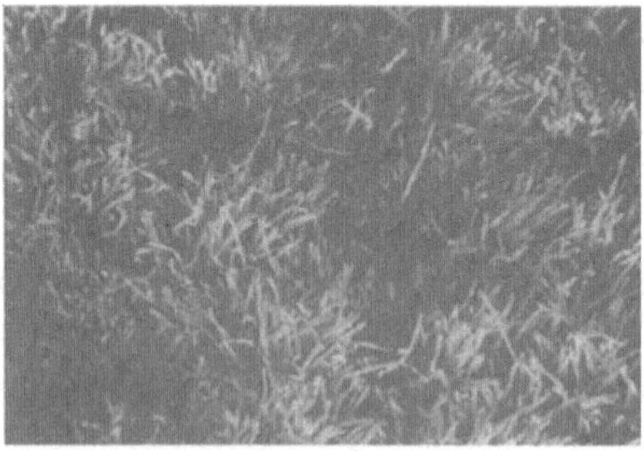

Fig. Sheared pile woven staple fabric

Very stable fabric structures can be produced by bringing two warp yarns into the loom at 60° angles to one another and inserting fill in the normal fashion to form a triaxial fabric.

KNITTING

Knitted fabrics are formed by use of hooked needles to interconnect loops of yarn to form a pattern. The needles used are one of three types, bearded, latch, or compound. The hooked needles pull the yarn through a previously formed loop and must close as they pass through the loop and use different closing mechanisms depending on needle type. The bearded needle uses a presser bar to close the needle, while the latch on a latch needle is closed by the loop of yarn it passes through. Compound needles are opened and closed by a Programmemed mechanism such as a cam mounted near the base (butt) of the needle. These types of needles are used for both warp and fill knitting. In warp and fill knitted structures the row of loops running horizontally across the fabric are referred to as courses, whereas the rows of loops running vertically up the fabric are called wales. Examples of a warp and fill knit are found in figures.

Fig. Warp knit

Fig. Fill knit.

Warp Knitting

Warp knitting is the knitting method most related to weaving. Warp knitting uses a warp beam prepared in the same way as for weaving. The warp is fed into the warp knitting machine, and finished warp fabric is taken up in much the same fashion found on a loom. The actual mechanism of fabric formation is quite different, however. The basic elements and motions of one element of a warp knitting machine are found in figure.

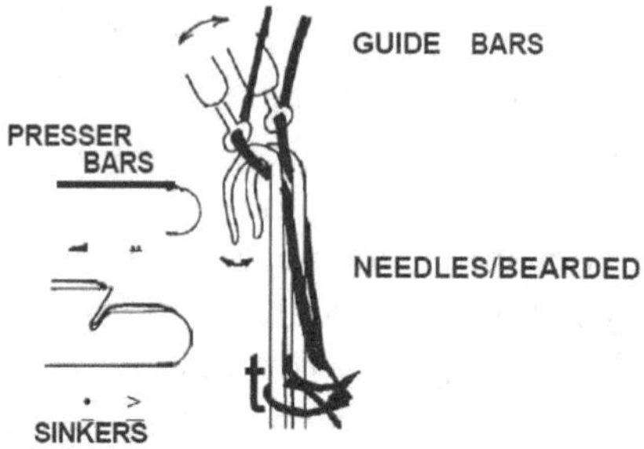

Fig. Basic elemen of a wap kintting machine.

In a warp knitting machine, there are one or more warp yarns per needle in the machine and each warp yarn is attached to and passes through the eye of an individual guide bar that is used to position the yarn for knitting and to move the yarn from needle to needle to give the characteristic zig-zag pattern of loops running vertically up the warp knitted fabric. The needles are bearded, latched or compound and individually mounted vertically in a horizontal row attached to a straight bar.

The guide bars are mounted in front of the horizontal row of needles and move back and forth in concert in predetermined patterns to present the yarns to the appropriate needles for knitting. The more warp yarns per available needle, the more complex a knitted structure that is possible. In forming fabric the guide bars wrap the warp yarns across the faces of the individual needles. The needles then close by the mechanism appropriate to that type needle and the yarns are drawn down through the loops formed in the previous cycle. The needles then move back up into their original position and open to receive the next yarn or yarns. Sinkers are used to hold the loops taut and in place, while the guide bars introduce the yarns to the needles. In general, the simplest warp knitting machines use bearded needles and limited guide bars per needle to form tricot fabrics.

In these machines a presser bar is mounted parallel to the needle bar to close the needles as the warp yarns are drawn through the set of loops formed in the previous cycle. The more complex Raschel, Milanese and Simplex machines are capable of more extensive patterning and use latched or compound needles. The production speed and efficiency of warp knitting machines is very high and represents the fastest means of converting filament yarn into fabric. Warp knitted fabrics can be stabilized by introduction of warp and/or fill yarns into the warp knit structure during knitting. Pile or terry fabrics are produced by introduction of tow into the knitted structure followed by cutting to form a pile or by use of multiple warp yarns to introduce terry loops, respectively.

Fill (Weft) Knitting

Fill (weft) knitting involves insertion of a filling yarn into the knitted structure by drawing the fill yarn as a successive row of loops through the previously knitted loop structure to form a new course of loops running across the knitted structure. The yarn may be fed back and forth across the knitted fabric to form a flat fabric or successive yarns may be laid in the same direction spaced behind each other to form a tubular fabric. The basic elements of a segment of a fill knitting machine are presented in figure.

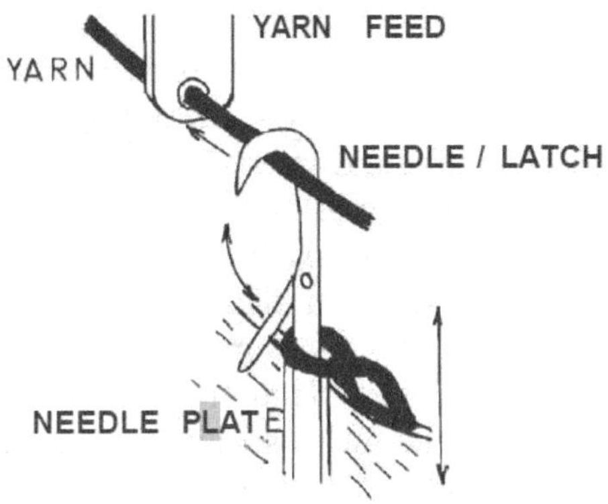

Fig. Basic elements of a fill knitting machine.

In fill knitting the needles operate in a sequenced sine-shaped wave. After the needle takes the yarn, it starts down through the previously made loop and the needle closes by the process appropriate to the needle type. After the need 1e with yarn passes completely through the previous loop, the needle opens, releases the newly made loop and moves upward through the loop to begin the process again.

It is possible to Programme individual needles within the fill knitter to go through the full knitting cycle, or go through a partial knitting cycle in which the knitting needle does not fully clear the previous loop (a tuck stitch), or does not go through the knitting cycle at all (float or miss stitch). Such Programmeming is carried out by use of cams or pattern wheels that interact with patterns of tabs or indentations on the butt of the knitting needle. More complex programming of patterns is possible when multiple needle systems are used in conjuction with slot, pin and blade, multiple disk, or punched card and tape mechanisms referred to as Jacquard mechanisms. Multiple needle systems are mounted in parallel rows 60 to go out of plane to one another. Such multiple systems also can be used without complex Jacquard programming to simultaneously form double sets of loops in knitting to form ribbed and interlock knitted fabrics depending on the gauging of the needles in relation to each other. Fashioning or shaping of the knitted tube formed is al so possible by programming of the number and which individual needles are used within each row of courses formed or through movement of loops from one needle to another before the next row of knitted loops are formed. As in the case of warp knitting, fill knitting can be used to form tufted or pile fabrics or additional warp or fill yarns can be laid into the structure to provide additional stability.

TUFTING AND PILE FORMATION

Tufting is a rapid process for formation of continuous rows of yarn loops across the face of a textile backing substrate. The yarn loops on the tufted substrate then can be cut or sheared to form a cut pile if desired. Woven and knitted pile substrates can be formed by introduction of additional yarns into woven and knitted structures. Because of its high efficiency, today 95per centof carpets are formed by the tufting process. In tufting, a woven or film primary backing material is passed under a row of tufting needles each carrying a yarn. The needles penetrate the backing to form a loop which is held in place as the needle is withdrawn through the backing. Subsequent loops are made in the same manner to form continuous rows of tufted loops on the substrate. After the loops are formed all or some of the tufted loops may be cut by special cutting devices. After the tufted yarn has been inserted into the backing, a layer of adhesive must be applied to the back side of the tufted structure to mechanically fix the tufted yarns in place. Often a secondary backing in the form of a textile substrate is placed and fixed by adhesive over the back side of the yarns or a rubber coating or a rubber or polyurethane foam backing is applied to the back of the tufted substrate. The basic elements of the tufting process are found in figure.

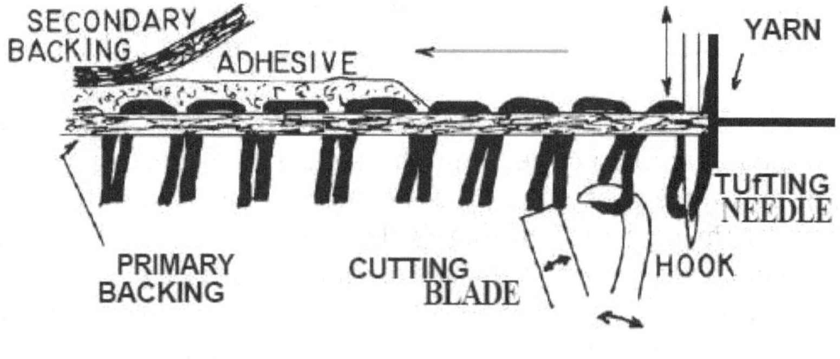

Fig. Basic elements of a tufting machine.

The primary and secondary backings used for carpets until recently were woven jute; however, polyolefin and polyester films and coarse woven polyolefin slit film filaments have been increasingly used as backing materials. Nylon yarns account for over 70per cent of the yarns used to form the tufted face of the substrate, with polyester, polypropylene, acrylic, modacrylic, and wool yarns being used to lesser extents. Nylon dominates the tufted carpet market due to its overal1 toughness and resiliency. Tufting leads to a wide range of looped pile and cut pile carpet substrates with the number of tufts per unit area, the length of loops or pile, and the Fibre type affecting the nature and performance of the carpet. The tufting process utilizes a large quantity of yarn that is not seen in the finished product, namely the segments of yarn on the backside of the primary backing. Several alternative processes have been developed to minimize this problem, although tufting remains the primary method used. In the alternative processes either premeasured lengths of yarn are forced through a tightly woven fabric backing and then fixed with adhesive or adhesive is applied to the face of the backing substrate and pile is fixed into the adhesive layer on the face of the substrate. Such techniques are limited to the production of cut pile substrates. Flocking is a related method for forming a pile or fuzzy surface on a fabric substrate. In flocking, adhesive is applied to a fabric substrate and then cut fibers or yarns are fixed to the adhesive using mechanical or electrostatic methods to orient the pile vertical to the fabric substrate.

NON-WOVEN FORMATION

Non-woven textile substrates can be formed through entanglement and/ or bonding of fibers in the form of webs or yarns by various chemical and mechanical means. Such methods used to form Non-wovens minimize greatly the number of steps required to go from the Fibre to the finished substrate and greatly reduce the cost of production of the substrate. The strength, flexibility, and utility of the resulting Non-woven substrate is generally less than that of a similar woven or knitted substrate. Therefore, Non-woven fabrics have been used little in apparel other than in interfacing or as felts. but have found extensive use in medical, industrial, and home furnishing applications. Examples of a non-woven felt and an adhesive bonded non-woven are found in figures.

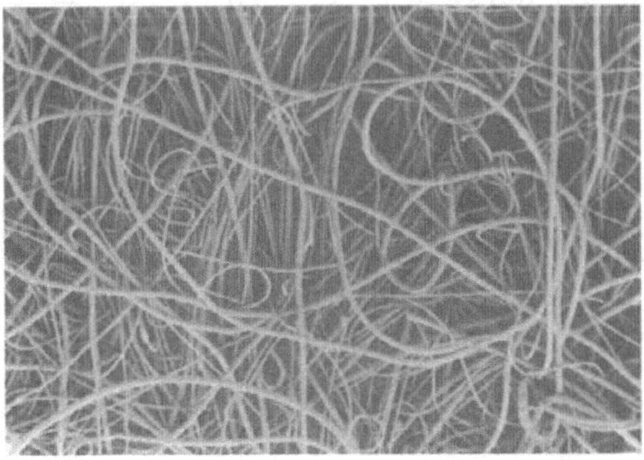

Fig. Non-woven felt.

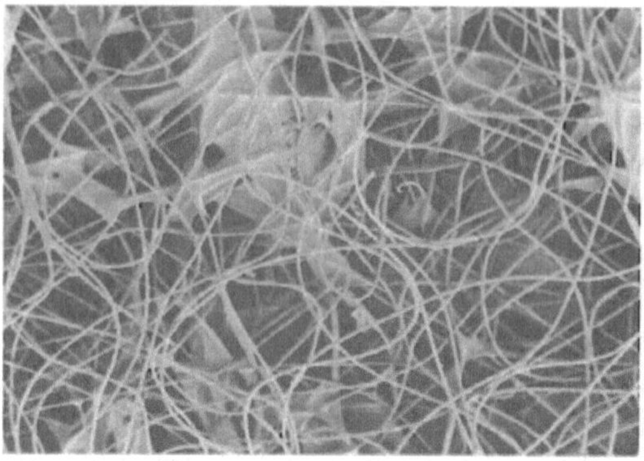

Fig. Non-woven adhesive bonded material

The Fibre web or in some cases films or yarns used in Non-woven formation are prepared by conventional processes. Web formation is carried out by carding, garnetting, air laying, or wet laying of staple fibers. The web can also be formed by direct laying of extruded filaments onto a moving belt or through extrusion of a film followed by slitting or embossing and stretching. The webs are laid in a random or oriented fashion with oriented webs being parallel, plated, or crosslaid, in one or more plies. The webs or in selected cases yarns are entangled or are bonded together by four basic processes:

- mechanical entanglement,
- stitching,
- self-bonding, and
- adhesive bonding.

Mechanical Bonding or Entanglement of Non-wovens

A number of processes exist for mechanically entangling or bonding a web to form a Non-woven substrate. One method that is mechanical in nature does not involve a web but rather splitting of a polyolefin film. In this process a film is extruded followed by either a discontinuous pattern cut by knife slitting or by embossing the film followed by bia xial orientation to form a network structure. Methods that involve mechanical bonding or entangling of Fibre web include wet laying, felting, needl ing, and air or water jet techniques. Wet laying essentially uses the process used for making paper. A slurry of short staple fibers in water or solvent is continuously laid onto a moving wire screen. The water or solvent remaining in the Fibre mass that does not fall through the screen is removed by squeeze rolls and subsequent drying to form a randomly oriented Non-woven substrate. The strength of the wet laid Non-woven can be increased by adding chemicals or binders to the slurry during the process. Felting of a Fibre web involves the entanglement of fibers in such a way that the fibers "ratchet" on one another and make the web more dense. This can be accomplished by use of the natural felting action of scaled keratin fibers such as wool in the presence of moisture and mechanical action or by heat-induced shrinkage of fibers like polyester accompanied by mechanical action. Needle punching of a web involves passing the web between bands of needles which continuously punch through and withdraw from the web. The needles have reverse barbs and cause reorientation of some fibers in the web and complex entanglement of the web in the vertical as well as horizontal direction. Needle punching from both sides of the web will provide a stronger substrate. Water jets or air jets can also be used to entangle a web. The Fibre web is passed under a series of high velocity water jets or air jets that are programmed to deliver short bursts of water or air at intervals. The jets disrupt the web and form bonds through mechanical entanglement of fibers in the web. In general, mechanical bonding and entanglement methods give soft drapable fabrics of low to moderate mechanical strength.

Stitching or Stitch Bonding

Yarns or Fibre webs are bonded together by stitching the webs together by a series of needles that operate very similarly to conventional warp knitting machines. Therefore guide bars must be used as part of the mechanism to interconnect the threads being sewed into the yarns or web being fed into the machine. Stitch-bonded fabric substrates are usually made using filament thread to form a reasonably strong but quite flexible structure. Yarns and webs can also be crosslaid in these stitch-bonded processes. The processes operate at very high operating speeds using inexpensive yarns or webs with fabric substrate being formed at 50 to 200 meters/ hr. Stitch bonding of yarns is usually carried out on Malimo machines, whereas web stitching is carried out on Maliwatt or Arachne machines. Pile stitching machines such as Malipol, Araloop, and Locstitch have also been developed for formation of pile fabrics.

Self Bonding

Techniques have been developed to bond thermoplastic fibers to each other by use of heat and/or solvent. In some cases, low levels of binder are added to the substrate prior to bond formation to assure a higher level of bonding. Spunbonding is the most widely used of these techniques. In spunbonding, a web of continuous filament thermoplastic fibers, such as polyester and polyolefin are extruded and randomly laid onto a moving conveyor belt and then subjected to heat and pressure between embossed rolls to form periodic bonds between fibers in the web. The flexibility of the web formed by spunbonding is determined by both the density of the web and the number of bonding points per unit area. Melding involves use of specially prepared web of sheath/core filaments that can be subsequently bonded by use of heat. The sheath/core fibers have a lower melting polymer substrate making up the sheath than contained in the core. Passing the web between embossed rolls under pressure held at a temperature above the softening point of the sheath polymer, causes the fibers in the web to be bonded at crossover points in the structure. Since the cores of the fibers are unaffected by the treatment, the resulting bonded web is quite strong. A third method, called gel bonding, involves use of a solvent applied to the Fibre web. Solvent is metered onto the Fibre web causing softening and gelation of the Fibre web, and the web is then passed through embossed pressure rollers. The gelled surfaces of the fibers merge at crossover points and on solvent removal by heating form a bond between fibers. The above processes form fabric substrates of moderate stiffness.

Adhesive Bonding

Application of adhesive to the Fibre web followed by pressure and heating to cure the binder and bond the fibers in the web to one another is another extensively used method of Non-woven formation particularly for Non-thermoplastic fibers such as cellulosics. The adhesive is applied as a solution, suspension, or emulsion to the Fibre web by padding, spraying, or printing or by appl ication as a foam. The adhesives used are usually thermosetting or thermoplastic (hot melt) adhesives. After application the web is subjected to heat to drive off solvent (usually water) and to melt and soften a thermoplastic adhesive or to cause reaction or curing of a thermoset adhesive. The adhesive, on curing, tends to concentrate at the Fibre crossover points due to capillary action. Non-woven formation by adhesive bonding tends to form moderate to stiff substrates with the degree of stiffness being dependent on the nature and density of the Fibre web, the type and concentration of adhesive used, and the number of bonds per unit area.

COMPOSITE FORMATION

Flexible composite te xtile substrates are formed by either coating te xtile substrates with a continuous polymer layer or lamination of two or more textile substrates together by use of an adhesive polymer layer. The polymer coating can be applied neat, from solution, as an emulsion, or in the form of a film or thin foam. Inflexible Fibre-polymer composite substrates are formed by imbedding fibers, Fibre webs, Fibre tows, or fabrics in a stiff polymer matrix. Inflexible composites have found extensive use in engineering and aerospace applications particularly where high performance properties are important. Owing to their inflexibility, they lose any properties characteristic of textile substrates. In order to get a good bond between the fabric substrate and the polymer coating in laminated and coated fabrics it is essential to get a good adhesive bond between the textile substrate and the coating or adhesive. This is particularly important since the substrate and polymer may have very different stretch and recovery properties. Adhesive failure under stress is the most probable cause of failure in these materials. In coated fabrics, elastomeric properties are necessary to achieve a serviceable coated substrate. In coated substrates, the polymer coating is metered and spread evenly onto the fabric surface as a controlled viscosity solution, aqueous emulsion, or hot melt polymer followed by drying and curing if needed. A urethane foam can be bonded to a te xtile substrate, by carefully melting the foam surface with a flame and joining the melted foam surface to the texti Ie substrate under pressure followed by cool ing. Laminated fabrics are bonded under similar conditions except that two fabric faces are brought together and bonded by the polymer layer used.

10

Preparation, Dyeing, and Finishing Processes

PREPARATION AND DRYING

Before dyeing and finishing a textile substrate, it must undergo preparation to assure complete wetting, uniform application of dye or finish, and/or to remove Colour or yellowness from the substrate that would interfere with coloration and the visual aesthetics of the textile. Fabric preparation includes wetting out of the fabric, scouring to remove natural waxes or spin finishes on the substrate, desiring to remove sizing from woven and warp knitted fabric, and bleaching to remove Colour from and to brighten the fabric. Natural fibers in the form of top, sliver, or yarn often undergo a degree of preparation prior to substrate formation.

Following fabric preparation and dyeing or finishing, the textile substrate must be dried. Drying processes are currently under intense examination, since they account for nearly two thirds of the energy consumed in textile wet processing.

PREPARATION

In order to conduct the processes associated with preparation, the textile substrate must effectively be wet out. Surfactants or surface active agents are used to lower the surface tension of water in relation to the Fibre substrate so the water will wet the substrate. Surfactants are organic molecules that effectively bridge the transition between water and the more hydrophilic Fibre surface through combination of a polar head and a hydrocarbon tail within a single molecule.

Surfactants are also necessary for scouring processes in which surface waxes or spinning oils must be lifted from the Fibre surface and suspended as micelles of oil surrounded by surfactant, followed by rinsing. Scouring processes are necessary to ensure complete removal of oil or wax from textiles so that dyes and finishes can uniformly penetrate the Fibre and fabric structure. Low concentrations of sodium hydroxide in addition to detergent are used to scour cotton fabrics to assist in removal of their natural wax coating. Sometimes additional scouring of wool substrates may be necessary to remove excess spin finishes or residual lanolin prior to dyeing and/or finishing. Use of solvent-based scouring techniques for synthetic Fibre substrates has become more popular in recent years. Sizing applied to warp yarns prior to weaving or warp knitting by the slashing process must be removed prior to dyeing or finishing assuring even and uniform application of the dye or finish. Sizes are not as readily removed as spin finishes and spinning oils, so special techniques must be used. Sizes can consist of starches, modified starches, and adhesives based on synthetic organic polymers. Starches and modified starches generally may be removed by dilute acid or enzyme treatment, whereas synthetic adhesive sizes can be removed by specialized short washing treatments. Many natural fibers and some synthetic fibers possess a degree of yellow coloration and must be bleached to brighten and remove yellow coloration from the textile substrate prior to dyeing and/or finishing.

DRYING

After textile processes including sizing and desiring, scouring, dyeing, and finishing, the textile Fibre substrate must be dried before further processing can take place or before the substrate can be delivered as the final product. Wet textile substrates can be considered an irregular but somewhat ordered array of interlocking fibers making up a wet porous fibrous structure. Water within the structure lies between individual fibers being held by capillary action or by cohesive interactions between water molecules. Moisture is also present within the Fibre depending on the moisture regain characteristics of that Fibre. In textile processes water not firmly held by capillary action or water within the Fibre is removed by passing the substrate between squeeze rolls or by passing it over a narrow slit from which sonic velocity steam blows water from the fabric. Such processes can reduce the water uptake of the fabric to approximately 30 to 120per cent wet pickup with sonic steam giving the lowest values. The water remaining then must be removed by convection, conduction, or radiation heating. Drying of textile substrates is carried out most often by convection heating. The textile fabric under tension is passed into a convection oven heated to a temperature well above the boiling point of water. Hot air passing over the wet textile evaporates water from the fabric surface, and internal water moves by capillary and related actions to the fabric surface and in turn is evaporated. The resulting moist air is continuously removed as dry heated air is introduced into the oven. The tension devices holding the edge of the fabric remove wrinkles and introduce a uniform width to and tension on the fabric. These drying units are referred to as stenter or tenter frame dryers. Conduction heating finds more limited use in drying of textiles and is used more in high temperature curing of finishes on textile substrates. In conduction heating, the textile fabric is brought in contact with and passed through a set of heated rolls. Since the fabric is in direct contact with the heated metal rolls, heat is transferred more quickly to the textile than possible by convection heating. The basic process of drying remains the same for both processes with heating of the outside of the textile structure, and drying from the outside to the interior of the structure. In radiation heating, the wetted textile structure is irradiated by thermal radiation using a heat source such as an infrared lamp. On striking the wetted surface the radiation is absorbed and the surface is heated. Radiation heating has limited use for drying textile structures and is used more often in curing or pre-drying printed textiles. Radio frequency and microwave energy can be used to heat wetted textile structures. The textile structure can be heated uniformly throughout, since water as a polar medium absorbs the energy. Radio frequency and microwave heating is efficient for initial drying, but as the moisture level decreases, there is less water to absorb the energy and the process is not as viable. Therefore, these processes are used in pre-drying fabrics and heating of wet packages. In recent years the efficiency of drying textiles has been improved through insulation of the heating units and through installation of heat exchangers to remove heat from the outflow of warm moist air from the heating units.

COLOUR, DYES, DYEING AND PRINTING

COLOUR THEORY

Colour is defined a s the net response of an observer to visual physical phenomena involving visible radiant energy of varying intensities over the wavelength range 400 to 700 nanometers. The net Colour seen by the observer is dependent on integration of three factors:

- the nature of the light source,
- the light absorption properties of the object observed, and
- the response of the eye to the 1ight reflected from the object.

The relative intensities of the various wavelengths of visible 1ight observed by the eye are translated by the mind of the observer resulting in the perception of Colour. In Colour measurement, the human eye is replaced by a photocell which detects the light energy present at various visible wavelengths. Visible 1ight is a narrow band of electromagnetic radiation from 400 to 700 nm detected by the human eye. Radiation falling below 400 nm is ultraviolet radiation, and that falling above 700 nm is infrared radiation; both are unseen by the human eye. If pure light of a given wavelength is ob served, it will have a Colour corresponding to that wavelength. Pure wavelengths of light are seen when white light is refracted by a prism into a "rainbow" spectrum of continuous Colour. Light sources such as sunlight, incandescent light, and fluorescent light are continuums of various wavelengths of light with the relative amounts of the various wavelengths of light being dependent on the overall intensity and type of light source. Sunlight at noon has very nearly the same intensity of each wavelength of light throughout the visible spectrum, whereas at dusk sunlight is of lower intensity and has greater quantities of the longer, red wavelengths than of shorter, blue wavelengths. Fluorescent lights generally contain large amounts of shorter, blue wavelengths, while incandescent tungsten lights contain a large component of longer, red wavelengths compared to noon sunlight. Differences in intensity and wavelength distribution between light sources has a profound effect on the Colour observed for a dyed textile, since the textile can absorb and reflect only that light available to it from the source. When a dyed fabric appears different in Colour or shade under two different light sources, the phenomenon is referred to as "flare." When two fabrics dyed with different dyes or dye combinations match under one light source but not under another, the effect is called "metamerism." When light from a source strikes a dyed textile surface, different portions of the light of the various wavelengths are absorbed by the dye, depending of the structure and light absorption characteristics of the dye. Light not absorbed by the dye on the textile is reflected from the surface as diffuse light. The Colour seen is a composite of all the wavelengths reflected from the fabric. If significant direct reflectance of light from the fabric occurs, the fabric exhibits a degree of a gloss. If little or no light throughout the visible range is absorbed by the fabric and the majority of light is reflected, the fabric appears white. If the fabric absorbs all of the light striking it, the fabric is black. If uniform light absorption and reflectance across the visible wavelengths occurs at some intermediate level, the fabric will be a shade of grey.

Table. Colors After Absorption/Reflectance

Wavelength of Light Absorbed (nm)	Light Absorbed by Dyed Textile	Color Seen by the Observer
400-435	Violet	Yellow-green
435-480	Blue	Yellow
480-490	Green-blue	Orange
490-500	Blue-green	Red
500-560	Green	Purple
560-580	Yellow-green	Violet
580-595	Yellow	Blue
595-605	Orange	Green-blue
605-700	Red	Blue-green

The dye absorbs discrete packages or quanta of light, and the dye molecule is excited to a higher energy state. This energy is normally harmlessly dissipated through increased vibration within the dye molecule as heat, and the dye is then ready to absorb another quantum of light. If the dye cannot effectively dissipate this energy, the dye will undergo chemical attack and Colour fading or Colour change will occur or the energy will be transferred to the Fibre causing chemical damage.

Organic molecules that contain unsaturated double bonds are capable of absorbing light within a given wavelength range. If these double bonds are conjugated and alternate within the molecule, they are able to mutually interact with one another as a cloud of n electrons. If sufficient conjugation exists, the molecule will partially absorb light in the lower energy visible wavelength range and will be considered a dye or a pigment. In general, dyes are colored molecules soluble or dispersible in water or solvent media which can penetrate the Fibre on coloration, whereas pigments are not dispersible and must be mechanically entrapped in or locked to the Fibre by a binding resin.

A series of colorless dyes exist which are called "fluorescent brighteners." These dyes absorb little light in the visible region, but absorb radiation at unseen ultraviolet wavelengths and then emit this radiation at long blue wavelengths to provide an optical bleaching effect on fabrics.

Fluorescent dyes combine fluorescence with visible light absorption characteristics to give extremely bright colors, since unseen ultraviolet light is also being made visible to the eye. The unsaturated groups which can be conjugated to make the molecule colored are referred to as "chromophores." Groups which enhance or alter the Colour within a conjugated system through alteration of the electron density are referred to as "auxochromes." A series of typical chromophores and auxochromes follows:

CHROMOPHORES -

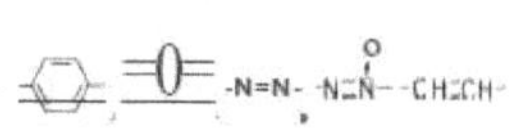

AUXOCHROMES — $-CH_3$, $-OCH_3$, $-OH$, $-NH_2$, $-NO_2$, $-SO_3Na$

The third component in Colour is the observer, which can be the human eye or a photodetector in a Colour instrument. Most Colour measurement systems are based on a standard light source and a "standard observer" for quantitative measurement of Colour. The human eye doesn't respond uniformly to Colour throughout the visible region, but gives maximum response in the middle visible wavelengths. In summary, then, the Colour observed is a composite of three factors:

- the light source,
- the object, and
- the observer.

A change in any one of these three factors will affect the net Colour observed.

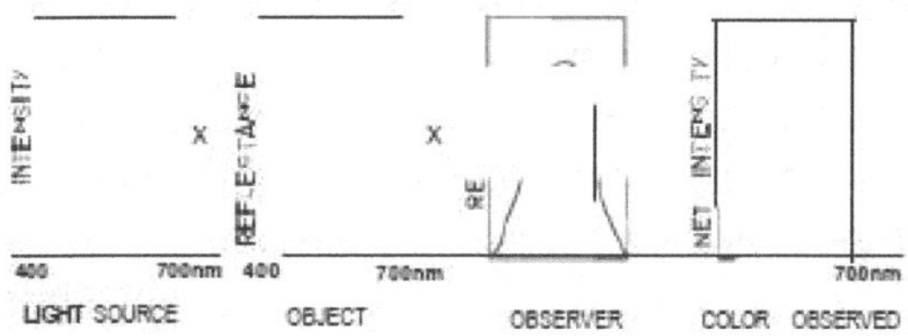

The Colour of a textile can be completely defined by the reflectance Colour spectrum of the fabric as seen above; however, since spectra are of limited value in defining given colors or shades, and numerical Colour systems have been developed to quantitatively or qualitatively define Colour. Knowing the relative position of two colors within the Colour space permits one to determine differences between the two colors. Colour is often defined by the following terms: hue, chroma or saturation, and lightness or value. These Colour terms are defined as follows:

- Chroma or saturation — basic difference between Colour and the closest primary hue.
- Lightness or value — relative degree of lightness.
- The Munsell system uses standard hues and numerical values for chroma and lightness to define Colour.

An array of chips of equal Colour difference as defined by many observers has been prepared that make up the Munsell Book of Colour. The primaries consist of all major hues plus purple arranged logically to form a horizontal circle. The value or lightness makes up the vertical axis varying from 0 to 10 with 0 being black, 10 being white, and the intermediate numbers being varying shades of grey. The chroma is the degree of deviation from the "true" primary and can vary from 0 to 14, with the larger numbers being closest to the "pure" Colour. To determine the Munsell value for a given Colour, the Munsell Colour book is consulted to determine the Colour designation. For example, if a Colour had a designation of 5BG *SIB,* the sample would be a blue green with a value or lightness of 5 and a chroma of 8. I

n such a way Colour can be characterized, but it is not possible to assign strict numerical values to Colour differences between two dyed samples by use of the Munsell system. Additive Colour systems used in Colour measurement determine the amount of three Colour primary lights necessary to define a given Colour using a standard light source and observer. Subtractive systems are used by dyers and colorists, since the net Colour reflected from a dye bath or dyed fabric is more meaningful in matching dyeings and depends on the amount of the three subtractive primaries present. Colour theory and Colour measurement are complex and are further complicated by an individual's response to physical, physiological, and psychological aspects of Colour. Nevertheless, Colour differences can be effectively measured using additive

Colour systems provided the light source, the observer, and angle of viewing and degree of field observed is defined. In order to clarify and standardize the additive Colour system and Colour difference measurement, the Commission Internationale de 1'Eclairage was formed in 1931. The CIE has provided the definitions and standards necessary for Colour measurement. The primaries defined by CIE are not real colors, but are imaginary primaries used to define all colors in the Colour space. The amount of each of these primaries in a given Colour are used to define the shade and depth of shade for that Colour. Since the tristimulus values are difficult to plot, the values are normalized and reduced to trichromatic coefficients according to the equations x = $X/X+Y+Z$ and y = YjX+ Y+Z. Then the shade of a sample is defined by x and y, and the relative lightness by the Y value which is equivalent to the total reflectance of the dyed textile as observed by the human eye. Colour differences between two samples are determined by the equation $E E > (E x^2 , E y^2 , E Y^2)1/2$ The Colour space can be mathematically presented in a number of way s and numerous Colour difference formulas and systems exist. Recently, the CIE has established a Colour and Colour difference system based on rectangular Colour coordinates called CIELAB. This system use s L rather than Y for lightness, since L is more closely related to the response of the human eye, with black being assigned 0 and white being assigned 100. A and B provide the shade value of the Colour in rectangular coordinates with A+ being red and A- being green, and with B+ being yellow and B- being blue. Then the equation $E E > (E L^2 , E A^2 , E B^2)^{1/2}$ will give the Colour difference. Al so cylindrical coordinates of L, C, and HO, C is the distance from neutral grey shade $(C\ A^2 , B^2)$, and H° is the arc tangent B/A expressed on a 360° scale where A , 0¬, B , > 90¬, A. > 180¬, B. > 270- may be used if desired. The concentration of dye on a fabric after a given dyeing time can be determined by three methods:

- measurement of the decrease in dye concentration in solution with time by ultraviolet-visible spectroscopy,
- determination of the dye concentration on the fabric dyed for a given time by dye extraction and ultraviolet-visible spectroscopy, or
- by measurement of the reflectance spectra of the dyed fabric followed by application of the Kubelka-Munk relationship in which

$$K/S > \frac{(1 . R)^2}{2R}.$$

DYES AND DYE CLASSIFICATION

Dyes as colored unsaturated organic molecules must have affinity for fibers to be effectively applied. The dye s on fibers are physically bound to the Fibre by one or more physical forces including hydrogen bonding, van der Waal s, or ionic forces and in certain cases chemically bound by covalent bonds. Dyes may be classified in a number of ways, including Colour, intended use, trade name, chemical constitution, and basis of application. Of these classification methods, chemical constitution and basis of application have been most widely used.

Chemical constitution indicates the major chromophores present in the dye but does not indicate more than such structural aspects of the dye. A classification scheme for dyes has been developed and evolved for use by dyers which is based on the method of application and to a lesser degree on the chemical constitution of the dye class. Dyes classified by this scheme are assigned standard designations according to dye class, Colour, and overall constitution by the Society of Dyers and Colorists in the Colour Index.

Dyes Containing Anionic Functional Groups

A number of dyes, including acid, direct, mordant, and reactive dyes, contain functional groups that are sodium salts of sulfonic or carboxylic acids. These functional groups provide water solubility to the dyestuff. The dyes differ in sub-classification in their affinity for fibers and/or the presence of special functional groups.

Acid Dyes

The acid dyes are large dyes containing one or more sulfonic or carboxylic acid salt functional groups. These dyes are dyed onto fibers from acid solution, since positive charge development within the Fibre in acid solutions acts as a driving force for dye diffusion and migration into the Fibre. Only fibers which develop a positive charge in the presence of acid, such as wool, silk, and other protein fibers, nylon, and certain modified synthetics, are readily dyed by acid dyes, Acid dyes on fibers are reasonably colorfast to light and laundering, but mordanting will improve the overall fastness properties of the dye. The Colour of the dye may be affected somewhat by mordanting; however, pre-metallized acid dyes are a special class of acid dyes which have been reacted with a mordant prior to dyeing and which have sufficient solubility to be dyed under conditions normally used for acid dyes.

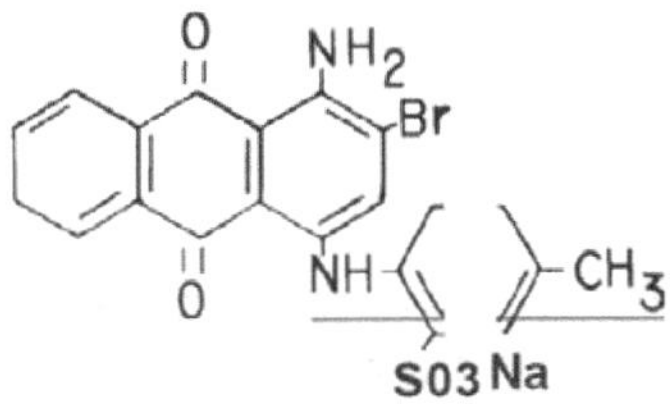

ACID BLUE 78

One method of classifying acid dyes involves dividing them into three groups according to their application and wetfastness. Leveling acid dyes dye evenly to a moderate degree of fastness. For wool and protein fibers they are small molecules requiring highly acidic dyebaths for good exhaustion. For nylon, leveling dyes are somewhat larger molecules and are applied from weakly acidic dyebaths. Milling dyes have better wash fastness, but generally give duller shades and lack good leveling characteristics. They are larger than leveling dyes and are applied from dilute acetic acid solutions. Acid milling or super milling dyes are applied from neutral solutions and have poor leveling characteristics due to their larger molecular size. They generally are applied to wool and have high wetfastness and light fastness.

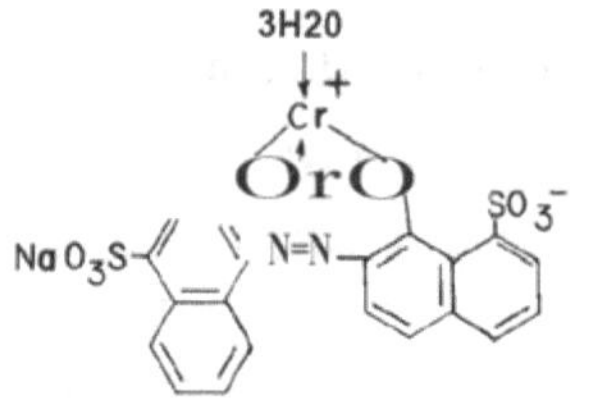

ACiD BLUE 158
PRE-METALLIZED ACID DYE

Direct Dyes

Direct or substantive dyes are a special class of dyes which penetrate cellulosic fibers readily and have good affinity for these fibers due to their size and shape. Whereas acid dyes are large and bulky, direct dyes are long, narrow, and flat in molecular structure, which allows them to readily enter the cellulose structure and interact with the cellulose in such a way as to provide good Fibre affinity. Direct dyes often contain one or more azo groups connecting aromatic chromophores, thereby providing a straight chain dye molecule. Since charge development is not a primary consideration in diffusion of direct dyes onto cellulosics, the dyes are usually applied from basic solutions in which cellulosics are more stable and more likely to swell.

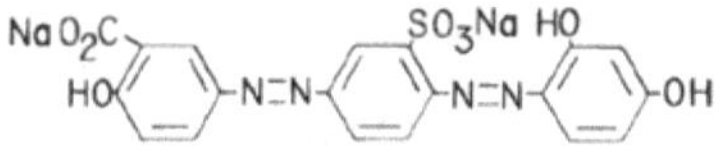

DIRECT RED 185

Direct dye s fall into three main categories. Class A direct dyes level well even in the presence of high salt concentrations, while Class B direct dyes have poor leveling properties which can be improved by proper adjustment of salt concentration. Class C dyes have poor leveling properties, but exhaust well on raising the temperature of the dyebath in the absence of salt. The direct dyes are reasonably colorfast on cellulosic fabrics, but fastness can be improved by mordanting with metal salts, cationic fixing agents, or formaldehyde or through reaction with diazonium compounds or diazotization of the dye and reaction with a coupling compound.

Mordant Dyes

Mordant dyes are acid dyes that have special sites other than acid salt anion groups that can react with a metal salt mordant. Mordant dyes are "tailor-made" to chelate with metal ions to form a strong organometallic complex of limited solubility and greater colorfastness. The Fibre may be dyed initially and then mordanted, dyed and mordanted simultaneously, or mordanted and then dyed.

Of the three methods, post-mordanting is preferred. Salts of chromium, aluminum, copper, iron, tin, and cobalt are commonly used as mordants. Since the mordant affects the electron distribution and density within the dye, the Colour of the dyed fabric tends to change.

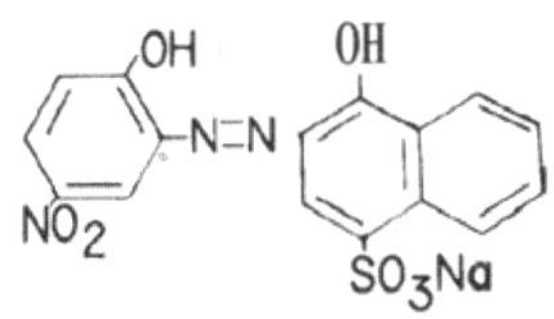

MORDANT BROWN 35

Reactive Dyes

Reactive dye s are dye s which usually have the basic structure of acid, direct, or mordant dyes but which in addition have a reactive group capable of covalent bond formation with the Fibre. Since the Fibre must have reasonable reactivity Towards the dye reactive group, application of these dyes has been limited to cellulosic, protein, and nylon fibers for the most part.

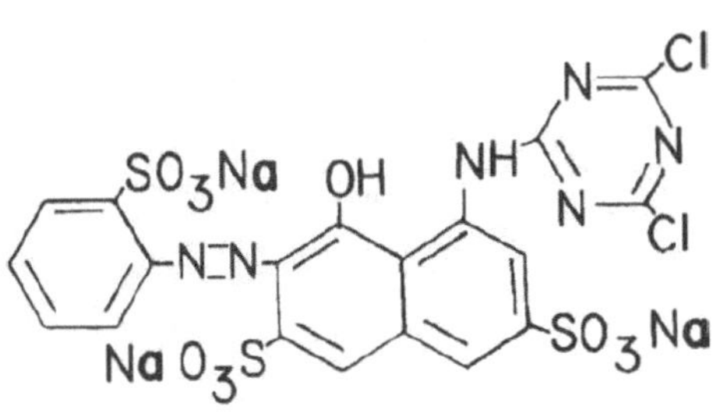

REACTIVE RED

The fastness of reactive dyes covalently bound to the Fibre is excellent. Reactive functional groups have been selected for incorporation into reactive dyes which will react readily with the Fibre after diffusion into the structure but which will not hydrolyze in the water solvent used in dye application.

Acidic or basic conditions are necessary for successful and rapid reaction of the reactive dye with the Fibre, so dye application is carried out at either slightly acid or basic pH. Procion dyes are the best known of the reactive dyes.

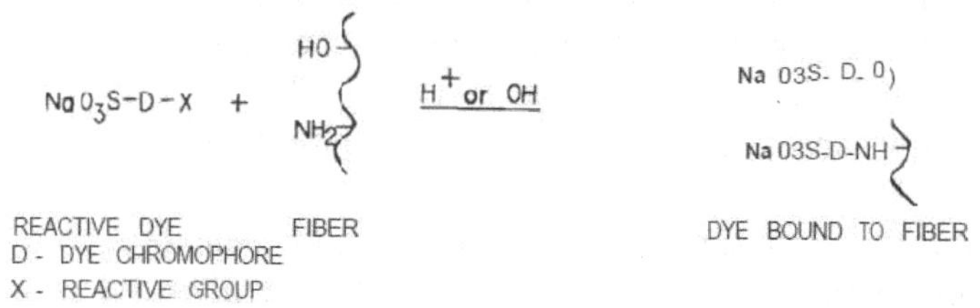

Dyes Containing Cationic Groups

Basic or cationic dyes are colored cationic salts of amine derivatives. Basic dye cations will migrate

Towards negative charges inside the Fibre. The dyes may be applied to cellulosic, protein, nylon, acrylic, and specially modified synthetic fibers. Although the dyes generally are of striking brilliance and intensity, the colorfastness of the dyes on cellulosic, protein, and nylon fibers is generally poor.

Colorfastness can be improved through mordanting with tannins or other complexing agents. Since the insoluble parent amine is regenerated from basic dyes at basic pH, the basic dyes are applied from mildly acid or neutral solutions. The dyes are also important in "one-shot dyeing" of paper products.

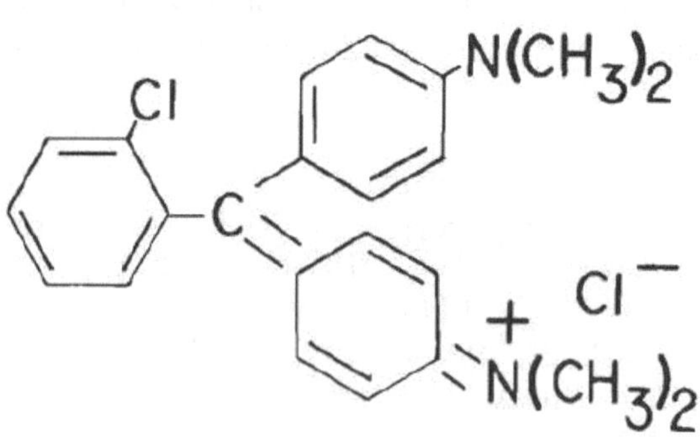

BASIC BLUE

Dyes Requiring Chemical Reaction before Application

Both vat and sulfur dyes must be chemically reduced before application to a Fibre, whereas azoic or naphthol dyes are formed through reaction of two separate dye components after application to the Fibre. These dyes tend to penetrate the Fibre less than other dyes, and care must be exercised in application to get reasonable fastness properties.

Vat Dyes

Vat dyes are usually water-insoluble dyes that can be chemically reduced in the presence of base to form a water-soluble and colorless leuco form of the dye, which is then applied to the Fibre. Vat dyes can be readily applied to cellulosic fibers and most synthetic fibers, but care must be taken in applying the dyes to protein fibers due to the high passivity of the leuco dye solution which can damage protein fibers.

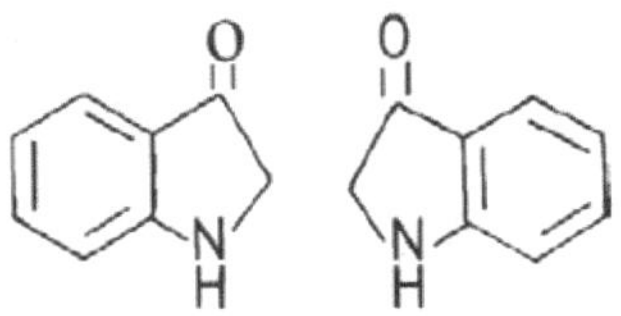

VAT BLUE I

INDIGO

The dyes are usually indigoids or anthraquinone derivatives and applied at low temperatures. After application of the leuco form of the vat dye, the dye is reoxidized on the fabric by oxygen in the air or through treatment of the dyed fabric with a mild oxidizing agent. The vat dyes are reasonably colorfast if poorly held surface dye has been removed:

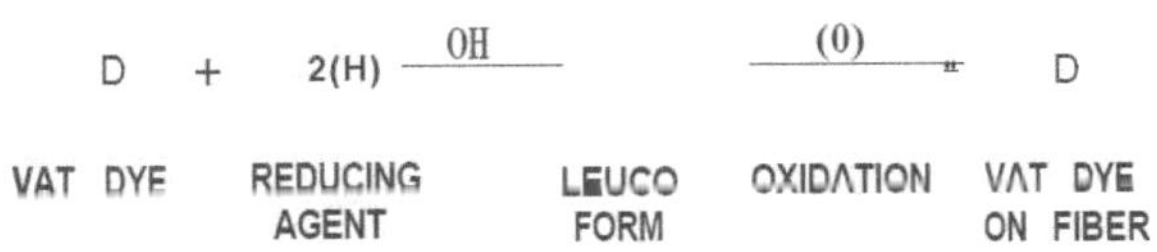

Stable sulfate esters of reduced leuco forms of vat dyes are available which do not require prior chemical reduction before application.

Sulfur Dyes

Sulfur dyes are inexpensive complex reaction mixtures of selected aromatic compounds with sodium polysulfide. The sulfur dyes are chemically reduced in the presence of base prior to application to the Fibre, and are reoxidized after dyeing on the Fibre by oxygen in the air or by application of a mild oxidizing agent such as hydrogen peroxide.

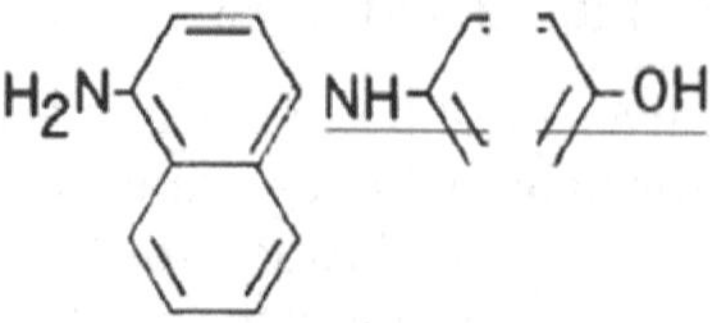

SULFUR GREEN 6

The sulfur dyes are similar in application and Fibre affinity to vat dyes, and also are available in a solubilized leuco form:

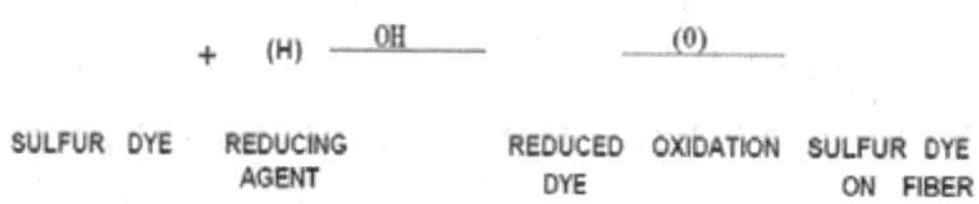

Sulfur dyes are inexpensive, and have adequate colorfastness properties for most applications.

Azoic Dyes

Azoic or naphthol dyes are formed in situ on the fabric through a coupling reaction of an aromatic alcohol or amine such as naphthol. The dye formed contains an azo group:

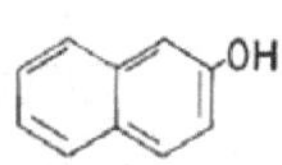

DIAZO COUPLING COMPONENT

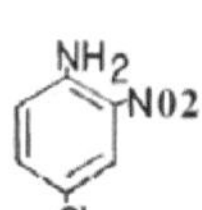

AZOIC DIAZO COMPONENT 9

The fabric is usually impregnated first with the aromatic coupling component followed by immersion of the fabric in a solution containing the diazo component, with the azoic dye being formed instantaneously. Owing to the instability of the components, the dyeings are carried out near room temperature. The dyes have moderate fastness, and dyeings may be mordanted to increase fastness.

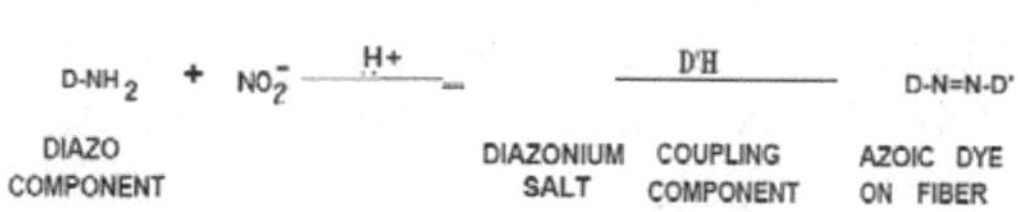

Special Colorant Classes

Several types of dyes or colorants do not "fit" logically into the other classifications and have been included in this special classification. Disperse dyes are small polar dye molecules which can be used to dye thermoplastic fibers such as triacetate, nylon, polyester, and other synthetic fibers. Solvent dyes are dyes which are specially formulated so that they can be applied from solvents other than water. Pigments are not dyes at all, but rather colorants that must be incorporated in the Fibre during spinning or fixed on the surface of the Fibre by use of a polymer adhesive. Natural dyes are complex mixtures derived from natural sources which can be placed in a number of the above classifications.

Disperse Dyes

Disperse dyes were formulated and introduced to permit dyeing of hydrophobic thermoplastic fibers including acetate, triacetate, nylon, polyester, acrylic, and other synthetics. The disperse dyes are small polar molecules, usually containing anthraquinone or azo groups, which do not have charged cationic or anionic groups within the structure. The disperse dyes are sparingly soluble in water and must be dispersed with aid of a surfactant in the dyebath. As the small amount of dissolved disperse dye diffuses into the Fibre, additional dye dispersed in solution is dissolved, until the disperse dye is nearly completely exhausted onto the Fibre. A special class of reactive disperse dyes has been introduced that can react with fibers like acetate and nylon after diffusion into the Fibre. The light- and wash fastness of these dyes is generally good, but difficulty has been encountered with fume fading with certain of the disperse dyes. Many disperse dyes have appreciable vapor pressures at elevated temperatures and can be "dyed" onto thermoplastic fibers by sublimation, which involves diffusion of the dye vapors into the Fibre.

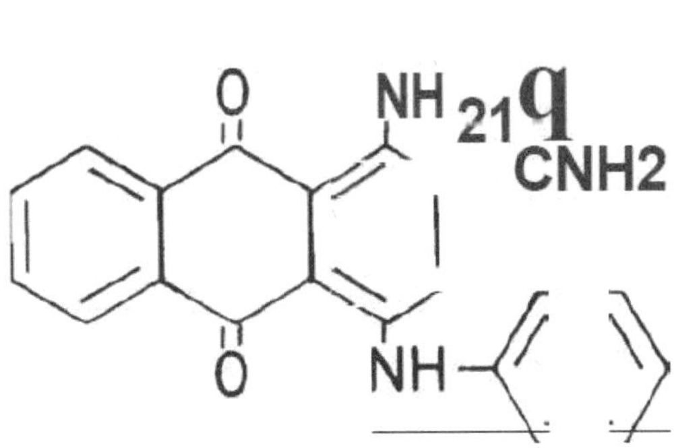

DISPERSE BLUE 6

Solvent Dyes

Solvent dyes often resemble dyes in other classes, except that these dyes contain groups that improve their solubility in solvents such as alcohols and chlorinated hydrocarbons. It is often possible with solvent dyes to dye certain fibers which would be poorly dyeable from aqueous solution.

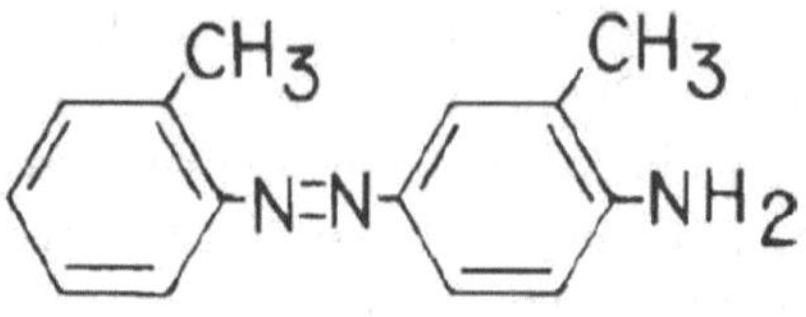

SOLVENT YELLOW 3

Pigments

Pigments as a class are colored materials that are insoluble in their medium of application. As a result, they cannot penetrate or become readily fixed to a Fibre and must be "locked" onto the Fibre surface by use of a polymeric adhesive binder that encapsulates and holds the pigment onto the fibers. Pigments include inorganic salts, insoluble azo or vat dyes, toners, lakes, metallic complexes, and organometallic complexes. Some binders are preformed water-insoluble polymers applied from solvents or as emulsions, whereas others are water-soluble or emulsifiable polymers which can be chemically crosslinked and insolubilized after application by drying and heating. The pigment-binder systems tend to stiffen textiles, and have moderate to poor fastness, since they are surface treatments.

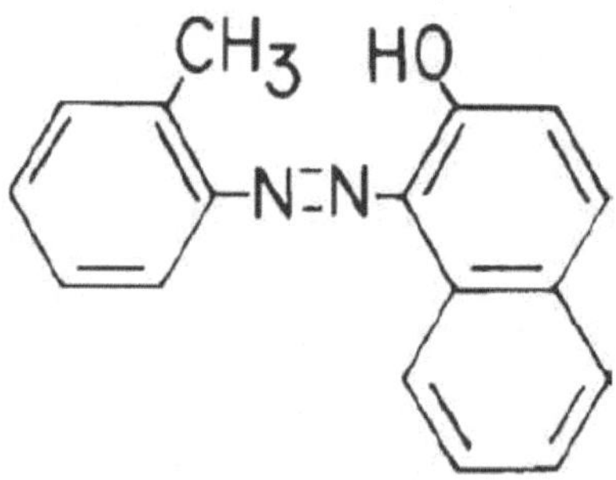

PIGMENT ORANGE 3

Natural Dyes

Natural dyes and pigments are derived from mineral, animal, or plant sources and generally are complex mixtures of materials. A number of natural dye s can be classified as acid or vat dyes and can be readily mordanted. In general, the natural dyes give more muted tones than synthetic dyes and are useful only on natural or regenerated fibers.

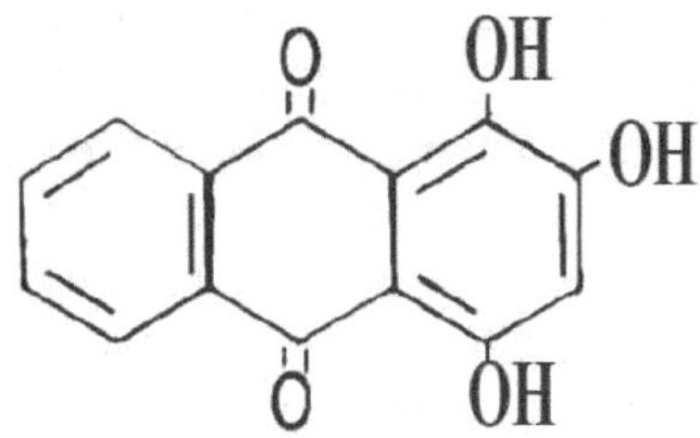

NATURAL RED 8

PURPURIN

Dyeing of Blends

Textile structures are often blends of more than one Fibre type. Blends containing similar fibers are easy to dye with dyes of similar structure and application characteristics. Care must be taken in dyeing blends containing fibers of highly different dye affinities, if the dyeing is carried out in the same dyebath. The dyes and their auxiliaries must be compatible with one another. When fibers in a blend are dyed the same Colour, the dyeing is referred to as "union dyeing," whereas dyeing fibers in a blend of different colors is referred to as "cross dyeing." In blends, interesting tone-on-tone, tone-on-white, and differential dyeings are possible by selection of appropriate dyes and dyeing conditions.

APPLICATION METHODSAND FACTORS AFFECTING DYEING

Dyes may be applied to textile structures in a number of ways and at a number of points within the textile construction process. Dyeing or printing techniques can be used. Dyeing methods involve application of dye solutions to the textile, whereas printing can be considered a specialized dyeing technique. In printing the concentration of dye is higher, and the dye medium is thick and viscous to limit dye migration on the fabric, permitting formation of a design or pattern. Fibre or stock dyeing involves dyeing of the loose fibers or Fibre top or sliver before yarn formation. Colorants can be added to man-made polymers before they are formed into fibers to ensure uniform coloration throughout the Fibre. This technique is referred to as "solution" or "dope dyeing." Yarns or skeins can be dyed, and piece dyeing of textile fabrics can be carried out. Resist techniques such as tie-dyeing or batik limit dye migration to certain parts of the fabric. Discharge processes use bleaches or other chemicals to remove dye from selected areas of a dyed or printed fabric. The above processes are used for application of dyes to textile substrates in the form of Fibre sliver or top, yarn, or fabric by batch or continuous methods. Fibre or stock dyeing of sliver or top is usually conducted in large vats with movement of the dye liquor through the fibers to assure intimate contact of the dye liquor with the fibers. Yarn or skein dyeing is conducted by suspending skeins of yarn in an agitated dyebath with possible additional movement of the skeins during dyeing. Yarns can also be wound as packages on perforated spindles or spools and immersed in the dyebath. The dye liquor is then circulated back and forth through the packages. Fabric rolls can also be dyed by this package dye technique. Package dyeing is often carried out in closed systems at elevated temperatures and pressures. Jig dyeing involves passage of a fabric piece back and forth from one spindle to another through a dyebath, whereas a dye beck containing a winch is used to move a continuous fabric piece through the dye liquor. The techniques are all batch processes.

The pad-batch method is a specialized batch technique for application of reactive dyes to cellulosic fibers. In the pad-batch method, cellulosic fabric is passed through a concentrated solution of reactive dye, followed by storage of the wet fabric in a vapor tight enclosure for 24 to 48 hours *at* room temperature, to permit time for diffusion and reaction of the dye with the fabric substrate prior to washing off unreacted dye. The process leads to significant energy savings compared to conventional processes. Various specialized techniques have been developed for application of disperse dyes to polyester. Unless the dyeing is carried out at 100 °C or above, the rate of dyeing is very slow. Dyeing with disperse dyes from aqueous solutions *at* 120°-130°C to achieve rapid dyeings is common but requires the use of closed high-pressure equipment. Recently, jet dyeing has been introduced, which permits not only high-temperature dyeing but also impingement of the dye onto the moving fabric through use of *a* venturi jet system. Also, carriers can be introduced to permit dyeing of polyester at atmospheric pressure *and* below 100°C. Carriers are usually aromatic organic compounds that can be emulsified in water *and* which have affinity for the polyester.

The carriers penetrate the polyester, open up the molecular structure of the Fibre, and aid in passage of the disperse dye across the dye solutionfiber interface and within the Fibre. Suitable carriers include aromatic hydrocarbons such as diphenyl *and* methyl naphthalene, phenolics such *as* the di- and trichlorobenzenes, aromatic esters including methyl salicylate, butyl benzoate, and diethylphthalate, and benzaldehydes. Carriers must be removed after dyeing, and the presence of carriers in mill effluents presents a problem because of their toxicity. Continuous dyeing is carried out on *a* dyeing range where fabric or carpet is continuously passed through *a* dye solution of sufficient length to achieve initial dye penetration. The dye on the fabric or carpet is fixed by subsequent s teaming of the substrate. Recently, foamed dye formulations have been applied to fabrics, thereby effectively reducing the dye liquor to fabric ratio and reducing energy and effluent treatment costs. A novel approach to continuously dyeing polyester with disperse dyes involves sublimation of disperse dye under heat and partial vacuum into polyester by the technique called "thermosol dyeing." Polyester containing disperse dye applied to the Fibre surface is heated near 200°C under partial vacuum for a short period of time. At this temperature, the molecular motion within the polyester is high, permitting the dye vapor to penetrate into the Fibre. On cooling, the disperse dye is permanently trapped *and* fixed within the Fibre. The above processes are graphically represented in figure

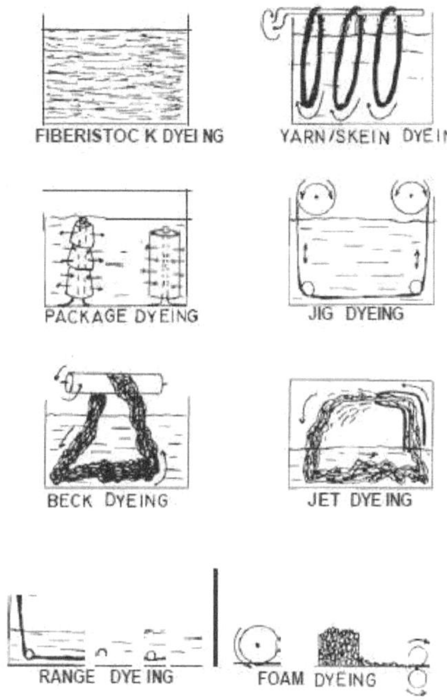

Fig. Dyeing processes

In printing, the printing paste is applied through use of direct transfer dye using a block or engraved roller or through application using a partially marked flat or rotary screen and squeegee system. Printing of polyester by disperse dyes can be accomplished by heat transfer printing, which is a modification of thermosol dyeing. In this process disperse dyes are printed onto paper followed by bringing the polyester fabric and printed paper together under pressure with sufficient heating to cause diffusion of disperse dyes into the polyester. Block, flat screen, and heat transfer processes are batch processes, whereas engraved roller and rotary screen printing are continuous processes. Special techniques using dyeing solutions which give printed style fabrics have been developed. In these processes, multiple jets of different dye solutions are sprayed in programmed sequence onto the fabric as it passes under the jets to form patterns with definition very nearly like that of prints. Specialized techniques for formation of patterns on carpets have been developed. In one process dye solution is metered and broken or cut into a pattern of drops which are allowed to drop on a dyed carpet passing underneath to give a diffuse overdyed pattern on the carpet. Representations of the printing and printed style processes are found in figure.

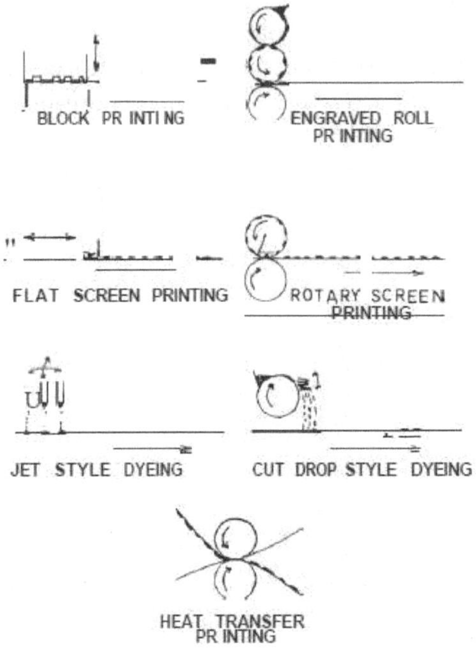

Fig. Printing and pridted style processes.

Physical factors such as temperature and agitation and auxiliary chemicals added to the dyebath or printing paste can alter the rate of dyeing and/or the total dye absorbed by the Fibre. In dyeing, the rate of dyeing of the Fibre is dependent on the rate of migration of dye in solution to the Fibre surface, the rate of diffusion of dye at the Fibre interface, and the rate of diffusion of dye in the Fibre matrix. Agitation of the dyebath effectively eliminates the effect of dye diffusion to the Fibre in the dyebath.

The rate of dye passage across the Fibre-liquid interface is rapid in most cases; therefore the rate of dyeing is solely determined by the rate of dye movement within the Fibre matrix. As the temperature of dyeing is raised, the rate of "strike" of dye onto the Fibre and diffusion in the Fibre increases, whereas the total amount of dye present in the Fibre at equilibrium decreases. In other words, heating a dyebath speeds dyeing but decreases the total dye exhausted on the fabric. Dyeing is usually carried out at a temperature above the glass transition temperature of the Fibre, since the molecular segments of the polymers within the Fibre have more mobility and permit more rapid dye diffusion above that temperature. A number of chemical reagents are added to the dyebath to alter in some manner the course of dyeing. Acids or bases may be added to the dyebath to induce charge formation in the Fibre in order to increase dye diffusion or to promote reaction of the dye with the Fibre as in the case of reactive dyes. In addition, bases may be used to open up the structure of cellulosics to improve dyeing or to aid in dissolving of vat and sulfur dyes. Common inorganic salts such as sodium chloride or sodium sulfate may be added to a dyebath and act as leveling agents to retard the rate of dyeing and give a more even dyeing. Surface-active agents such as anionic detergents can act as leveling agents and also permit the rapid and complete Fibre wetting needed to give even dyeing.

Water softening chemicals such as sodium hexametaphosphate are added to bind or chelate hardwater calcium or magnesium ions that may interfere with dyeing. Reducing agents are added to vat and sulfur dyes to react with the dye to give the leuco form. Added organic solvents and/or certain chemical reagents may increase the dye solubility and rate of penetration into the Fibre. Carriers are aromatic derivatives added in disperse dyeings to open up the Fibre structure of polyester and other thermoplastic fibers and to increase the rate of dye diffusion. Urea has a similar effect in speeding dyeing of cellulosic and protein fibers. In printing pastes, natural and synthetic gums and thickeners are necessary to provide viscosity and thickness to the dye paste and to limit dye migration across the fabric face.

Heating of fabrics following printing is necessary to achieve good dye penetration into the Fibre, and superheated steam is preferred, since wet fixation causes a faster rate of dye migration than dry heating methods. The fastness of the dyeing will be dependent on the location of the dye within the Fibre, chemical and physical forces holding the dye to the Fibre, and the stability of the dye-Fibre combination to environmental factors. Dye distributed evenly throughout the Fibre in small aggregates is preferred, whereas surface or ring dyeing leads to poor washfastness.

DYES APPLIED TO FIBRE CLASSES

Dyes for Cellulosic Fibers

Cellulosic fibers can be dyed readily using a number of dye classes. The less crystalline regenerated cellulosics dye more readily than the more crystalline natural cellulosic fibers such as cotton. Cellulosics can be dyed or printed using direct, reactive, basic, vat, sulfur, or azoic dyes. Cellulosics can also be colored using pigment-binder systems. Mordanting of direct and basic dyes on cellulosics improves overall fastness. Since the hydroxyl groups in cellulosics are less reactive than many functional groups found in protein fibers, reactive dyes such as the dichlorotriazine or dichloroquinoxaline dyes are often used on cellulosic fibers due to their higher reactivity. The general stability of cellulosic fibers to basic solutions permits them to be effectively dyed by vat, sulfur, and azoic dyes without significant damage.

Dyes for Cellulose Ester Fibers

Acetate and triacetate fibers can be effectively dyed using disperse dyes. The rate of dyeing is more rapid with the more hydrophobic triacetate fibers than with acetate. Under special conditions, azoic and vat dyes may be used to dye these fibers. Acetate fibers also have affinity for selected acid and direct dyes. Since acetate loses its luster above 85°C, dyeings must be carried out at or below this temperature. Addition of pigments or solvent-soluble dyes to the acetate or triacetate spinning "dope" prior to Fibre spinning leads to colored fibers possessing excellent colorfastness, although the colors available are limited.

Dyes for Protein Fibers

Protein fibers are the most readily dyed fibers due to the numerous reactive functional groups present. They can be dyed with a wide range of dyes under acid, neutral, or slightly basic conditions. Since the keratin fibers are less crystalline and oriented than secreted fibers such as silk, they tend to dye more rapidly and more readily to deeper shades. Under acid conditions amino groups in the protein fibers are protonated to form NH; groups. In this form, they are able to attract dyes containing acid anions including acid, direct, mordant, and reactive dyes. Special premetallized acid dyes of sufficient solubility are used to dye protein fibers to fast colors. The functional groups available in protein fibers are more reactive than hydroxyl groups in cellulosic fibers. Reactive dyes of more 1imited reactivity have been developed especially for protein fibers. Protein fibers complex very readily with multivalent metalcations. Acid dyes and mordant dyes may be rendered very fast by mordanting with metal salts, and chromium salts are especially effective as mordants. At neutral or slightly acid pH, protein fibers may be dyed with cationic or basic dyes; however, the fastness of the dyed Fibre is poor without mordanting with tannic acid or other mordants for cationic dyes. Azoic and vat dyes find only limited use on protein fibers due to the damaging effect that basic solutions of these dyes have on protein fibers. Sulfate esters of reduced vat dyes can be used effectively on protein fibers, however. Natural dyes from many sources have good affinity for protein fibers and are used extensively in the crafts area and often in conjunction with mordants.

Dyes for Polyamide Fibers

With the exception of the aramid fibers, the polyamides dye readily with a wide variety of dyes. Since the polyamides contain both acid carboxylic and basic amino end groups and have a reasonably high moisture regain, the fibers tend to dye like protein fibers such as wool and silk. Since the molecular structure is somewhat more hydrophobic, more regular, and more densely packed in the polyamides than in protein fibers, they also exhibit to some degree the dyeing characteristics of other synthetic fibers such as polyesters and acrylics. Due to their highly regular molecular structure and dense chain packing, the aramid fibers resemble polyester and are dyed only by small dye molecules such as disperse dyes.

Polyamides such as nylon 6, 6,6, and Qiana can be readily dyed with dyes containing anionic groups, such as acid, metallized acid, mordant dyes, and reactive dyes and with dyes containing cationic groups such as basic dyes. Acid dyes on nylon can be mordanted effectively for additional fastness; however, the colorfastness of basic dyes is poorer and more difficult to stabilize by mordanting. Vat and azoic dyes can be applied to nylons by modified techniques, and polyamides can be readily dyed by disperse dyes at temperatures above 80°C. Aramids can only be dyed effectively with disperse dyes under rigorous dyeing conditions. The biconstituent Fibre of nylon and polyester can be effectively dyed by several dye types due to the nylon component, but for deep dyeings disperse dyes are preferred. Nylon 6 and 6,6 are produced in modifications that are light, medium, or deep dyeable by acid dyes or specially dyeable by cationic dyes.

Dyes for Polyester Fibers

Owing to their high crystallinity and hydrophobicity, the polyester fibers are extremely difficult to dye by normal dyeing techniques unless the Fibre has been modified, as in the case of modified terephthalate polyesters. A limited amount of polyester is solution dyed through incorporation of dye or pigment into the polymer melt prior to spinning. It is more common to use this technique to incorporate fluorescent brightening agents into polyester. Only smaller, relatively Non-polar dye molecules can effectively penetrate polyester; therefore disperse dyes have been the dye class of choice for the Fibre. Azoic dyes and pigment-binder systems have also found limited use on polyesters. Polyester modified with appropriate comonomers can be dyed at lower temperatures or with acid or basic dyes depending on the nature of the modifying groups.

Dyes for Acrylic Fibers

The nature and distribution of acrylonitrile and comonomer or comonomers in the acrylic fibers affect the overall dyeability and the classes of dyes that may be used in dyeing these fibers. Both acrylic and modacrylic fibers can be dyed using disperse dyes, with the more hydrophobic and less crystalline modacrylic being more dyeable with this dye class. The polar cyanide groups in the acrylonitrile unit of these fibers have some affinity for acid dyes and particularly mordanted systems containing copper or chromium ions. Addition of an acid or basic comonomer such as acrylic acid or vinyl pyridine as comonomer imparts improved dyeability with basic and acid dyes, respectively, for these fibers. Vat dyes can be used on acrylic fibers to a limited extent.

Dyes for Polyolefin Fibers

Polyolefin fibers are hydrophobic, and the molecular chains within the Fibre are tightly packed. Therefore it is extremely difficult to dye polyolefin fibers or to increase their affinity to dyes.

Colored inorganic salts or stable organometallic pigments have been added to the polymer melt prior to Fibre spinning to Colour the fibers. Also, Non-volatile acids or bases or materials such as polyethylene oxides or metal salts have been added to the polymer prior to Fibre formation to increase the affinity of the Fibre for disperse, cationic, acid, or mordant dyes. Polyolefin fibers can be chemically grafted with appropriate monomers after Fibre formation to improve their dyeability.

Dyes for Vinyl Fibers

The vinyl fibers, with the exception of vinal and vinyon-vinal matrix fibers, are extremely hydrophobic and difficult to dye, and consequently they can be dyed only through pigmentation of the polymer melt before Fibre formation or through dyeing with disperse dyes. Vinal and vinyon-vinal matrix fibers dye readily with dyes used on cellulosics including direct, mordant, reactive, vat, and sulfur dyes.

Dyes for Elastomeric Fibers

Since the elastomeric fibers are often a component in the core of blended yarns, coloration is not important in all applications. Rubber fibers cannot be dyed readily and are colored through mixing of pigments into the rubber prior to extrusion into fibers. Spandex fibers are more dyeable and can be dyed with acid, reactive, basic, or vat dyes. Anidex can be dyed with disperse or basic dyes. The nylon component of spandexnylon fibers can readily be dyed with acid, basic, disperse, or vat dyes.

Dyes for Mineral and Metallic Fibers

The mineral and metallic fibers are essentially undyeable, and special techniques must be used to impart Colour to the fibers. Thermally stable ceramic pigments can be added to molten glass prior to Fibre formation, or pigment-binder systems may be applied to the surface of the mineral and metallic fibers. Glass fibers can also be sized with a protein which then can be insolubilized and dyed with conventional protein dyes. Glass fibers are colored by coronizing, which involves preheating of the glass substrate to high temperatures to remove all organic materials followed by coloration with a pigment-binder system. The metallic fibers may also be colored through anodizing the metal filament present or through pigmentation- of the plastic layer coating the metal. The nature of the metal in the organometallic fibers determines their ultimate Colour.

FINISHES AND FINISHING

Often fibers in textile substrates are deficient in one or more properties, or improved properties are desired for the substrate. Textile finishing provides a method whereby deficiencies in the textile can be corrected or specific properties can be introduced. Physical finishing techniques or chemical finishing methods are used. Physical finishing is usually carried out on the yarn or formed textile substrate, whereas chemical finishes can be added to the spinning bath prior to Fibre formation for man-made fibers or applied to individual fibers, yarns, or completed textile structures.

PHYSICAL FINISHES AND FINISHING

Physical finishing methods for textiles include optical finishing, brushing and napping, softening, shearing, and compacting of the textile structure.

Optical Finishes

Luster may be imparted to a fabric by physical means. The techniques basically involve flattening or smoothing of the surface yarns using pressure. Beating of the fabric surface or passing the fabric between hard calendering rolls under pressure and with some friction will tend to flatten out the yarns and lower light scattering by the fabric surface, thereby improving reflectance and luster. Luster may be improved further if the calendering rolls are scribed with closely spaced lines which will be imprinted on the fabric to reinforce light striking and reflecting from the Fibre surface. Similar techniques can be used to impart optical light interference patterns to the fabric. Thermoplastic fibers which can deform under heat and pressure can most readily be modified to impart luster.

Brushing and Napping

Physical delustering of a fabric as well as bulking and lofting of the fabric can be achieved by treatments which roughen the Fibre surface or raise fibers to the surface. Fibre raising processes such as brushing and napping involve use of wires or brushes which catch yarns in the textile structure and pull individual fibers partly from the yarn structure. The resulting fabric is warmer, more comfortable, and softer.

Softening and Shearing

During calendering or beating of a fabric interaction between individual fibers within yarns may be lessened and the textile structure softened. Also, when a smooth textile structure free of raised surface fibers or hairiness is desired, the fabric may be sheared by passing the fabric over sharp moving cutting blades or by passing the fabric over a series of small gas jets which singe and burn away raised fibers.

Compacting

During fabric formation processes, stresses often are introduced into a textile. Such stresses can be controlled by drying the finished fabric on a stenter frame, which controls the width of the fabric and the tension on the fabric during the drying process. A second method involves compression of the fabric structure, as in the Sanforizing process. In this process, the fabric and backing blanket is fed between a feed roller and a curved braking shoe, with the blanket being under some tension.

The tension on the blanket is released after passing the fabric and blanket between the roller and braking shoe. The net result is compaction of the fabric being carried along in the system. Such a simple technique permits fabrication of the fabric of finished textile goods without fear of excessive shrinkage on laundering. Protein hair fibers such as wool, and thermoplastic fibers such as polyester, can be compacted by felting action. The scale structures on protein fibers entangle and stick on agitation, particularly in the presence of moisture. The resulting "ratcheting" effect causes the fibers to compact and felt. Many processes for wool take advantage of this effect, and Non-woven felt structures are produced by this method. Compaction of thermoplastic structure occurs when the fibers are raised to near their softening point. At a sufficiently high temperature the fibers shrink and contract, causing compaction of the textile structure.

CHEMICAL FINISHES AND FINISHING

Chemical finishes are chemical reagents or polymeric materials applied to textile structures by a number of methods. The major types of chemical finishes used on textile structures are listed in table.

Table. Major Chemical Finishes.

Finishes Affecting Aesthetics, Comfort and Service	Protective Finishes
Optical finishes	Photoprotective agents and antioxidants
Hydrophilic and soil release finishes	Oil and water repellants
Softeners and abrasion resistant fibers	Antistatic agents
Stiffening and weighting agents	Biologically protective finishes
Laminating agents	Flame retardants
Crease resistant and stabilizing finishes	

Chemical finishes can be applied by a number of methods including padding, spraying, printing, foam application, or vapor techniques. In addition, the finish can be added to the spinning bath prior to formation of man-made fibers. Of these methods, padding is most important. Many finished fabrics must be dried and cured before chemical finishing is complete. Thorough wetting of the Fibre by the finish solution and spreading of the finish evenly over the Fibre surface is critical in most cases to get the desired effect. The location of the finish on the surface or within the Fibre is important, depending on the finish and its function.

Finishes Affecting Aesthetics, Comfort, and Service

Optical Finishes

Optical finishes do little to affect the Colour of a textile substrate, but rather act to destroy or mask Colour centers. They may either brighten the textile, making it more reflective, or deluster the textile, making it less reflective, depending on the treatment. Bleaches are usually chemical oxidizing or reducing agents that brighten the textile by attacking unsaturated molecules that make the textile appear off-Colour. Chlorine bleaches such as sodium hypochlorite are strong oxidizing agents capable of destruction of Colour centers on a textile substrate. Unfortunately, sodium hypochlorite is fairly Non-selective and attacks many dyes and finishes and certain fibers, causing loss or change in Colour and a deterioration in Fibre properties. Sodium chlorite and peracetic acid also are used as strong oxidizing bleaches on some synthetic fibers at the mill to achieve desired whiteness. Oxygen bleaches such as hydrogen peroxide and sodium perborate are milder in oxidizing action and can be used on sensitive fibers such as wool. With hydrogen peroxide, pH adjustment is critical in getting the desired bleaching. Sodium perborate bleaching must be carried out at elevated temperatures, although chemical activators may be added to lower the effective bleaching temperature.

Oxygen bleaches are less likely than chlorine bleaches to damage the Fibre and dyes present on the Fibre. Reducing agents chemically reduce and saturate double bonds within Colour centers and find limited use as bleaches in mill applications. Reagents such as sodium dithionate and sodium formaldehyde sulfoxylate are commonly used as reducing bleaches and in compositions for stripping dyes from fibers. Fluorescent brightening agents are colorless fluorescing dyes which mask yellow coloration on fabrics. Delustering of fibers may be carried out through alteration of the Fibre surface or through addition of alight scattering and/or absorbing agent to the Fibre substrate. Chemical attack or etching of the Fibre surface leads to a more irregular surface morphology, thereby increasing light scattering and making the Fibre dull. With man-made fibers, titanium dioxide is added to the spinning solution before Fibre formation. Titanium dioxide is an excellent light scattering agent, and this delusterant is effective when added at 0.5per cent-2.0per cent levels to the Fibre. Titanium dioxide can also be applied to the surface of natural fibers, if a binder such as urea-formaldehyde resin is used to fix the delusterant to the Fibre surface.

Hydrophilic and Soil Release Finishes

Hydrophilic finishes that promote absorption or transport of water and aid in Fibre wetting and soil removal in a textile construction are useful in many Fibre applications. In man-made fibers surface-active agents such as nonionic polyethylene glycol derivatives can be added to the spinning solution prior to spinning, which will make the Fibre more wet table and hydrophilic. Hydrophilic polymers such as copolymers containing acrylic acid can be fixed to Fibre surfaces to provide improved water absorbency and to limit penetration of soils. Hydrolytic attack of the surface of hydrophobic man-made fibers such as polyester improves Fibre wetting, moisture transport, and soil removal. None of the above absorbent techniques, however, are sufficiently effective to alter the absorbency properties of hydrophobic fibers like polyester enough to totally resemble naturally absorbent fibers like cotton and wool.

Softeners and Abrasion Resistant Finishes

Softeners and abrasion resistant finishes are added to textile structure:

- to improve aesthetics,
- to correct for harshness and stiffness caused by other finishes on the textile substrate, and
- to improve the ability of the fibers to resist abrasion and tearing forces.

The softeners and abrasion resistant finishes are generally emulsions of oils or waxes, surface-active agents, or polymers that lubricate the surface of individual fibers in the textile substrate to reduce friction between fibers and permit them to pass over one another more readily. Emulsions of oils and waxes and related derivatives have been used in the past as softeners. Non-ionic and cationic detergents act as softeners, but lack permanence. Emulsions of polyacrylates, polyethylene, or organosilicones impart softening properties and possess reasonable fastness. In addition, they may impart a more full hand to the textile.

Stiffening and Weighting Agents

Textile auxiliaries that stiffen and weight fabrics have included temporary and permanent sizes and metal salts applied alone or with a binding agent. The sizes stiffen the fabric through formation of bonds between fibers, particularly at Fibre crossover points. Temporary sizes include starch, naturally derived gums, carboxy methylcellulose, and polyvinyl alcohol. Acrylic binders, polyvinyl chloride, and polyvinyl acetate emulsions act as permanent sizing agents. Metal salts may be used to weight certain fibers such as silk. If the metal salt has affinity for the Fibre, as is the case for stannic chloride and silk, no binder is needed. Urea-formaldehyde or acrylic resins can act as suitable binders for fixation of Non-permanent metal salts onto the Fibre surface. High intakes of metal salts used to weight textile structures may cause fibers to become brittle and to be sensitive to photochemically induced damage.

Laminating Agents

In recent years lamination of two textile structures together to form a composite structure has become very important. This technique requires use of adhesive materials that will not delaminate under normal use, including flexing and bending, shearing forces, and cleaning. The acrylics, polyurethanes, and a number of hot melt thermoplastic polymers are used as adhesives. Some adhesives have reactive groups which on curing lead to strong textile-adhesive covalent bonds in addition to the physical forces normally responsible for a good adhesive Fibre bond.

Crease Resistant and Stabilizing Finishes

When fibers are bent or deformed under various environmental conditions and then allowed to recover, the degree of recovery will depend on the morphology and inherent structure of the Fibre. Most synthetic fibers show reasonable recovery from such deformation, whereas the cellulosics and, to a lesser extent, the protein fibers have poor recoveries, particularly under moist conditions. Crease resistant finishes for cotton and rayon have been developed that give much improved wrinkle recovery properties, whereas crease resistant finishes for wool are still under development, since the process is complicated by felting shrinkage.

Crease resistant finishes for cellulosics are chemical crosslinking agents and include wash-and-wear and durable press finishes. The wash-and wear finishes are generally urea-formaldehyde or melamine-formaldehyde resins and are cured on cellulosics as flat goods in the mill, whereas durable press treatments consist of cyclic urea-aldehyde derivatives which are partially cured at the mill, after which curing is completed after the fabric is made into a garment in apparel manufacture. Other dysfunctional finishes, including epoxides, isocyanates, vinylsulfones, and aziridines, have been introduced as crease resistant finishes for cellulose. These have met with only limited success owing to their higher cost and other deficiencies. Since wool is already heavily crosslinked, a different approach to crease resistant finishing must be taken. The disulfide crosslinks in the textile structures must be chemically reduced followed by a setting treatment with bifunctional reagents such as capped diisocyanate derivatives. Chemical treatments for setting of thermoplastic man-made fibers are generally not used, since heat setting of the s e thermoplastic fabrics is an effective technique for imparting crease re s istance. Ureaformaldehyde resins also are often used on cellulosics to impart dimensional stability to the textile structure and to prevent yarn slippage within the structure. Finishes capable of causing interfiber bonding can act as effective stabilizing finishes, although they may stiffen the overall structure.

Protective Finishes

Photoprotective Agents and Antioxidants

Oxidative attack of fibers by oxygen in the air is particularly enhanced by the presence of moisture and light and leads to overall deterioration of the Fibre, its properties, and useful lifetime. Weathering of natural cellulosic and protein fibers is well known, and many man-made fibers rapidly deteriorate unless a protective agent is added to the Fibre. Protective agents are applied topically to natural fibers or added to the spinning solution prior to man-made Fibre formation. They are basically of two types,

- photoprotective agents that interact with light and
- antioxidants that quench oxidative free radical attack.

Photoprotective agents interact with ultraviolet and visible light absorbing and/or scattering the light and often releasing the absorbed energy harmlessly as heat. Photoprotective agents include metal oxides such as titanium dioxide and tin oxide, various metal salts, and ultraviolet absorbing organic compounds. Ortho-substituted benzotriazoles, triazines, and benzophenones are aromatic organic compounds that can harmlessly transfer the absorbed energy within the aromatic ring structure and thereby dissipate the energy eventually as molecular vibration and heat. Antioxidants interfere with oxidative free radical reactions formed through quenching and/or removal or the free radical from the Fibre matrix so that it will not react further. Antioxidants are usually heavy metal salts that interfere with oxidation or aromatic organic compounds such as phenols and thiol derivatives which can readily donate a hydrogen to the radical, forming a stable free radical species that will not react further. Such antioxidants present in fibers are eventually exhausted and thereafter can provide no further protection.

Oil and Water Repellents

Several classes of chemical agents exist that impart water and/or oil repellency when applied to textile substrates. Some finishes give water repellency only, whereas other finishes impart both water and oil repellency. Water repellent finishes are those which permit the fabric to continue to breathe after treatment, whereas waterproof treatments completely seal the spaces between individual yarns, as in the case of rubberized or unsaturated fatty acid-cured fabrics. Older water repellent treatments used derivatives of soaps and fatty acids to impart water repellency. Owing to their hydrocarbon nature they exhibit no oil repellency and actually are somewhat oleophilic. Soaps combined with zirconium, aluminum, or rare earth salts and methylol or pyridium salt derivatives of fatty acids were the major water repellent finishes used prior to World War II. With the development of organosilicon polymers after World War II, the methylated polysiloxanes were introduced as water repellent finishes. Also, fluorocarbon finishes generally are preformed organic polymers that have fluorine groups substituted on the side chains. It is important that the fluorine constituents be present at the end of the polymer side chains, with terminal trifluoromethyl groupings being most effective. The unique properties of fluorine that enable fluorocarbon polymers to repel water, oil, and waterborne soils have contributed greatly to their extensive use on consumer goods under trade names such as Scotchgard and Zepel.

Antistats

Antistats or antistatic agents are finishes that can be applied to a fabric to aid in the dissipation of static charge buildup on the fibers. Antistats can be applied to the Fibre as a temporary finish or added in the spinning bath prior to Fibre formation to give a more permanent finish. Chemical crossl inking of an antistat applied to a textile structure will provide a permanent finish, also. Most natural fibers and regenerated natural fibers are hydrophilic and possess charged or polar groups on the Fibre surface that can dissipate static charge to the atmosphere and prevent static buildup. Therefore antistat treatments are confined to the synthetic fibers such as nylon, polyester, etc. The antistats are surface-active agents related to detergents, ethylene oxide derivatives, silicones, or polar polymers such as polyamine resins. Because of their polar nature, they are able to bleed static charge from the Fibre and dissipate it into the air.

Biologically Active Finishes

Natural fibers and regenerated natural fibers are subject to attack by various biological agents including bacteria, fungi, and insects. While synthetic fibers are not normally attacked by these biological agents, substrates of these fibers can act as a support for growth of bacteria and fungi on the Fibre surface. Several metallic and organometallic salts, phenolic and anilide derivatives, and quaternary amine salts can inhibit growth of bacteria and fungi on fibers. Cellulosics most often require such treatment to prevent mildew and rot from feeding on and attacking the Fibre substrate. Introduction of cyanide groups in cotton by graft polymerization of acrylonitrile on the substrate also prevents attack by rot. In the past, commercial insecticides such as DDT and Dieldrin were used to prevent insect attacks on cellulosic and protein fibers. Since many of the insecticides previously applied can no longer be used for such treatments, new insecticides which are not environmentally persistent or damaging have been developed and are related to the natural products, the pyretherins.

Flame Retardants

All textile fibers with the exception of glass are flammable. The degree of flammability is dependent on the chemical structure of the Fibre, the construction of the textile substrate, and the environmental conditions present at the time of Fibre ignition. A Fibre is flame retardant when it self-extinguishes on removal of the flaming source. Certain fibers, including wool, modacrylic, aramid, and vinyon, are flame retardant by virtue of their inherent chemical structure and combustion characteristics. Thermoplastic synthetic fibers such as nylon and polyester are not self-extinguishing and continue to burn after ignition; however, owing to their melt-drip characteristics, the molten flaming polymer drops away from the fabric, causing the fabric to stop burning. Other fibers, such as the cellulosics, including cotton and rayon, burn readily and completely on ignition, leaving an ash which continues to oxidize and glow even after the flame is out. Blends of different types of fibers will show differing flammabilities which may not be directly related to an average flammability of the fibers in the blend.

Flame retardants can be used on all fibers to improve their flammability characteristics further, with the most flammable showing the greatest degree of improvement on application of a flame retardant. Flame retardants for man-made fibers are generally introduced to the spinning solution prior to Fibre formation, whereas natural fibers must be topically treated. Since the mechanism of flaming combustion of fibers varies with the Fibre type, different retardants must be used for the various Fibre types. The more flammable cellulosic fibers have received the greatest attention. Flame retardants can act in the gas phase or condensed phase of the burning Fibre to interrupt oxidation and flaming and/or smoldering combustion. Organic compounds containing halogens and/or phosphorus, sometimes in conjunction with inorganic salts, have been found to be effective in many applications. Often different compounds in combination give a synergistic effect. Water-soluble inorganic salts containing phosphorus, boron, and ammonium have been found to be effective in "one-shot" Non-durable applications, particularly on cellulosics.

FINISHES APPLIED TO FIBRE CLASSES

Finishes for Cellulosics

Cellulosics have certain deficiencies in properties that require finishing to improve these properties and maximize consumer usage and acceptance. The most important cellulosic finishes include crease resistant and stabilizing finishes, soil release and softening finishes, oil and water repellant finishes, biologically protective finishes, and flame retardant finishes.

Crease Resistant and Auxiliary Finishes

Cellulosics characteristically do not recover well from bending deformation, particularly in the presence of moisture, and crease resistant finishes have been developed for cellulosics to improve the wrinkle recovery of the cellulosic fabrics in the wet and/or dry states. Such finishes also stabilize cellulosics against relaxation shrinkage induced by mechanical forces during fabric formation. The crease resistant finishes used for cellulosics are for the most part derivatives of urea and various aldehydes.

These resins chemically crosslink adjacent cellulose chains and provide a chemical memory within the cellulose which aids in recovery from bending deformation or wrinkling. The degree of dry and wet wrinkle recovery of the resin-treated cellulosics will differ depending on whether curing to achieve crosslinking was conducted in the dry state or in the swollen wet state. Although crease resistant finishes for cellulosics were developed shortly after World War I, it was not until after World War II that they appeared extensively on consumer goods. The finishes introduced in the late 1940s were generally referred to as drip dry or wash-and-wear finishes.

The resin-treated cellulosics were cured flat while they were still moist at the mill. This required that the treated fabric be permitted to drip dry to achieve the maximum desired recovery effect. Since the fabric was cured in the flat state before being made into a finished textile, the garment did not retain creases placed in the finished garment. Flat set items such as sheets and table cloths can be effectively made crease resistant by this process, however. In the early 1960s durable press finishes were introduced. After resin application and partial curing at the textile mill, the treated fabrics were sent to the apparel manufacturer.

The apparel manufacturer fabricated the treated fabric into garments which were then fully cured in their finished state to give a textile product that retained its creases and recovered to its finished form after washing and tumble drying. Although it is difficult to make a clear distinction between the resins used for the wash-and-wear and for durable press, certain generalizations can be made.

Both types of resins are urea-aldehyde derivatives used in conjunction with possible other coreactants. The reactive functional groups in the resins are multiple in methyl groups which undergo acid catalyzed reaction with hydroxyl groups in adjacent cellulose molecular chains to form bridging crosslinks to its unwrinkled state. Representatives of the major types of crease resistant resins for cellulosics a re shown below:

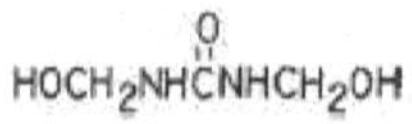

DIMETHYLOL UREA

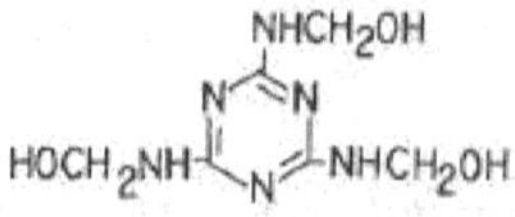

TRIMETHYLOL MELAMINE

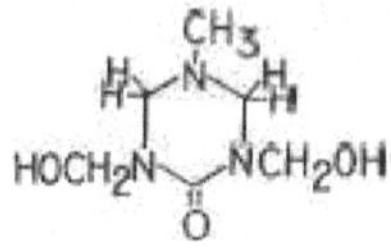

DIMETHYLOL-N-METHYL-TRIAZONE

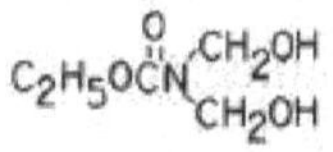

DIMETHYLOLETHYL-CARBAMATE

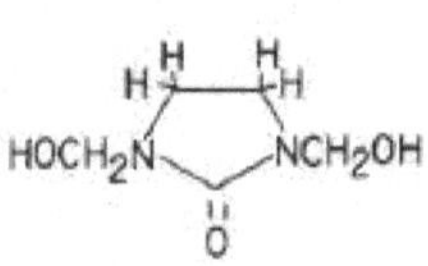

DIMETHYLOLETHYLENEUREA (DMEU)

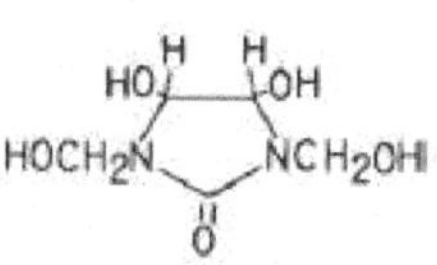

DIMETHYLOLDIHYDROXY-ETHYLENEUREA (DMDHEU)

In general, urea-formaldehyde condensates, methylolmelamines, and triazone-formaldehyde resins are used in wash-and-wear finishing. All suffer to varying degrees from chlorine retention and yellowing when in contact with sodium hypochlorite bleaches, which can cause damage to the cellulosic, In addition, the triazone resins give off a fishy Odour under moist conditions. The ethyleneurea resins and the carbamates are used extensively in durable press treatments. In general, these resins are less susceptible to chlorine retention and undergo partial cure readily followed by a final cure. Latent or mild acid catalysts including inorganic and Lewis acid metal salts are added to the resin solution to achieve curing of all of these resins applied to cellulosics. A gaseous formaldehyde-sulfur dioxide vapor treatment for 100per centcellulosics has been developed as well as a radiation curing treatment using methylolacrylamide. These resin treatments tend to lower the strength and overall abrasion resistance of the cellulosic fibers. Careful control of curing conditions and selection of catalysts as well as addition of softeners to the finishing solution can reduce this effect to some degree. Nevertheless, it has been necessary to blend thermoplastic heat settable fibers such as polyester with the treated cellulose to achieve acceptable wear characteristics. As a result polyester-cellulosic blends have made large inroads into markets where 100per cent cellulosic constructions were found previously. Soil retention on polyester -

cellulosic blends treated with crease resistant resins has been a problem, and a series of polar hydrophilic soil release finishes have been developed to meet this need. In recent years, low free formaldehyde durable press finishes have become important to protect workers from exposure to unreacted formaldehyde. In these formulations the methylol groups are capped with methyl groups, or a formaldehyde scavenger is added to the finishing solution.

Biologically Protective Finishes

Cellulosics particularly used in outdoor conditions are susceptible to attack by microorganisms. Treatment of cellulosics with chlorinated phenolic derivatives, organometallics such as copper 8-hydroxyquinolate, cationic surfactants, or grafting with polyacrylonitrile can reduce biological attack, although reductions in certain other properties may be expected.

Flame Retardant Finishes

Cellulosic fibers are among the more flammable fibers and are subject to rapid flaming combustion as well as slow smoldering combustion. Flammability legislation and accompanying demand for flame retardant cellulosics over the last 25 years has resulted in development of a large number of flame retardant treatments for cellulosics, while only limited development of agents inhibiting smoldering combustion has taken place. A number of temporary water-soluble flame retardant treatments for cellulosics are used. For the most part, these treatments use mixtures of inorganic salts, including borates, phosphates, silicates, and ammonium, and aluminum, antimony, and other metal ion salts. Borax and boric acid salt combinations are the most commonly used. Temporary flame retardants find only limited use on "one shot" items that will not be laundered. Permanent treatments are necessary for most cellulosic apparel and other cellulosics which must be laundered. Permanent treatments for Non-apparel textile cellulosic substrates have involved antimony oxide in conjunction with chlorinated hydrocarbons and crease resistant nitrogenous resins or other resin binders. For apparel purposes, a number of resin treatments based on tetrakis phosphonium chloride or on THPC plus sodium hydroxide have been used. THPC-amide based finishes include THPC reacted with trimethylolamine and urea or THPC reacted with urea and sodium hydrogen phosphate to form a complex polymer chemically fixed to the cellulose. THPOH-amide finishes include THPOH-urea-trimethylolmelamine and can be mixed with chlorinated hydrocarbons. THPC reacted in situ on the cellulosic with gaseous ammonia followed by mild oxidation has become a popular flame retardant finish in recent years. A dialkylphosphonamide, Pyrovatex CP, can be effectively cured on cellulosics when used with a triazine melamine resin to provide flame retardant properties. All of these flame retardant treatments result in some loss to the wear characteristics and aesthetics.

Finishes for Cellulose Ester Fibers

A limited number of finishes are used to correct deficiencies in acetate and triacetate. Delusterants, antioxidants, and/or antistat finishes added to the spinning solution prior to Fibre formation. Plasticizers can also be added to the spinning solution to improve the flexibility of the Fibre. When flame retardant treatment is required, haloalkylphosphonates are added to the spinning solution or padded onto the yarns to effectively lower the flammability of these fibers.

Finishes for Protein Fibers

Chemical Setting

Only limited finishing is carried out on protein fibers. Keratin fibers such as wool can be chemically set by chemical reduction of the disulfide crosslinks in the Fibre, mechanically arranging the Fibre in the shape desired, and then reoxidizing and joining the cleaved disulfide bonds in the new configuration. Reducing agents commonly used are adducts of organic compounds with sodium dithionate or sodium bisulfite. The reduced disulfide bonds are oxidized in time by oxygen in the air, but oxidation can be speeded up by application of a mild agent such as sodium bromate. This same process is used for permanent waving of human hair, which is also a keratin.

Shrinkproofing and Wrinkle Resistance Finishes

Wool and other keratin fabrics felt in the presence of moisture, agitation, and heat because of the surface scales on the fibers and resultant ratcheting action between individual fibers. Wool can be made resistant to felting shrinkage through modification of the scale structure by oxidizing the scales and/or by fixation of polymers on the scales. Oxidative treatments have included treatment by dry chlorination, wet chlorination, dichlorocyanuric acid, and permonosulfuric acid, as well as other oxidizing agents. The oxidizing agents attack the scales and partially destroy them, limiting their ability to cause felting. Although attack is mostly limited to the surface, significant damage and weakening occurs within the Fibre. Polymer finishes that effectively can be spread and fixed on the Fibre surface render wool shrink resistant even at uptakes as low as 1per cent-3per cent. Polymer is formed in situ on the Fibre surface through reaction of step growth monomers such as diacid chlorides or diisocyanates with diamines or preformed polymers in solution or as an emulsion are applied to the wool and cured to chemically bond and fix the polymer to the wool. The presence of polymer on the Fibre surface effectively prevents felting in a way that is not completely understood. The most satisfactory and effective treatments involve combined mild surface oxidation followed by polymer application. The usual shrinkproofing techniques have little effect on the wrinkle resistant characteristics of the wool. Finishes that contain chemically protected isocyanate groups have been developed that show some promise as durable press treatments for wool and have some ability to impart shrink resistant properties, too. Such treatments must be used in conjunction with chemical reduction of disulfide bonds in the Fibre to set the Fibre in the desired conformation in order to fully develop durable press properties.

Mothproofing Treatments

In the past, wool and other keratin fibers were protected from attack by moth larvae through use of mothballs or chlorinated insecticides. In recent years, the chlorinated insecticides used have been banned, and less toxic organic insecticides have been developed for use on wool.

Weighting Treatments

Silk is characteristically weighted through chemical treatment to give a heavier fabric. Silk is treated with metal salts such as tin chloride in acidic aqueous solution followed by washing with sodium phosphate and sodium silicate to form a complex tin phosphosilicate. High levels of salt will impart a harsh hand to the silk and further lower its resistance to sunlight-initiated oxidative attack.

Flame Retardant Treatments

Although the protein fibers are naturally flame retardant, additional flame retardancy is required in some applications. Wool can be effectively rendered flame retardant through treatment with halogenated organic acids, organophosphonates, and complex inorganic salts of zirconium, titanium, tungsten, etc.

Finishes for Polyamide Fibers

Although the polyamides are strong, tough fibers, they possess deficiencies with regard to oxidative attack by oxygen in the presence of light and/or heat which necessitates finishing to limit such attack. Static charge buildup on nylon is also a problem, and antistatic treatments are often applied to the Fibre. Improved flame retardancy is desirable in some applications, and flame retardant treatments are often used. Selection of appropriate comonomers can introduce groups into polyamides which alter their dyeing characteristics.

Photoprotective Agents and Antioxidants

Several approaches have been used to stabilize the polyamides against oxidative attack in the presence of heat and/or light. Application or addition of antioxidant compounds such as stannic chloride or aluminum chloride plus base, copper and manganese salts, tung states and molybdates, substituted phenols, catechols, organic phosphites, and combinations of these materials to the polymer melt or as finishes have been found to be effective in preventing such attack. Ultraviolet absorbers such as benzotriazoles and hydroxyphenones are effective in preventing light-induced oxidative attack. Titanium dioxide added to nylon as a delusterant will lower the absorption of light by the Fibre but will al so contribute to light-induced photodegradation through peroxide formation.

Antistatic Agents

Polar hydrophilic polymers applied as finishes or added to the polymer melt prior to Fibre formation improve the static buildup characteristics of polyamides. Polyethylene glycols and acidcontaining vinyl and acrylic copolymers have been used, as have insoluble salts of amphoteric and cationic detergents.

Flame Retardant Finishes

Polyamide fibers can be most effectively flame retarded by thiourea-aldehyde condensate treatments in presence and absence of durable press finishes. Amino phosphorazines, ammonium bromidebinder combinations, and antimony oxide with halogen-containing polymers have also been found to be effective. Tris phosphate is effective as a halogenated organophosphorous compound flame retardant, but due to its mutagenic and possible carcinogenic characteristics it has been excluded from use.

Finishes for Polyester Fibers

The major finishes for polyester fibers include photoprotective agents and antioxidants, antistatic agents, soil release finishes, antipilling finishes, and flame retardant finishes. Many of these chemicals can be added to the polymer melt prior to spinning. The finishes can also be added topically during dyeing and finishing.

Photoprotective Finishes and Antioxidants

Compounds that are used to stabilize polyesters against heat- and/or light-induced oxidative attack are similar to those used for nylon and other synthetics. They include metal salts as well as benzotrizoles, benophenones, and hindered phenols.

Antistatic Finishes

Owing to its hydrophobicity, polyester builds up static charge readily. Finishes have been developed for polyester that increase the hydrophilicity and ionic character of the Fibre and permit more ready dissipation of static charge from the Fibre surface. These treatments include lauryl phosphate, morpholine, various polyethylene glycols, organosilicones, and polyamine resins.

Soil Release Finishes

Owing to its combined hydrophobic and oleophilic character, polyester is soiled readily by oily soils. If these soils penetrate into the Fibre, they are difficult to remove by laundering. In order to improve the removal of oily soils from polyester, oil repelling soil release finishes have been developed to limit the penetration of oily soils into the Fibre, thereby making their removal easier during laundering. Perfluoroalkylamides and other fluorocarbon derivatives have been used for this purpose, but more inexpensive hydrophilic finishes such as copolymers of acrylic acid, polyhydroxy compounds, and carboxymethylcellulose have been used more extensive1y. Since po1yester is often used with cellulosics in blends in which the cellulosic must be treated with a durable press finish, combined durable press-soil release finishes have been prepared and are useful and very effective on such blends.

Antipilling Finishes

Because of its strength, polyester is subject to pill formation during wear. To make polyester more pill resistant, weaker and therefore less pillable, polyester fibers have been prepared by using lower molecular weight polymers or through partial hydrolysis of the polyester after textile substrate formation.

Flame Retardant Finishes

With the implementation of various flammability regulations in the U.S., improved flame retardant properties for polyester were required particularly with regard to the continued burning of melted polyester falling from burning fabric. In order to impart self-extinguishing characteristics to the polyester, two approaches have been used— introduction of ha 1ogen- and/or phosphorus-containing co monomers to the polymer structure or introduction of a halogenated phosphonate or related compounds to the melt prior to spinning. In both cases, antimony oxide may be added to improve the overall flame retardant characteristics. Tris phosphate was an ideal flame retardant for polyester and was used extensively for this purpose. Unfortunately, Tris was found to be mutagenic and a potential cancerproducing agent and had to be removed from the marketplace. As a result of the ensuing confusion, all topical flame retardant finishes have become suspect, and certain flammability requirements have been relaxed to allow polyester to pass the flammability test for children's sleepwear without further treatment.

Other Finishes

Comfort and the moisture related properties of polyester are of great interest, and finishes and treatments that improve these characteristics should be of great importance in coming years. Polyester fibers can be rendered more water wet table by surface treatment with basic solutions or through application of a hydrophilic finish to the polyester surface. Such finishes improve the water wicking and water transport characteristics of the polyester. Base hydrolysis has the added benefit of giving the polyester a crisp silk-like hand. Fine denier base-treated polyesters are produced in Japan that is effectively used as a silk substitute.

Finishes for Acrylic Fibers

Antistatic treatments for acrylic fibers include polyglycol esters, fatty acids, and hydroxyl and amino derivatives that can be crosslinked with aldehydes or epoxy groups. Certain of these antistats as well as cationic surfactants can be used as softening agents to impart a softer hand to acrylic fibers. Modacrylics are flame retardant and self-extinguishing and usually do not require an additional flame retardant treatment. Acrylic fibers can be made flame retardant by use of a halogenated comonomer with acrylonitrile or through addition of an organic phosphorus derivative or halogenated material to the polymer solution prior to spinning.

Finishes for Polyolefin Fibers

Antistatic finishes and antioxidants normally used on synthetic fibers are added to the polymer melt prior to polyolefin Fibre formation. The most common antioxidants used include hindered substituted phenols, organometallic antioxidants, and substituted phenols. Blending of comonomers with polyolefin prior to Fibre formation is used not only to improve dyeability but al so to plasticize the Fibre and improve other Fibre properties. Polyolefins may be effectively made flame retardant through incorporation of a metal oxide such as antimony oxide in conjunction with a brominated hydrocarbon or brominated organophosphate.

Finishes for Vinyl Fibers

Vinyl fibers except for vinal and vinal-vinyon does not generally undergo specialized finishing, although antioxidants and antistatic agents may be used. Vinal and vinal-vinyon matrix fibers can be treated with ureaaldehyde based durable press finishes to improve their wrinkle recovery, and vinal can be made flame retardant with phosphorus-containing finishes used for cellulosics.

Finishes for Elastomeric Fibers

The elastomeric fibers often have antioxidants and antistatic agents of the types outlined previously incorporated into them to protect the Fibre against oxidative attack and to reduce static buildup, respectively.

Finishes for Mineral and Metallic Fibers

A limited number of finishes are used on mineral and metallic fibers. Glass fibers are often surface treated with agents such as chromium salts or silanes to improve their ability to adhere to organic adhesive materials used in polymer-glass Fibre composites. Various organic starches, proteins, and synthetic polymer sizings are applied to glass fabrics to protect the individual fibers and to lower abrasion between individual fibers.

11

Effect of Radiation on Textile Dyeing

INTRODUCTION

Love for colours is a natural instinct and every individual has his own choice and liking for colour. The icy appearance of Hamaliyan ranges or lush green forests or fields of agriculture or trees laden with colorful fruits or butterflies moving from flower to flower presents the beauty of nature, generation after generations are being attracted. The choice of beautiful fascinating colours reflects the aesthetic sense of humans that varies. Colour is visual perceptual property corresponding in humans to the categories called red, yellow, blue and others. It is a sensation that arises from the activity of retina of the eye and its attached nervous mechanism, and results in a specific response to the radiate energy of certain wavelength and intensity. Thus it is a quality of an object with respect to light. Colorants may be either pigment or a dye which are characterized by their ability to absorb or emit light in the visible range 400-700nm.They may be organic or inorganic depending upon their structure and method of production.

Dyes are the coloured substances which are capable of imparting their colours to the matrix which may be Fibre, paper or any object. They must have fixing tendency on a fabric that is impregnated with their solution and the coloured fixed dyes must be fast to light as well as resistant to action of water, dilute acids, alkalies, various organic solvents used in dry cleaning, soap solutions, detergent, etc. A pigment generally is a substance which is insoluble in the medium in contrast to dye in which it is applied and has to be attached to a substrate by additional compounds *e.g.* polymer in paints and plastics. A compound looks coloured because it has absorbed certain electromagnetic radiation from the visible region. The moieties, present in colouring substance, responsible for the absorption of electromagnetic radiation and reflect in the visible region are called chromophores. Ultraviolet radiation constitutes to 5per cent of the total incident sunlight on earth surface. Even though, its proportion is quite less, it has the highest quantum energy compared to other radiations. Light is electromagnetic in nature. Within the electromagnetic spectrum, human eye captures visible light in the range between about 380 nm and 700 nm. Dyes absorb electromagnetic radiation of varying wavelength in the visible range of spectrum. Human eyes detect the visible radiations only for the respective complementary colours.

CLASSIFICATION OF DYES: NATURAL AND SYNTHETIC DYES

All colourants obtained from animals, plants and minerals without any chemical processing are called natural dyes.*e.g.*Alizarin a pigment extracted from madder, tyrian purple from snail and ochre which is a mineral of Fe_2O_3. Natural dyes may be vat dyes, substantive or mordant dyes as they require the inclusions of one or more metallic salts of tin, chromium, iron, copper, aluminum and other for ensuring reasonable fastness of the colours to sun light and washing. The natural dyes have several advantages such as: these dyes need no special care, wonderful and rich in tones, act as health cure, have no disposal problems, have no carcinogenic effect,easily biodegradable, require simple dye house to apply on matrix and mild reactions conditions are involved in their extraction and application. There are some limitations of natural dyes which includes, lesser availability of colours, poor colour yield, complex dyeing processing, poor fastness properties and difficulty in blending dyes.

Table shows the classification of dyes based upon both colours and structures.

Colours	Chemical Classification	Common Names
Yellow and Brown	Flavone Dyes	Quercitron, Tesu
Yellow	Iso-quinoline Dyes	Barberry
Orange- Yellow	Chromene Dyes	Kamala
Brown	Naphthoquinone Dyes	Henna, Alkanet
Black	Benzophyrone Dyes	Cochineal, Madder
Blue	Indigoid Dyes	Logwood
Red	Anthraquinone Dyes	Indigo

Neutrals Tannins Pomegranate, Eucalyptus

Commercialization of natural dyes can be done successfully by a systematic and scientific approach to extraction, purification and use. Optimization of extraction condition is a must to minimize the investment cost and to avoid discrepancy in the dye shade quality. Natural dyes occur in many plant parts in small quantities and as complex mixtures with many chemical compounds of similar or different structures. These compounds vary considerably with change in general, same genus but different species and ecological conditions of the plant source. So when natural dyes extracted from these sources are used for dyeing and printing, variation in shade, depth and tone, among others, may arise. Further, chemical components of plants change with age and maturity of the parts. Extraction may include drying, pounding, soaking, skimming, crystallizing, condensing, caking and liquidifying, among others, depending on the quality and species of the dye yielding plant, mineral and insect. *Synthetic dyes* are a class of highly coloured organic substances, primarily utilized tinting textiles that attach themselves through chemical bonding between the molecules of dye and that of Fibre. The use of natural dyes in textiles was eliminated since synthetic dyes give variety of reproducible shades and colours. Synthetic dyes are classified on the basis of chemical structure or on the basis of methods of application to the material. Dyes are synthesized in many ways by using different chemicals. On the basis of methods of application dyes are categorized as:-

- *Acid dyes:* These dyes are anionic and form ionic bonds with fibers that are cationic in acid solutions. These dyes are applied onto the acrylic, wool, nylon and nylon/cotton blends. These are called acidic because they are normally applied to nitrogenous fibers in inorganic or organic acid solutions.
- *Azoic dyes:* These dyes contain azo component used for dyeing of cotton fabrics. In the dyeing process Fibre is first treated with coupler followed by application of azo dye. This type of dye is extremely fast to light.
- *Basic dye*: These dyes are cationic and form ionic bonds with anionic fibers such as acrylic, cationic dyeable polyester and cationic dyeable nylon. These are amino derivatives used mainly used for application on paper
- *Disperse dyes:* These dyes are colloidal and are soluble in hydrophobic fibers. Mostly these dyes are used for coloring polyester, nylon, and acetate and triacetate fibers. They are usually applied from a dye bath as dispersion by direct colloidal absorption method
- *Direct dyes:* These are also azo dyes applied generally on cotton-silk combination from neutral or slightly alkaline baths containing additional electrolyte. These dyes predominantly interact and attach themselves with the Matrix through electrostatic interactions. These dyes are used to Colour cellulose, wool, nylon, silk etc.
- *Reactive dyes:* Reactive dyes are the best choice and other cellulose Fibre at home or in the art studio. Fixation of dye occur onto the Fibre under alkaline conditions by forming a covalent bond between reactive group of dye molecule and OH, NH, SH etc groups present in the fibers.
- *Mordant dyes:* Applied in conjunction with chelating salts of Al, Cr and Fe. Metallic salts or lake formed directly on the Fibre by the use Al, Cr or Fe salts which cause precipitation in situ.
- *Sulfur dyes:* These dyes are used for dyeing cotton and rayon. The application of this dye requires careful process due to its water-soluble reduced form and insoluble oxidized form. These dyes are fast to washing but poorly fast to chlorine and give dark and dull colors.
- *Vat dyes:* These dyes are insoluble in water and cannot be directly applied to textiles. These dyes require oxidation as well as reduction step for its application onto matrix.

- *Acetate rayon dyes:* Developed for cellulose acetate and some synthetic fibers

Dyes are synthesized in a reactor, filtered, dried, and blended with other additives to produce the final product. The synthesis step involves reactions such as sulfonation, halogenation, amination, diazotization, and coupling, followed by separation processes that may include distillation, precipitation, and crystallization. In general, organic compounds such as naphthalene are reacted with an acid or an alkali along with an intermediate and a solvent to form a dye mixture. The dye is then separated from the mixture and purified. On completion of the manufacture of actual colour, finishing operations, including drying, grinding, and standardization, are performed. These steps are important for maintaining consistent product quality.

CHEMISTRY OF FIBERS

Cotton the most abundant of all naturally occurring substrates and is widely used. For the fabric strength, absorbency quality, capacity to be washed and dyed, cotton has become the principal clothing fabric of the world. The materials characteristically exhibit excellent physical and chemical properties in terms of water absorbency, dye ability and stability and can not be entirely substituted by artificial polymer fibers. The cellulose consists of glucose units linked together through oxygen atoms, 30 to several hundred chains from micro fibrils. By dry weight 94per cent of cotton is made up of cellulose. The remaining constituents include 1.3per cent protein, 1.2per cent pectic substances, 0.6per cent waxes, 1.2per cent ash, and 4per cent of other components. Of three hydroxyl groups on the cellulose ring, two are secondary, and one is primary. Most of the reactions with cellulose occur at the primary hydroxyl groups.

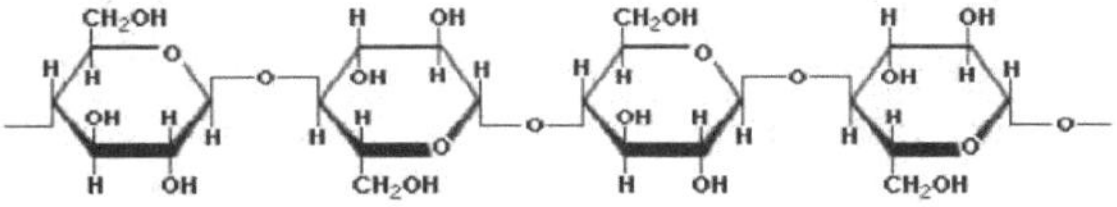

When cellulose is chemically modified with the compounds containing cationic and anionic groups, the molecular chains are modified. In the modified Fibre surface, the chemical and physical properties of cellulose Fibre are changed. Through chemical modification, the reactivity of the cellulose Fibre is enhanced. And several classes of dyes such as direct, azo, reactive etc can be successfully applied. The application of the cationic dyes has not gained widespread success. Our study comprises of the treatment method such as high energy radiation treatment which may create the anionic centre in the fabric to transfer the cationic dye onto the physically or chemically modified fabric. The reports of modified cellulose with the compounds containing multifiber cationic and anionic groups are scarce. *Wool* is different to other fibers because of its chemical structure that influences its texture, elasticity, staple and crimp formation. It is composed of keratin-type protein having more than 20 amino acids and very small amount of fat, calcium and sodium. The amino acids in wool linked together in ladder-like polypeptide chain to form a protein/polymer type structure.

Wool polymer contains some important chemical groups that able to form inter-polymer forces of attraction. These groups are: the polar peptide groups and the carbonyl groups, which forms hydrogen bonds with the slightly positively charged hydrogen of the amino groups of another peptide groups. There are also carboxylate groups and amino groups present in wool as side groups, between these two groups salt linkages or ionic bonds may be formed. Finally, the existence of the inter-polymer forces tends to make the van der Waals' forces rather significant.

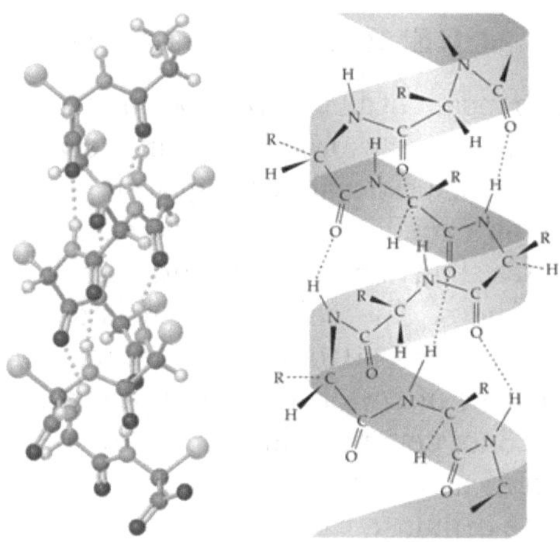

Wool is easy to dye since the surface of the wool Fibre diffuses light giving less reflection and a softer colour. The proteins in the core of the Fibre absorb and combine with a wide variety of dyes and allow the wool to hold its colour. *Silk* is an insect Fibre comes from the silkworm that spins around itself to form its cocoon.

A single filament from a cocoon can be as long as 1600 meters. It is considered an animal Fibre because it has a protein structure. Like other animal fibers silk does not conduct heat, and acts as an excellent insulator to keep our bodies warm in the cold weather and cool in the hot weather. The flat surfaces of the fibrils reflect light at many angles, giving silk a natural shine. Natural and synthetic silk is known to manifest piezoelectric properties in proteins, probably due to its molecular structure. Silk emitted by the silkworm consists of two main proteins, sericin and fibroin, fibroin being the structural center of the silk, and sericin being the sticky material surrounding it. Fibroin is made up of the amino acids Gly-Ser-Gly-Ala- Gly-Ala and forms beta pleated sheets. Hydrogen bonds form between chains, and side chains form above and below the plane of the hydrogen bond network.

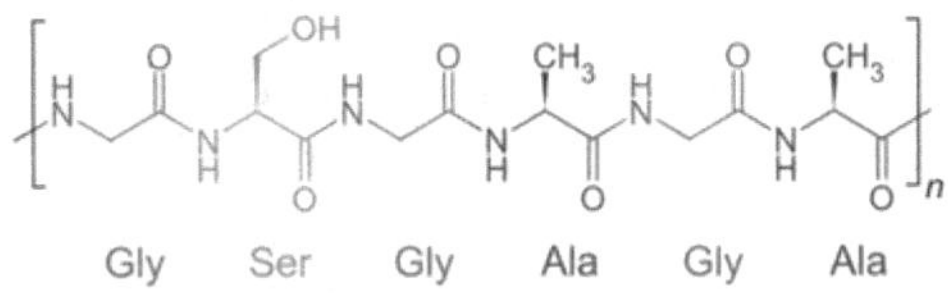

Silk polymer is composed of sixteen different amino acids where as wool polymer contains twenty amino acids of wool polymer. Three of these sixteen amino acids namely, alanine, glycine and serine, make up about four-fifth of the complete polymer chain The important chemical groupings of the silk polymer are the peptide groups which give rise to hydrogen bonds, the carboxyl and amine groups give rise to the salt linkages. The high proportion of glycine, which is a small amino acid, allows tight packing and the fibers are strong and resistant to breaking. The tensile strength is due to the many interceded hydrogen bonds, and when stretched the force is applied to these numerous bonds and they do not break *Polyester* was first introduced to the American public in 1951by W.H. Carothers Laboratory. It was advertised as a miracle Fibre that could be worn for 68 days without ironing and still look presentable. Polyester was once hailed as a magic Fibre capable of being washed, scrunched and pulled on without showing any signs of water or wrinkles.

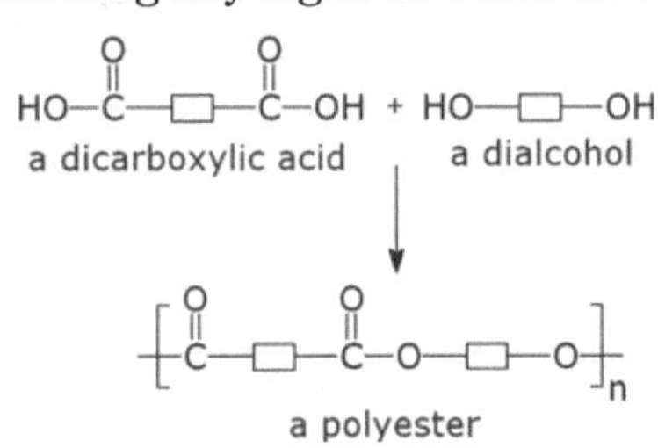

Now it is remembered for its bright double knit fabrics and comfortable texture. The name "polyester" refers to the linkage of several monomers within the Fibre. Polyester is long chain polymer chemically composed of at leas 85per cent by weight of an ester and a dihydric alcohol and a terephthalic acid. *Polyester Cotton* (PC) is a blend of polyester and cotton in varied proportions. This particular fabric is well received by customers around the world. The yarn is available in single and twisted form. The polyester cotton (PC) fabric yarn commonly has a blend ratio of 50per cent polyester to 50per cent cotton. In polyester cotton fabric (PC), polyester provides wrinkle resistance and shape retention while cotton provides absorbency and consequent comfort.

IRRADIATION IN TEXTILES

Irradiation processes have several commercial applications, in the coating of metals, plastics and glass, in printing, wood finishing, film and plastic cross linking and in the fields of adhesive and electrical insulations. The advantages of this technology are well known energy saving, low environmental impact, simple, economical and high treatment speed. Despite these advantages, there have been few applications of radiation curing in the textile industry, such as Non- woven fabric bonding, fabric coating and pigment printing. Radiation treatment on fabric and garments can add value in colouration. Modification of the surface Fibre can allow more dye uptake; its fixation at low temperature and increase wettability. Cotton knitwear pilling can be eliminated from the surface of the fabric by radiation treatment without affecting the strength of the Fibre. Effect of UV radiation in natural as well as synthetic dyeing using irradiated cotton fabric has given significant results.

EFFECT OF UV AND GAMMA RADIATION ON THE FABRIC DYED WITH NATURAL DYES

There is a remarkable difference in colour strength when different extracts of irradiated and un-irradiated turmeric powder were used to dye the irradiated and un-irradiated fabric. The methanol solubilized extract gave more colour strength than aqueous solubilized and alkali solubilized extract as displayed in figure.

Fig. Effect of UV radiation on the colour strength of the irradiated and un-irradiated cotton dyed with heat solubilized, alkali solubilized and methanol solubilized extract of irradiated and un-irradiated turmeric powder (Where URP- un-irradiated powder, RP – irradiated powder, RC- irradiated cotton fabric, URC- un-irradiated cotton fabric)

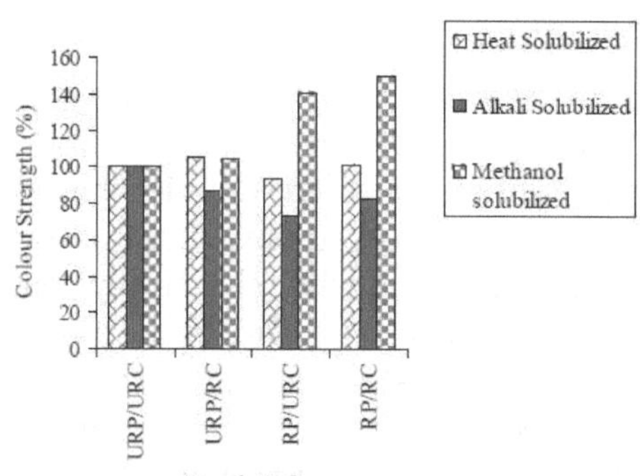

The low colour strength using alkali solubilized extract is due to alkaline degradation of curcumin into products like vaniline, vanilic acid, feruloylmethane, ferulic acid and other fission products, which sorb on the fabric along with colourant and impart dull redder shades.

While using (heat) aqueous solubilized extract, the colourant being insoluble in water may undergo hydrolytic degradation and the actual colourant concentration becomes low onto the fabric as a result low colour strength is observed. By using methanol solubilized extract, the actual colourant get significant chance to sorb onto fabric and impart yellow colour with dark shades.

The irradiation of fabric is also another factor which affects the colour strength of the fabric. UV irradiation adds value to colouration and also increases the dye uptake ability of the cotton fabrics through oxidation of surface fibers of cellulose.

The colourants from Methanol solubilized extract reach the vicinities of fibres and upon investigation of colour strength using spectraflash SF 650, dark yellow shade was observed. In the case of un-irradiated fabric, the insoluble impurities get significant chance to sorb on the matrix along with colourant which showed the dull redder shades. Gamma rays are ionizing radiations that interact with the material by colliding with the electrons in the shells of atoms. They lose their energy slowly in material being able to travel through significant distances before stopping.

The free radicals formed are extremely reactive, and they will combine with the material in their vicinity. Upon irradiation the cross linking changes the crystal structure of the cellulose, which can add value in colouration process and causes photo modification of surface fibers.

The irradiated modified fabrics can allow: more dye or pigment to become fixed, producing deeper shades, more rapid fixation of dyes at low temperature and increases wet ability of hydrophobic fibers to improve depth of shade in printing and dyeing. The influence of gamma radiation on the colour strength values of the fabric dyed with natural dyes extracted from eucalyptus bark has been shown in figure.

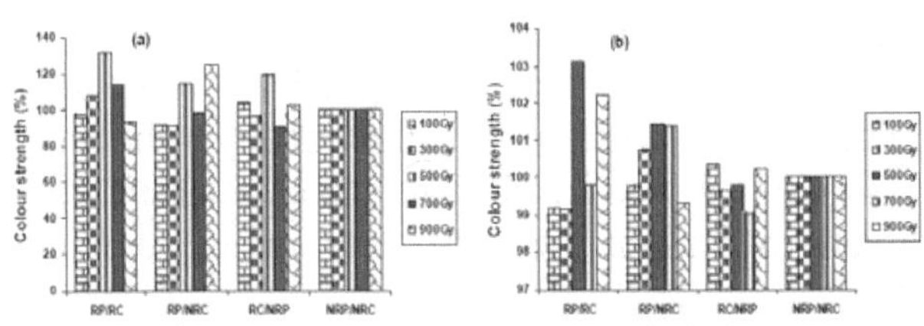

Fig. Effect of gamma radiation on the colour strength of the cotton dyed with (a) ethanolic (b) aqueous extracts obtained from irradiated and un-irradiated Eucalyptus, NRC-un irradiated cotton fabric

High colour strengths and dark brown shades of the fabric dyed in ethanolic extract were obtained as compared to aqueous extracts.

The low colour strength and un-evenness in shade in aqueous extract is due to presence of insoluble impurities that might come on the fabric along with colourant.

The results shown in figure, demonstrate that irradiated fabric dyed using alcoholic extract gave more colour strength than un-irradiated fabric. Gamma irradiation causes dislocation and fragmentation of fabric fibers however, only soluble colourant free from impurities get maximum chances to sorb on the fabric. But un-irradiated fabric contained less dye and yielded greener shade.

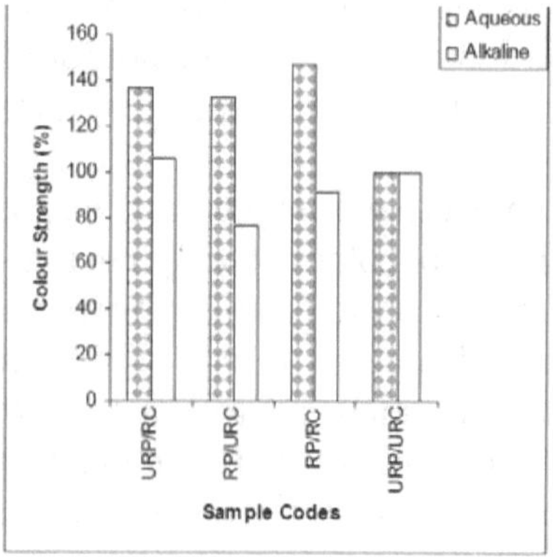

Fig. Effect of gamma radiation on the colour strength of the cotton dyed with extracts obtained form irradiated and un-irradiated turmeric powder usign aqueous and alkaline media (Where URP-un irradiated powder, RP - irradiated powder, RC-irradiated cotton fabric, URC-un-irradiated cotton fabric)

The colour strength changes significantly in aqueous than in alkaline media. The fabrics dyed in aqueous extract of turmeric powder were darker yellow in shades than that of fabrics dyed in alkaline extract.

The low colour strength was due to alkaline degradation of curcumin into water-soluble products like vaniline, vanilic acid, feruloylmethane, ferulic acid and other fission products, which gave dull redder shades.

Tonessen and Karlsen reported that below pH 7, curcumin existed in yellow colour and is insoluble in water.

Due to insolubility, the colourant might have tendency to get absorbed completely on the fabric without passing through the medium and shows darker yellow shades. Hence irradiated fabrics dyed in aqueous media gave more colour strength than un-irradiated fabrics due to oxidative degradation of cellulose fibres.

Treatment of fabric by high-energy radiation causes either dislocation and fragmentation or slight loss in mass of fabric.

However, only colourants get maximum chance to sorb on fabric than insoluble impurities. So more colour strength is obtained in case of irradiated fabric dyed using aqueous extract of irradiated turmeric powder. Thus it is found that if irradiated fabric dyed with aqueous extract of irradiated turmeric powder, maximum colour strength and darker yellow shade was obtained.

EFFECT OF UV AND GAMMA RADIATION ON THE FABRIC DYED WITH SYNTHETIC DYES

UV irradiation effects the colour strength values and shades of fabric dyed with synthetic dyes. Using suitable dye and fabric, the process of irradiation can produce large variation in shades.

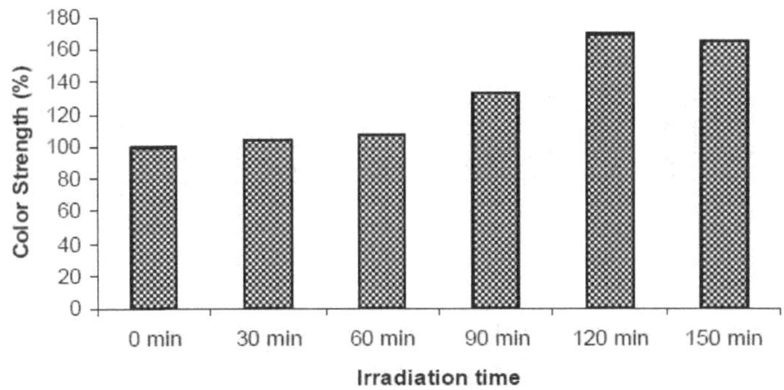

Fig. Effect of UV irradiation time on the colour strength of the irradiated cotton fabric dyed with irradiated stilbene based reactive dye.

The data displayed in figure shows that irradiated fabric for 120 min. gave maximum colour strength as compared to un-irradiated fabric.

The fabric irradiated for 120 min. showed even shade with better colour strength. The reason might be the oxidation of cellulose upon exposure to UV radiation. Michael and EL-Zaher in 2005 reported that the UV treatment of cellulose fibre created spaces between fibres which imbibed more dye and as a result the interaction between dye and cellulose fabric becomes more significant. The dye molecules rush rapidly onto the fabric and as a result darker shades were obtained.

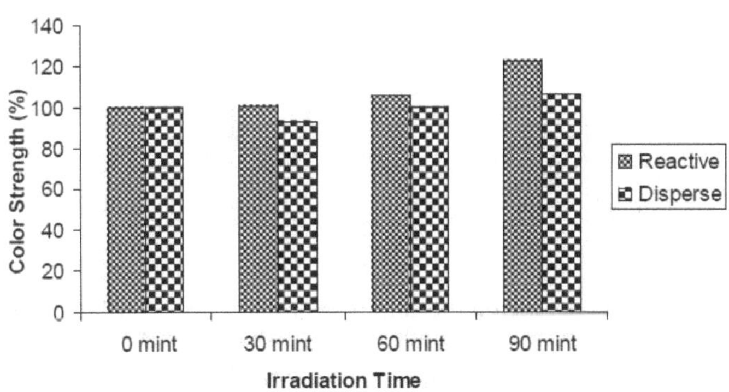

Fig. Effect of UV irradiation time on the colour strength of the irradiated cotton fabric dyed with un-irradiated reactive and disperse dye.

The figure shows that the fabric irradiated for 90 min. has maximum affinity for dye substrate to attach.

The fabric irradiated for 30 and 60 minutes show even shades having good colour strength. This improvement might be due to the oxidation of cellulose in to carboxylic acid group upon exposure of cellulose to UV radiation which interacts more towards the dye material to form covalent bond.

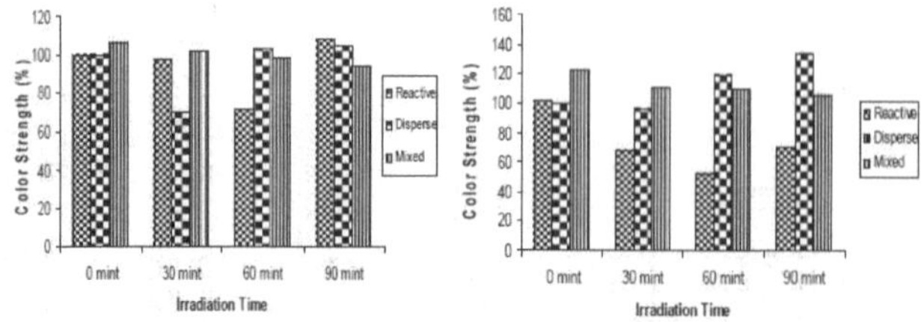

(a) (b)

Fig. Effect of UV irradiation time on the colour strength of the irradiated cotton fabric dyed with un-irradiated multifunctional triazine and irradiated reactiove and dianix disperse and mixed dye.

The un-irradiated and irradiated cotton fabric for the period of 30, 60 and 90 min was dyed, the results of fabrics have been shown in figure. shows that irradiated fabric and irradiated dyes for 90 min has maximum affinity for dye to attach on it. Oxidation of cellulose upon UV radiation significantly increases the dye uptake in the substrate due to the interstices available in the case of irradiated fabric surface. The dye molecules rush rapidly on to the fabric and as a result darker shades were obtained. K.R. Millington suggested that photo modification of surface Fibre may attain more dye or pigment to become fixed producing deeper shades. UV radiation causes more rapid fixation of dyes increases wettability of hydrophobic fibers to improve depth and shade in printing. For knitted wool and cotton fabrics, the problem of pilling can be eliminated.

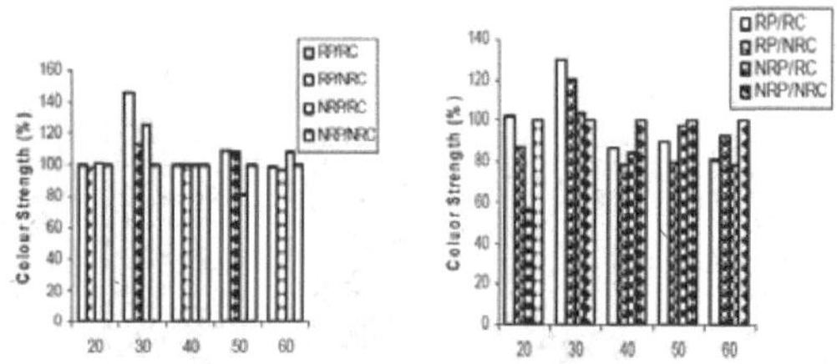

Fig. Effect of UV irradiation time on the colour strenght of the irradiated cotton fab dyed with un-irradiated and irradiated 5per cent (a) and 1per cent (b) Reactive dye.

The result shown in figure (a and b) indicate that colour strength values of 5per cent solution of dye powder are more as compared to colour strength values obtained in case of 1 per cent solution. The optimized time for irradiating cotton fabric is 30 minutes as shown in figure (a and b). At this time, oxidation of cellulose generates carboxylic acid group which helps in significant interaction of dye with oxidized a surface and show darker shades. Irradiation for less time does not activate the surface to interact with dye molecules to such an extent. While irradiation for long time may either facilitate insoluble impurities to rush onto modified fabric due to availability of wide interstices/gaps among the fibers which may cause dull and uneven shades having low colour strength.

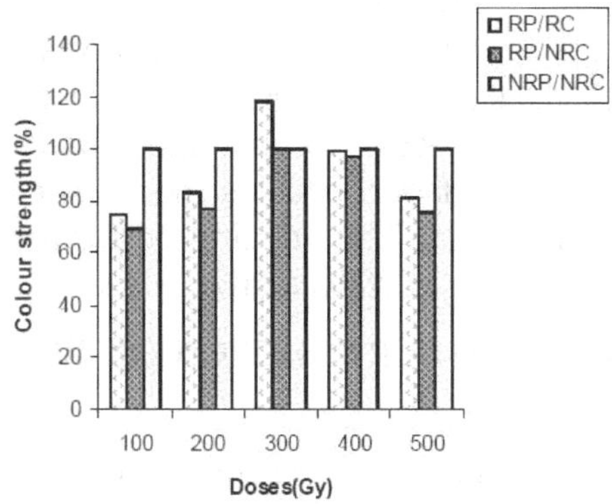

Fig. Effect of gamma irradiation on the colour strength of the irradiated and un-irradiated cotton fabric dyed with Irradiated and un-irradiated reactive blue dye powder.

Gamma radiation shows a promising influence in textile dyeing since irradiated fabric dyed with synthetic dye gave a prominent difference The result shown in figure indicate that colour strength values change remarkably using irradiated fabric which results in darker colour strength and more bluer shades than that of un-irradiated fabrics. This low colour strength is due the stuffing of insoluble impurities present in the dye solution on to the fabric. The dyeing performed using irradiated fabric treated with 300Gy absorbed dose gave maximum colour strength with darker bluer shades. At higher doses, low colour strength is obtained, which may be due to the degradation or dislocation of crystal moieties on cellulosic material. While at low dose, fabric surface does not activate enough to fix dye onto it and does not able to make firm interaction with dye material.

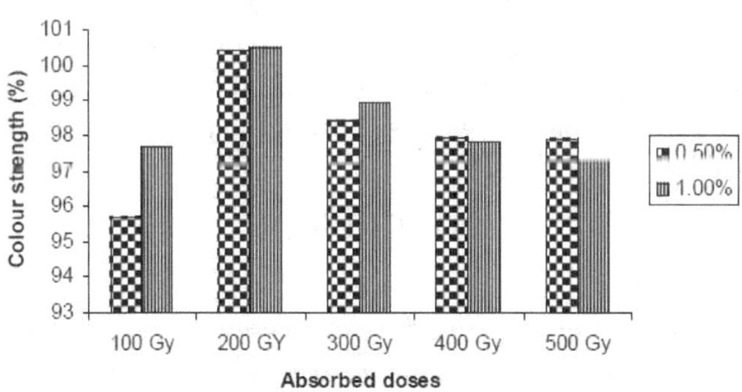

Fig. Effect of gamma irradiation on the colour strength of irradiated cotton fabric dyed with Irradiated reactive black dye powder

The result displayed in figure reveal that colour strength values decrease with increase in absorbed doses. The more colour strength is because of photo modified surface of cellulose which may have more affinity for dye substrate. The results show that the dyeing performed using 200Gy dose gives maximum colour strength with darker shades. At sufficient higher dose insoluble impurities along with dye molecule become fixed and causes uneven shades, while below optimum dose, surface of cellulose do not stimulate much to interact significantly with dye material. Thus dyeing performed using cotton fabric irradiated to an absorbed dose of 200Gy gave better colour strength.

EFFECT OF UV AND GAMMA RADIATION ON WOOL AND SILK AND POLYESTER

Studies on wool keratin have been previously performed in order to evaluate the effect of UV radiation. Chemical changes induced by short term UV radiation are confined to fibers at surface where as it is unable to penetrate into the fabric. The colour changes *i.e.* green followed by yellow in wool keratin due to UV radiation have been observed also. There are several processes to reduce the pilling yet no process can guarantee the zero pilling in wear. But Millington reported that it is only UV radiation which can reduce the pilling through siro flash technology followed by oxidation with hydrogen peroxide in germicidal UV Tubes. After using such techniques and then dyeing with UV irradiated wool fabric, the characterization of wool fabric meets standard marks by ISO. Thus the continuous UV reduction of the fabric followed by batch oxidation is of great commercial value. When wool fabric is exposed to UV radiation, it exhibits some physical and chemical changes on its surface. This interaction not only modifies the fabric of wool but also improve the shades particularly grey and black. It also helps in even dyeing, deeper shades, chlorine free printing and improve the photo bleaching of wool. Now a days, UV curing technology is being used for the modification of wool surface that helps in finishing as well as deepening the shades of wool when dyed using reactive dyes. By using UV curing technology, there are no risks involved to any loss of fabrics fibers in weight as well as in its physical appearance. This technology also do not cause any hazardous use of chemicals, smoothness of surface, unpilling as well as deep hues.

CONCLUSION

Radiation processes has several commercial applications starting from curing of fabrics, finishing, improvement of shades and characterization of dyed fabrics. The advantage of this technology are well known such as improvement in shades, enhanciong colour fastness, colour stregnth, low cost effective and reduction of the concentration of the dye. All these results have been seen from our above experiments. Radiation curing of silk, wool and cotton fabric to reduce pilling, their finishing and mercerization processes has also been improved. Thus both UV and gamma radiation has improved the textile sttuff according to standards of ISO, EPA and FAO. The use of eco-friendly technnology giving eco–label products under the influence of high energy radiations that may give new orientation for other dyes such as vat, reactrive azo and other brands. Similarly improvement of fibers of wool, silk, nylonn, Polyester cotton (P.C). etc., for dyeing to get good shades, even and lavelled dyeing, accepatable fastness properties yet are underway.So the dyers and colourists should try such techniques inorder to get better results and alternating methods for any risks related to human health.

Bibliography

Alfred Jenks & Son: *Illustrated Catalogue of Machines Built by Alfred Jenks & Son, Manufacturers of Every Variety of Cotton and Wool Carding, Spinning, and Weaving Machinery, in All Its Departments*, Bridesburg, Pa.: Alfred Jenks & Son, 1853.

Allen, Frederick James: *The shoe industry*, Boston: The Vocation bureau of Boston, 1916.

Allen, Nellie B: *Cotton and other useful fibers*, Boston: Ginn, 1929.

Allington, Sara May: *Practical sewing and dressmaking*, Boston: Page Co., 1913.

Allinson, May: *Dressmaking as a trade for women in Massachusetts*, Washington: Govt. print. off., 1916.

Atkinson, Edward: *Cotton and Cotton Manufactures in the United States*, Boston: Franklin Press, 1880.

Baird, Robert H: *The American Cotton Spinner, and Managers' and Carders' Guide*: A Practical Treatise on Cotton Spinning, Philadelphia: H. C. Baird, 1869.

Batchelder, Samuel: *Introduction and Early Progress of the Cotton Manufacture in the United States*, Boston: Little, Brown and Company, 1863.

Brown Brothers & Co: Brown Brothers & Co., Successors to Butler, Brown & Co.: *Manufacturers of Shaw's U. S. Standard Ring Travelers, Belt Hooks, Wire Goods, Etc., and Dealers in Supplies for Cotton, Woolen, Silk, Jute & Flax Mills, Providence*, R.I.: Brown Brothers & Co., c1881.

Deepankar Chakrabarti and Suchandra Choudhury, Sujit Choudhury: *An Introduction to Geographic Information Technology*, I K Pub, 2009.

E. Jenckes Manufacturing Co: *Illustrated Price List: E. Jenckes Mfg. Co., Sole Manufacturers of Hicks' U. S. Standard Ring Travelers*, Manufacturers' Supplies of Every Description, Pawtucket, R.I.: E. Jenckes Mfg. Co., 1883.

G Praveen Babu and M V Narayana: *Basic Concepts of Information Technology Workshop*, BS Pub, 2007.

G. Praveen Babu and M.V. Narayana: *Basic Concepts of Information Technology Workshop*, BS Publications, Delhi, 2008.

K.V. Kale, S.C. Mehrotra and R.R. Manza: *Computer Vision and Information Technology: Advances and Applications*, I.K. International Pub, 2010.

Mason Machine Works: *The Mason Machine Works, Taunton Massachusetts, U.S.A.: Inventors and Builders of Cotton Machinery*, Taunton, Mass.: The Company, 1898.

Merrill, Gilbert R: *American Cotton Handbook: A Practical Reference Book for the Entire Cotton Industry*, New York: American Cotton Handbook Co., c1941.

Raj Kumar Shrivastav: *A Textbook of Information Technology*, Dominant Publications, Delhi, 2013.

Rajnikant P. Patel: *Curriculum Design and Programming in Physical Education*, K.S.K. Publishsers, Delhi, 2011.

Sharad Sarin: *Strategic Brand Management for B2b Markets : A Road Map for Organizational Transformation*, Response Books, Delhi, 2010.

Shyama Balakrishnan and P.K. Paliwal: *Contemporary Information Technology Scenario*, Anmol, 2001.

Swarup K. Das: *A Textbook of Information Technology*, Dominant Publications, Delhi, 2010.

T.M. Samuel and Maria Samuel: *Advances in Information Technology*, Commonwealth Publications, Delhi, 2008.

Webber, Samuel: *Manual of Power for Machines, Shafts, and Belts, with the History of Cotton Manufacture in the United States*, New York: D. Appleton and Company, 1879

Index